Geoenvironmental Sustainability

RELATED TITLES

Forests at the Wildland-Urban Interface: Conservation and Management
by Susan W. Vince, Mary L. Duryea, Eward A. Macie, Annie Hermansen, and Robert L. France
(ISBN: 1566706025)

Restoration of Boreal and Temperate Forests
by John A. Stanturf, Palle Madsen, and Robert L. France
(ISBN: 1566706351)

Handbook of Water Sensitive Planning and Design
by Robert L France
(ISBN: 1566705622)

Geoenvironmental Engineering: Contaminated Soils, Pollutant Fate, and Mitigation
by Raymond N. Yong
(ISBN: 0849382890)

Geoenvironmental Sustainability

Raymond N. Yong
North Saanich, B.C., Canada

Catherine N. Mulligan
Concordia University, Montreal, Canada

Masaharu Fukue
Tokai University, Shizuoka, Japan

Taylor & Francis
Taylor & Francis Group
Boca Raton London New York

CRC is an imprint of the Taylor & Francis Group,
an informa business

CRC Press
Taylor & Francis Group
6000 Broken Sound Parkway NW, Suite 300
Boca Raton, FL 33487-2742

© 2007 by Taylor & Francis Group, LLC
CRC Press is an imprint of Taylor & Francis Group, an Informa business

No claim to original U.S. Government works
Printed in the United States of America on acid-free paper
10 9 8 7 6 5 4 3 2 1

International Standard Book Number-10: 0-8493-2841-1 (Hardcover)
International Standard Book Number-13: 978-0-8493-2841-1 (Hardcover)

Library of Congress Cataloging-in-Publication Data

Yong, R.N. (Raymond Nen)
   Geoenvironmental sustainability / Raymond N. Yong, Catherine N. Mulligan, Masaharu Fukue.
      p. cm.
   Includes bibliographical references and index.
   ISBN 0-8493-2841-1 (alk. paper)
   1. Environmental geotechnology. 2. Sustainable development. I. Mulligan, Catherine N. II. Fukue, Masaharu. III. Title.

TD171.9.Y66 2006
333.7--dc22   2006040225

Visit the Taylor & Francis Web site at
http://www.taylorandfrancis.com

and the CRC Press Web site at
http://www.crcpress.com

Preface

There are many who would argue that *sustainability* is a fashionable word that has lost its meaning when used in the context of society and the environment. They would further say that the word has been overused and that it never had any cachet because of one's inability to properly define what sustainability means. There are also those who will maintain that it is indeed foolhardy to attempt to write on the subject chosen for this book — simply because of the very amorphous nature of the subject.

The above notwithstanding, we have chosen to focus our attention on the geoenvironment and the need to protect the health and quality of the natural capital items that comprise the geoenvironment. We fully recognize that as long as continued depletion of the nonrenewable natural resources contained within the geoenvironment occurs, sustainability of the geoenvironment cannot be attained. We also recognize that the physical, chemical, and biological impacts to the geoenvironment from the various stress generators (humans, industry, agriculture, forestry, mining, cities, etc.) create situations that do not allow for sustainability goals to be achieved. If one combines all of the preceding impacts and their outcome with the recent spate of natural catastrophic disasters, such as the earthquakes, floods, hurricanes, slides, famine, etc., it will become all the more evident that geoenvironmental sustainability is an impossible dream. Faced with all of these, one has two simple choices: (1) to admit defeat and prepare to face the inevitable or (2) to correct, ameliorate, mitigate, and even prevent those detrimental elements that can be corrected, mitigated, and eliminated, and to find substitutes and alternatives that will replace the depleting nonrenewable resources. The material in this book is a first step in adoption of the second choice.

The subject addressed in this book is both an ambitious one and also a very difficult one, not only from the viewpoint of the basic science–engineering relationships involved in dealing with the various kinds of impacts on the geoenvironment, but also as much or more so from the fact that many crucial elements contributing to the generation of these same impacts cannot be properly addressed. One part of the problem is that many of these elements are not within the purview of this book (especially the critical subject of biological diversity). The other part of the problem is that there are many elements that are dictated by prominent forces. Among these are (1) socioeconomic factors and business–industrial relationships, (2) public attitudes, awareness, sensitivity, and commitment, and (3) political awareness and will.

The terrestrial environment, which is a major part of the geoenvironment, is the engine that provides the base or platform for human sustenance — food, shelter, and clothing. By all accounts, there is universal consensus that the stresses and demands imposed by society on the ecosphere have far surpassed the capability of the world's natural resources to regenerate and replenish themselves to meet sustainability requirements. It is contended that there are insufficient resources to meet the continued demands of the present world population, and that collapse of society will inevitably occur. Some of the major factors involved include (1) loss of biological diversity, (2) increasing discharge of greenhouse gases to the atmosphere, (3) loss of soil quality, (4) increasing generation of wastes and pollutants, (5) depletion of nonrenewable natural resources such as fossil fuels, and (6) depletion of natural living resources such as fisheries and forests. Considerable attention is being paid to many of these issues by researchers, policymakers, and other professionals well versed in engineering, scientific, and socioeconomic disciplines to alleviate the stresses to the geosphere and to seek sustainability and ways for society to live in harmony with the environment.

The primary focus of this book is on the geoenvironment and its importance as a resource base for life support systems. Mankind depends on the ecosphere, and in particular on the geoenvironment, to provide the raw materials to support life. Because of the many threats and negative impacts on the various life support systems within the geoenvironment, there is a pressing need for one to: (1) develop a better appreciation of the stresses imposed on the geoenvironment by mankind and (2) determine the requirements for sensible and proper management of our environmental resources to achieve a sustainable society.

The basic elements that define the geoenvironment will be developed in systematic detail and fashion, particularly in respect to their relationship to the five thematic areas, known as WEHAB: (1) water and sanitation, (2) energy, (3) health, (4) agriculture, and (5) biodiversity. These were identified as key areas of concern by the Johannesburg World Summit on Sustainable Development (WSSD) in 2002. Industrialization, urbanization, agriculture (food production), and resource exploitation (including energy) are basic activities associated with a living and vibrant society. We consider these basic elemental activities to be necessary to sustain life and also to be integral to *development*. In general terms, we consider development to (1) embody the many sets of activities associated with the production of goods and services, (2) reflect the economic growth of a nation, state, city, or society in general, and (3) serve as an indication of the output or result of activities associated with these four main elemental activities. Questions often arise as to how these activities accord with the aims of sustainability (of a society), and how one structures and manages programs and activities that would provide for a sustainable society.

The first two chapters provide the basic background needed to address the assimilative capacity of soils, particularly in the light of management of pollutants in the ground, and also in the light of sustainable development and land use. The intent of Chapter 1 is to provide an introduction to many of the basic issues that arise in respect to impacts and assaults on the geoenvironment as a result of anthropogenic activities associated with the production of goods and services. In Chapter 2 we will focus on contamination of the land environment as one of the key issues in the need to protect the natural capital and assets of the land environment. We will be paying particular attention to the various aspects of ground contamination and land management requirements to meet sustainability goals in this chapter.

In Chapter 3, the importance of water is also highlighted. The quality of the water can be and is significantly affected by all four components within sustenance and development, industrialization, urbanization, and resource exploitation and agriculture. Adequate quantities of good-quality water are also essential for health, agriculture, energy, and biodiversity. We will examine the sources and impacts so that they can be controlled to maintain the water quality and supply for future generations.

Chapter 4 examines the built environment. Populations within cities require clean water, sewage and waste management systems, housing, and transportation. They consume significant resources while polluting the air, land, and water. The increasing urban population will increase pressures on the geoenvironment in the years to come. The discussion in Chapter 5 will be confined to industrial activities associated with the extraction of nonrenewable mineral, nonmineral, and energy mineral natural resources (uranium and tar sands). Activities associated with the mining, extraction, and on-site processing of the extracted natural resource material (mineral and nonmineral) contribute significantly to the inventory of potential impacts to the terrestrial ecosystem.

In Chapter 6, we are concerned with the land environment and sustainability of the land ecosystem in relation to food production. We do not focus on food production from the agriculture engineering or soils science point of view since all the subjects and aspects of food production are well covered by soil science and agriculture engineering. Instead, the focus is from a geoenvironmental perspective on the results of activities in food production and in agro-industry on the geoenvironment itself. Chapter 7 directs its attention to the impacts on the geoenvironment in relation to industrial ecology. We consider the interactions on the geoenvironment by activities associated with manufacturing and service industries. Insofar as geoenvironmental resources are concerned, and in respect to sustainability goals, the primary concerns are (1) use of natural resources as both raw materials and energy supply and (2) emissions and waste discharges. Since the purview of this book addresses resource use from the geoenvironment framework and not from the industry perspective, we acknowledge the fundamental fact that the consequences of depletion of nonrenewable natural resources to society are a problem that must be confronted, and that the solution is not within the scope of this book. Accordingly, from the geoenvironmental protection point of view, we concentrate our attention on the impacts resulting from the discharge of liquid and solid wastes and waste products into the environment.

In Chapter 8, we discuss (1) the threats to the health of the coastal sediments realized from discharge of pollutants and other hazardous substances from anthropogenic activities, (2) the impacts already observed, and (3) the necessary remediation techniques developed to restore the health of the coastal sediments. A healthy coastal marine ecosystem ensures that aquatic plants and animals are healthy and that these do not pose risks to human health when they form part of the food chain.

Chapter 9 addresses the subject of land environment sustainability as it pertains to its interaction with the various waste discharges originating from industrial and urban activities. We focus our attention on developing concepts that involve the natural capital of the land environment. Chapter 10 discusses the magnitude of the problem of urbanization and industries. Particular attention is paid to the example of sites contaminated with hazardous wastes and other material discards. The discussion in this chapter recognizes that the impact from the presence of pollutants in the ground needs to be mitigated and managed — as a beginning step toward protection of the resources in the environment, and also as a first step toward achievement of a sustainable geoenvironment. The emphasis will be on using the properties and characteristics of the natural soil–water system as the primary agent for such purposes. Finally, in Chapter 11, we (1) discuss the case of nonrenewable nonliving renewable natural resources, (2) look at some typical case histories and examples of sustainability actions, and (3) present the geoenvironmental perspective of the present status of where we are in the geoenvironmental sustainability framework, with a view that points toward where we need to go.

Given the nature and scope of the multidisciplinary material covered in this book, the limitations, and given the need to present the information to highlight the importance of the land environment and sustainability of the land ecosystem in relation to food production, etc., we have had to make some difficult decisions as to the amount and level of *basic theory* needed to support the discussions presented. It was not our intent to develop or present extensive basic theories in any one discipline area of this multidisciplinary problem — except as is necessary to support the discussion from the sustainability viewpoint. There are basic textbooks that will provide the background theories for the various parts of the multidisciplinary problem treated in this book.

In the preparation of this book, the authors have benefited from the many interactions and discussions with their colleagues and research students, and most certainly with the professionals in the field who face the very daunting task of educating the public, industry, and political bodies on the need for conservation and protection of our natural resources. We have identified the sources of various kinds of noxious emissions in our discussions in the various chapters and have discussed the serious impact and consequences of such discharges. We have made mention in many chapters on the excess consumption of renewable resources and the significant problems of depleting nonrenewable resources — especially the energy resources. We have not embarked on detailed discussions on the kinds of alternate or substitute energy sources and the very pressing need for such sources to be found (researched and developed). That the need exists is eminently obvious. It was felt that the subjects of depleting energy resources and climate change deserve full attention from books dedicated specifically to them.

It is well understood that there is considerable effort directed toward alleviating many of the impacts described by industry, consumers, legislative bodies, the general public, and the professionals responsible for developing and implementing solutions. We wish to acknowledge these efforts and to remind all that much greater effort is needed. Finally, the first author wishes to acknowledge the very significant support and encouragement given by his wife, Florence, in this endeavor.

Raymond N. Yong
Catherine N. Mulligan
Masaharu Fukue
November 2005

About the Authors

Raymond N. Yong received his education in both the United States and Canada. He received his B.A. in math–physics from Washington and Jefferson College, his B.Sc. in civil engineering from the Massachusetts Institute of Technology, his M.Sc. in civil engineering from Purdue University, and his M.Eng. and Ph.D. from McGill University, Montreal, Canada.

He has authored and coauthored 7 other textbooks and more than 500 refereed papers in various journals. He also holds 52 patents. Of the many prizes he has received, notable among these are the Killam Prize from the Canada Council (Canada's highest scientific prize), the Legget Prize from the Canadian Geotechnical Society, the Dudley Award from ASTM, and the Canadian Achievement Award from Environment Canada.

Professor Yong is a fellow of the Royal Society of Canada, a chevalier of the Order of Québec, and a fellow of the Engineering Institute of Canada.

Catherine N. Mulligan has B.Eng. and M.Eng. degrees in chemical engineering from McGill University, and a Ph.D. specializing in geoenvironmental engineering, also from McGill University, Montreal, Canada. She has gained more than 20 years of research experience in government, industrial, and academic environments. She was a research associate for the Biotechnology Research Institute of the National Research Council and then worked as a research engineer for SNC Research Corp., a subsidiary of SNC-Lavalin, Montreal, Canada. She then joined Concordia University, Montreal, Canada, in the Department of Building, Civil and Environmental Engineering, where she currently holds the Concordia Research Chair in Environmental Engineering.

Professor Mulligan has taught courses in site remediation, environmental engineering, fate and transport of contaminants, and geoenvironmental engineering and conducts research in remediation of contaminated soils and water. She is the author of a textbook on biological treatment technologies for air, water, waste, and soil, coauthor of a book on natural attenuation of contaminants in soils, and has authored more than 45 refereed papers in various journals.

Professor Mulligan holds three patents. She is a member of the Order of Engineers of Québec, the Canadian Society of Chemical Engineering, the American Institute of Chemical Engineering, the American Society for Civil Engineering, the Air and Waste Management Association, the Association for the Environmental Health of Soils, the American Chemistry Society, the Canadian Society for Civil Engineering, and the Canadian Geotechnical Society.

Masaharu Fukue has B.Eng. and M.Eng. degrees in civil engineering from Tokai University, Japan, and a Ph.D. in geotechnical engineering from McGill University, Montreal, Canada. He joined a consultant firm for a short period after his Ph.D. and then joined Tokai University, where he is now the professor of marine science and technology.

Professor Fukue has given courses in geoenvironmental engineering, hydrospheric environment, shipboard oceanographic laboratory, and submarine geotechnology. He has invented a filtration system for seawater using a 2500-tonne barge, and has performed full-scale experiments in Kasaoka Bay, Japan, with this system for seawater purification. He has published more than 300 papers regarding the quality of seawater, sediments, and soils. He is also an advocate and a promoter of the Annual Symposium on Sea and Living Things and Rehabilitation of Coastal Environment, Japan, and is responsible for the organization of the Marine Geoenvironmental Research Association in Japan.

Professor Fukue is a member of the International Society for Soil Mechanics and Geotechnical Engineering, the International Society for Terrain-Vehicle System, the Japanese Society for Civil Engineers, the Japanese Geotechnical Society, and the Japan Society of Waste Management Experts. He has served as a chief editor for the *Japanese Standards for Soil Testing Methods* and for the *Japanese Standards for Geotechnical and Geoenvironmental Investigation Methods.* He has also served as a director of the standards division and is a member of the board of directors of the Japanese Geotechnical Society. He has been a member of the editorial board of the American journal *Marine Georesources and Geotechnology* since 1996, and is currently the co-chair of the Organizing Committee of the Third International Symposium on Contaminated Sediments, sponsored by the ASTM, ISCS2006-Shizuoka, Japan.

Contents

1

Sustainable Society and Geoenvironmental Management

1.1 Introduction

By all accounts, there is universal consensus that the stresses and demands imposed by society on the ecosphere have far surpassed the capability of the world's natural resources to regenerate and replenish themselves to meet sustainability requirements. It is contended that there are insufficient resources to meet the continued demands of the present world population, and that collapse of society will inevitably occur. Some of the major factors involved include: (1) loss of biological diversity, (2) increasing discharge of greenhouse gases to the atmosphere, (3) loss of soil quality, (4) increasing generation of wastes and pollutants, (5) depletion of nonrenewable natural resources such as fossil fuels, (6) increasing global population, and (7) depletion of natural living resources such as fisheries and forests. Considerable attention is being paid to many of these issues by researchers, policymakers, and other professionals well versed in engineering, scientific, and socioeconomic disciplines to alleviate the stresses to the geosphere and to seek sustainability and ways for society to live in harmony with the environment.

The primary focus of this book is on the geoenvironment and its importance as a resource base for life support systems. Mankind depends on the ecosphere, and in particular on the geoenvironment, to provide the raw materials to support life. Because of the many threats and negative impacts on the various life support systems within the geoenvironment, there is pressing need for one to: (1) develop a better appreciation of the stresses imposed on the geoenvironment by mankind and (2) determine the requirements for sensible and proper management of our environmental resources to achieve a sustainable society.

Many of the terms used in this book will have slightly different meanings depending on one's background, perspective, and scientific engineering discipline. It would be useful, at the outset, to define the terms that will be used in this book. The geoenvironment is a significant part of the ecosphere. Figure 1.1 shows the various components of the ecosphere and their relationship to the geoenvironment. The ecosphere shown in the simple schematic diagram in Figure 1.1 consists of the (1) atmosphere, (2) geosphere, which is also known as the lithosphere, (3) hydrosphere, and (4) biosphere. There are some that would add the anthrosphere as a significant component of the ecosphere. We will discuss this further in the next section.

Industrialization, urbanization, agriculture (food production), and natural resource exploitation (including energy) are basic activities associated with a living and vibrant society. We consider these basic elemental activities to be necessary to sustain life and also to be integral to *development*. In general terms, we consider development to (1) embody

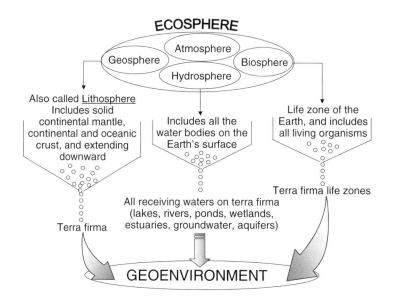

FIGURE 1.1
The various constituents of the ecosphere and their relationship to the geoenvironment.

the many sets of activities associated with the production of goods and services, (2) reflect the economic growth of a nation, state, city, or society in general, and (3) serve as an indication of the output or result of activities associated with these four main elemental activities. Questions often arise as to how these activities are compatible with, or are in conflict with, the aims of sustainability (of a society), and how one structures and manages programs and activities that would provide for a sustainable society.

1.1.1   Impacts on the Geoenvironment

Almost any physical event that happens in the ecosphere will have an impact on the geoenvironment. The question of whether these impacts will add value to or detract from the functionality of the geoenvironment and its ecosystems is among the issues at hand. It is difficult to catalog or list all of the impacts, not only because this cannot be done, but also because we need to arrive at some sets of criteria that will tell us *what constitutes an impact*. In many instances, we may not readily recognize or be aware of the impacts from many sets of activities or events — natural or man-made. To some extent, this is because (1) the effects of the impacts will not be immediately evident, as for example in the case of long-term health effects, and (2) the effects or results of the impacts cannot be recognized, i.e., we have yet to learn or recognize the results or effects of the impacts on the various biotic receptors and the environment.

1.1.1.1   *Geoenvironmental Impacts from Natural Events and Disasters*

The impacts on the geoenvironment from natural events (disasters), as for example earthquakes, tsunamis, hurricanes, typhoons, floods, and landslides, to name a few, are obvious — inasmuch as they are well reported in the daily newspapers. Recent events such as (1) the late December 2004 tsunami in Southeast Asia, (2) the various hurricanes in late summer 2005 that battered the gulf region of North America, (3) the typhoons and floods in East Asia and floods in Central America in the same period, and (4) the earthquake in October 2005 in the Pakistan–Afghanistan–India region are reminders that natural events can have considerable impact not only on the geoenvironment and its landscape, but also on human life and other life-forms (domestic animals and other wildlife).

Some of the dramatic impacts on the geoenvironmental landscape due to these natural catastrophic events include (1) collapse of man-made and natural structures and other infrastructure facilities, such as roads, pipelines, transmission towers, etc., (2) floods, (3) landslides, and (4) fires. Displacement of thousands and even millions of people due to loss of dwellings, and loss of life due to collapsing structures, floods and landslides, and ingestion of polluted waters are some of the impacts to humans. All of these events and their impacts on human life and other life-forms, and local and global societal response to such events, merit serious and proper consideration and attention in books and treatises devoted to the various aspects of these catastrophic events. They are not within the purview of this book. What is of direct concern in this book is the impact of anthropogenic activities mounted in support of production of goods and services to serve the needs of society.

1.1.1.2   Anthropogenic Forces and Impacts on Geoenvironment

The intent of this chapter is to provide an introduction to many of the basic issues that arise from impacts and assaults on the geoenvironment as a result of anthropogenic activities associated with the production of goods and services — in support of the focus of this book. We consider the geoenvironment to consist of the terrestrial (land surface) ecosystem — including the aquatic ecosystems contained within and contiguous to the land mass. Many of the driving forces responsible for these impacts and assaults can be attributed to actions of *production* and *implementation* technology. A large proportion of these actions are not willful. The negative impact on the environment and the geoenvironment in particular, resulting from many of these actions, are the result of a lack of appreciation of the fragility of the environmental systems. A pertinent example of this is the presence of historic and orphan toxic and nontoxic waste-polluted sites populating the land surface in many parts of the world. These are the legacy of our historic lack of appreciation of the damage done to the land environment by the many activities in support of production of *goods and services*. These kinds of goods and activities are necessary items in support of industrial development and a vibrant society. Not all of the kinds or types of natural and man-made impacts on the geoenvironment resulting from these activities can be considered. The geoenvironmental impacts that are health threatening constitute the major focus of this book. By and large, these (impacts) result from discharges from industrial operations and urban activities. A more detailed description and discussion of the anthropogenically derived impacts on the geoenvironment will be considered in the next chapter.

1.2   Geoenvironment, Ecosystems, and Resources

The *geoenvironment* is a specific compartment of the environment and, as such, concerns itself with the various elements and interactions occurring in the domain defined by the dry solid land mass identified as terra firma. These include a significant portion of the geosphere and portions of both the hydrosphere and the biosphere.

- We consider the *geosphere* to include the inorganic mantle and crust of the earth, including the land mass and the oceanic crust. Also included in this category are the solid layers (soil and rock mass) stretching downward from the mantle and crust. One could say that one part of the geoenvironment is the *terra firma* component of the geosphere, as seen in Figure 1.1.

- The *hydrosphere* refers to all the forms of water on earth — i.e., oceans, rivers, lakes, ponds, wetlands, estuaries, inlets, aquifers, groundwater, coastal waters, snow, ice, etc. The geoenvironment includes all the receiving waters contained within the terra firma in the hydrosphere. This excludes oceans and seas, but is meant to include rivers, lakes, ponds, inlets, wetlands, estuaries, coastal marine waters, groundwater, and aquifers. The inclusion of coastal (marine) environment in the geoenvironment is predicated on the fact that these waters are affected by the discharge of pollutants in the coastal regions via runoffs on land, and discharge of polluted waters from rivers or streams. Chapter 8 examines sustainability issues of the coastal marine environment in greater detail.

- The *biosphere* is the zone that includes all living organisms, and the *environment* is the biophysical system wherein all the biotic and abiotic organisms in the geosphere, hydrosphere, and atmosphere interact. The geoenvironment includes the life zones in or on terra firma from the biosphere.

The term *ecosystem* refers to a system where the various individual elements and organisms interact singly or collectively to the advantage or detriment of the whole. The relationships formed between these elements and organisms can be symbiotic or antagonistic. For an ecosystem to be self-sustaining, the relationships between the communities in the ecosystem need to be symbiotic, and furthermore, the interactions must be mutually beneficial. The various ecosystems that exist in the geoenvironment have functions, uses, resources, and habitats that are crucial to the production of goods and the means for ensuring life support.

The biological component within the ecosystems that comprise the terrestrial ecosystem do not fall within the purview of this book, except insofar as they contribute to the persistence, transformation, and fate of pollutants in the ground. In particular, we will be concentrating on the land aspects of the terrestrial ecosystem, and more specifically with the land surface (landscape) and subsurface systems.

1.2.1 Ecozones and Ecosystems

Ecology is generally defined as the study of the relationship between living and nonliving organisms and their environment. The study of these relationships is facilitated by establishing ecozones or ecosystems. Strictly speaking, *ecozones* are zones that are delineated according to some set of established ecological characteristics. They are essentially basic units of the land or marine environment that are distinctly characterized by the living and nonliving organisms within that region. Ecozones are geographical units that are usually several hundreds of square kilometers in spatial extent The ecosystem or ecosystems bounded or resident within the ecozone deals with the mutual interactions between the living and nonliving organisms in this zone. An *ecosystem* is defined herein as a discrete system that (1) contains all physical (i.e., material) entities and biological organisms and (2) includes all the results or products of the interactions and processes of all the entities and organisms in this system. With this classification scheme, one can distinguish between the two primary ecosystems constituting the ecosphere, namely, the land and aquatic ecosystems. The delineation of ecozones and ecosystems is somewhat arbitrary and can be performed or undertaken according to several guidelines. The boundaries demarking the ecozones are not fixed.

Classification and characterization of these land ecosystems can be performed according to various standards or guidelines. Classification according to the physiographic nature of the land is one of the more popular schemes available. Under such a scheme, one therefore has such ecosystems as alpine, desert, plains, coastal, arctic, boreal, prairie, etc. Another popular scheme for classification is the resource-based method of classification. This approach is based on identifying the sets of activities or the nature of the primary or significant resources constituting the specific land environment under consideration — such as agro-ecosystem and forest ecosystem. Within each ecosystem, there exist numerous elements and activities that can be examined and documented in respect to *before* and *after* ecosystem impact. One of the primary reasons for classification of any of the ecosystems is to define, bound, or document the *sphere of influence or examination wherein all the elements of the ecosystem interact and are dependent on the welfare of each individual element for the overall state, benefit, and function of the ecosystem.*

1.2.2   Natural Resources and Biodiversity in the Geoenvironment

Natural resources in the geoenvironment are defined as commodities that have intrinsic value in their natural state. They are the natural capital of the geoenvironment. In describing the natural resources in the geoenvironment, the most obvious ones are often cited immediately. These include water, forests, minerals, coal and hydrocarbon resources (oil, gas, and tar sands, for example), and soil. Other not so obvious resources are the developed resources, such as agricultural products, and alternative energy generation (resources), such as solar, geothermal, wind, tidal, and nuclear.

One natural resource that is often overlooked is biological diversity (biodiversity). This is one of the most significant of the natural resources in the ecosystem. We use the term *biodiversity* to mean the diversity of living organisms such as plants, animals, and microbial species in a specific ecosystem. They play significant roles in the development of the many resources that we have identified above, through mediation of the flow of energy (photosynthesis) and materials such as carbon, nitrogen, and phosphorous. According to Naeem et al. (1999), ecosystems consist of plants, animals, and microbes and their associated activities, the results of which impact on their immediate environment. They point out that a functioning ecosystem is one that exhibits biological and chemical activities characteristic for its type, and give the example of a functioning forest ecosystem that exhibits rates of plant production, carbon storage, and nutrient cycling that are characteristic of most forests. It follows that if the trees in the forest are harvested or if the forest ecosystem is converted to another type of ecosystem, the specific characteristics of a functioning forest ecosystem will no longer exist.

For the purpose of this book, we define a *functioning ecosystem* to include not only the biological and chemical activities, but also the physicochemical activities and physical interactions characteristic of the type of ecosystem under consideration. Trevors (2003) has enumerated a noteworthy list for consideration in respect to the role of biodiversity as part of our life support system. Included in the detailed list are such considerations as:

- Maintenance of atmospheric composition and especially the production of oxygen by photosynthesis and the fixation of carbon dioxide
- Water cycle via evaporation and plant transpiration
- Interconnected nutrient cycles (e.g., C, N, P, S) driven by microorganisms in soils, sediments, and aquatic environments
- Carbon sources and sinks

- Pollination of agricultural crops and wild plants
- Natural biocontrol agents, as for example in microbial degradation of pollutants in soil, water, sediments, wastewater, and sewage treatment facilities

1.3   Geoenvironment as a Natural Resource Base

Figure 1.2 shows a schematic of the various elements and interactions that contribute to the well-being of a sustainable society. The three basic components shown at the top of the schematic (energy, water and soil, and natural resources) are primary resource constituents of the geoenvironment. Figure 1.3 shows some of the primary natural resources in combination with various land and aquatic ecosystems. The health and accessibility of these are essential rudiments of life support systems. The beneficial interaction between all the elements shown in Figure 1.3 is needed to produce the necessary ingredients required for production of goods and services to sustain the population in a society. The ultimate goal for all of humankind is to obtain both a sustainable society and sustainable development. It therefore follows that until a sustainable society is obtained, sustainable development will not be achieved. The term *sustainable development* is used herein to mean that *all the activities associated with development in support of human needs and aspirations must not compromise or reduce the chances of future generations to exploit the same resource base to obtain similar or greater levels of yield.* There are some who contend that since the beginning of the industrial revolution, "mankind's occupation of this planet has been markedly unsustainable" (Glasby, 2002), and that the concept of sustainable development as defined in the World Commission on Environment and Development 1987 report is a chimera. There is very little doubt that with the present rate of exploitation of the renewable and nonrenewable resources, sustainable development per se is an illusion — a goal that is not sustainable as long as depletion of nonrenewable resources occurs, and as long as excessive or renewable resources outstrip the replenishment rate of these resources. Many of these issues will be addressed in the other chapters of this book from a geoenvironmental perspective.

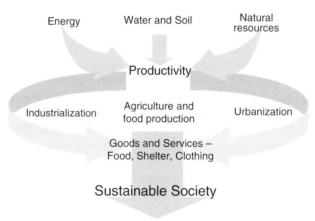

FIGURE 1.2
Basic elements and interactions contributing to a sustainable society and to sustainable development.

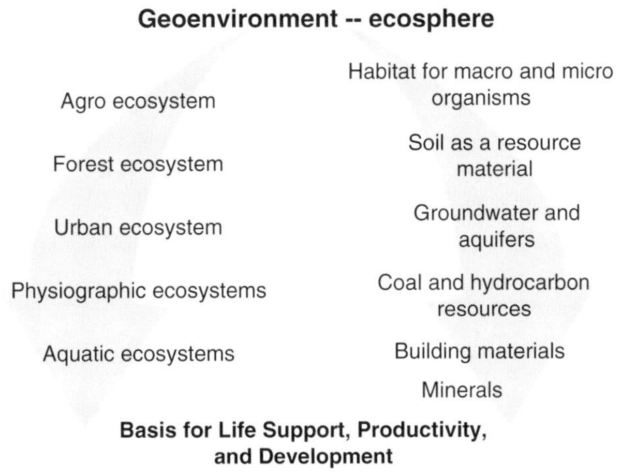

Geoenvironment -- ecosphere

	Habitat for macro and micro organisms
Agro ecosystem	
	Soil as a resource material
Forest ecosystem	
	Groundwater and aquifers
Urban ecosystem	
	Coal and hydrocarbon resources
Physiographic ecosystems	
	Building materials
Aquatic ecosystems	
	Minerals

Basis for Life Support, Productivity, and Development

FIGURE 1.3
Some of the major ecosystems, resources, and features of the geoenvironment. Note that the physiographic ecosystems include, for example, coastal, alpine, desert, and arctic ecosystems.

Specifically, recognizing that the land environment houses the terrestrial and aquatic ecosystems, and that these are the fundamental components in the engine responsible for life support, one needs to determine:

- Whether the economic growth and prosperity obtained as a result of all these activities contribute to the production of beneficial or detrimental impacts on the land environment
- Whether such impacts improve, increase, reduce, or degrade the functional capability of the land (terrestrial) and aquatic ecosystems that comprise the land environment

It is often argued that the technological tools used in implementation of development, and to provide for advances in development, are quite often in conflict with the ideals of sound environmental management. The accusation is frequently made that production technology is not always utilized in the interest of preservation of the natural environment. Historic examples of this are evident in the resource exploitation field, e.g., mining and forestry industries. Evidences of environmental mismanagement from past activities associated with development abound. Commoner (1971) has suggested, for example, that the wealth gained by the modern technology-based society has been obtained by short-term exploitation of the environmental system. While there is a just basis for this observation, we should note that present-day practices in most instances are supposed to be conducted in accord with strict environment protection guidelines. That being said, it needs to be understood that much remains to be done in determining the many factors and facets that go to make up environment protection. More specifically, while protection of the geoenvironment may lead to a preservation of many of the features and assets of the geoenvironment, it is clear from the various resource and habitat features shown in the right-hand compartment of Figure 1.3 that protection of the geoenvironment does not necessarily lead to sustainability of the geoenvironment. We recognize that techniques, procedures, and management programs should be structured to permit exploitation of the geoenvironment to occur with minimal negative impact on the geoenvironment. To that end, the intent of this book is to provide one with an appreciation and understanding of (1) the geoenvironmental impacts that result from activities associated with the means to maintain a sustainable society and implementation of *sustainable development* and (2) measures, requirements, and procedures needed to avoid, minimize, or mitigate these impacts.

We have previously defined the *geoenvironment* to include the land environment. This includes all the geophysical (geological and geomorphological) features, together with the aquatic elements classified as receiving waters. Figure 1.3 shows a simple schematic of the various major ecosystems, habitats, and resources that constitute the geoenvironment. The major ecosystems that constitute the land ecosystem, shown in the left compartment of Figure 1.3, consist of a mixture of physiographic and resource-type ecosystems. The mixed method of presentation of ecosystems shown in the diagram has been chosen deliberately because it does not require one to detail every single physiographic unit and every single resource-type unit.

The geoenvironment contains all the elements that are vital for the sustenance and well-being of the human population. Commoner (1971) states that the ecosphere, together with the earth's mineral resources, is the source of all goods produced by human labor or wealth. It is obvious that any degradation of the ecosphere will have a negative impact on the capability of the ecosphere to provide the various goods produced by human labor or wealth.

1.3.1 Survival and Limits to Population Growth

The posit that the geoenvironment is in itself a natural resource is founded on the fact that it provides the various elements necessary for life support, such as food, energy, and resources. Degradation of any of the physical and biogeochemical features that permit life support systems to function well will be a detriment to the requirements for a sustainable society. The Malthusian model (Malthus, 1798), for example, links availability of food with population growth (or reduction). Postulating that food is necessary for human existence, and that rate of food production increases linearly, the Malthusian model contends that since the rate of human population increase is geometric, there will come a time when food production will not be sufficient to meet population needs. While no account was given to availability of resources and industrial output in the original model, one presumes that these were accounted for in the *ceretis partibus* condition.

Almost 180 years after Malthus's publication, Meadows et al. (1972) utilized a systems dynamics model that considered five specific quantities: industrialization, population, food production, pollution, and consumption of nonrenewable natural resources. One of the conclusions of the Meadows et al. (1972) report on the phase I study entitled "The Project on the Predicament of Mankind" was that "if present growth trends in world population, industrialization, pollution, food production, and resource depletion continue unchanged, the limits to growth on this planet will be reached sometime within the next hundred years." The report also concludes that "it is possible to alter these growth trends and to establish a condition of ecological and economic stability that is sustainable far into the future."

While this initial report created a maelstrom of discussion, there has been almost unanimous agreement in the relevance of the primary message. There have been some that have argued that the conclusions are perhaps too pessimistic, since the system dynamics model used in the Meadows et al. (1972) report, by their own admission, was, at the time of publication of the report, somewhat "imperfect, oversimplified and unfinished." However, it cannot be denied that the essence of the model and the analyses were fundamentally sound. The five specific quantities examined in the model were deemed significant in view of the global concern on (Meadows et al., 1972) "accelerating industrialization, rapid population growth, widespread malnutrition, depletion of non-renewable resources, and a deteriorating environment."

The subsequent analyses by Meadows et al. (1992) using data gained from the 20-year period following their first publication showed that "in spite of the world's improved technologies, the greater awareness, the stronger environment policies, many resource and pollution flows had grown beyond their sustainable limits." They concluded that not only were the initial conclusions in Meadows et al. (1972) valid, but that these conclusions needed to be strengthened. In particular, their first conclusion reflects the growing concern we have regarding the capability of our geoenvironmental resources to provide the long-term necessities to sustain life at the pace we now enjoy. "Human use of many essential resources and generation of many kinds of pollutants have already surpassed rates that are physically sustainable. Without significant reductions in material and energy flows, there will be in the coming decades an uncontrolled decline in per capita food output, energy use, and industrial production" (Meadows et al., 1992). Glasby (2002) has offered the suggestion that a marked decrease in world population (to 1.2 billion) is needed if sustainability is to be achieved.

Arguments against consideration of the environment, and specifically the geoenviron-ment, as a limited natural resource are generally based on a very limited appreciation of the totality of the geoenvironment as an ecosphere, and also on negligent attention to the many negative impacts attributable to anthropogenic activities. Not all the geoenviron-mental resources are nonrenewable (e.g., forest resource). However, for those resources that are renewable, overuse or overexploitation will surpass their recharge rate, thus creating a negative imbalance. Is geoenvironmental deterioration a threat to human sur-vival? Commoner (1971) has examined the overall environmental problem and has posed the question in terms of ecological stresses: "Are present ecological stresses so strong that — if not relieved — they will sufficiently degrade the ecosystem to make the earth uninhabitable by man?" His judgment: "Based on the evidence now on hand, … the present course of environmental degradation, at least in industrialized countries, repre-sents a challenge to essential ecological systems that is so serious that, if continued, it will destroy the capability of the environment to support a reasonably civilized human society."

1.3.2 Geoenvironmental Crisis and Development

The *declaration* issued at the beginning of the U.S. National Environmental Policy Act (NEPA) of 1969 recognizes "the profound impact of man's activity on the interrelations of all components of the natural environment, particularly the profound influences of population growth, high-density urbanization, industrial expansion, resource exploita-tion, and new and expanding technological advances," and further recognizes "the critical importance of restoring and maintaining environmental quality to the overall welfare and development of man." The significance of this declaration cannot be over-looked. We can easily appreciate the need for environment protection and geoenviron-mental sustainability.

The report issued by the World Commission on Environment and Development (WCED) in 1987, commonly referred to as the Brundtland report, listed two major challenges as follows:

- To propose long-term environmental strategies for achieving sustainable devel-opment by the year 2000 and beyond;
- To define shared perceptions of long-term environmental issues and appropriate efforts needed to deal successfully with the problems of protecting and enhancing the environment.

The concept and definition of *sustainable development* differs considerably between individuals and specific interest groups. The WCED report states that "humanity has the ability to make development sustainable — to ensure that it meets the needs of the present without compromising the ability of future generations to meet their own needs." Without entering into a protracted debate about specification of indicators for determination of sustainability, and whether sustainability can ever be achieved when nonrenewable resources are being depleted, we can agree with the general ideas embedded in the WCED statement on sustainable development. Following from this, it is evident that it is proper to view sustainable development as a process and as a path toward *sustainability* — where sustainability in itself may be an elusive goal. From the geoenvironmental perspective, it is recognized that the needs of future generations should not be compromised by the environmental crisis brought about by the stresses resulting from the elemental activities described in the beginning of this chapter. The geoenvironment is the resource base that serves as the engine that provides for the various elements necessary for human sustenance. Through resource exploitation and industrial activities, it is the source for everything that is necessary for the production of food, shelter, and clothing. It is also the habitat for various land and aquatic biota. Impacts to the geoenvironment and its ecosystem need to be minimized and mitigated if one wishes to undertake the necessary steps toward sustainable development. Management of the geoenvironment is required if a sustainable geoenvironment is to be obtained — for without a sustainable geoenvironment, sustainable development will not be realized.

1.3.3   Sustainable Development and the Geoenvironment

The sets of forces needed to sustain a forceful economic climate and provide for a dynamic population base or population growth can be gathered into two main groups. These are defined by some very clear factors:

- Urban–industrial: This grouping includes those efforts and industries associated with the production of food, shelter, clothing, and economic health.
- Socioeconomic–political: The grouping of factors that include the social, economic, and political dimensions of a society.

The subjects covered in this book will deal with geoenvironmental stresses resulting from the various activities associated with a sustainable society and development. The framework within which these will be examined will be confined to the one determined by the urban–industrial factors defined above. This by no means diminishes the significance of the socioeconomic–political grouping of factors. These are important, but are not within the purview of this book.

It is well recognized that to meet the present and future needs of the ever-increasing global population, there will be increased requirements for adequate supplies of goods and services to feed, shelter, and clothe the population. The 27 principles articulated in the 1992 Rio Declaration show the need for protection and maintenance of environmental quality while meeting the needs of the global population. These were reinforced in the 2002 World Summit on Sustainable Development (WSSD) held in Johannesburg. Principles 1, 3, and 4 of the Rio Declaration state that:

- Human beings are at the center of concerns for sustainable development. They are entitled to a healthy and productive life nature. (Principle 1)

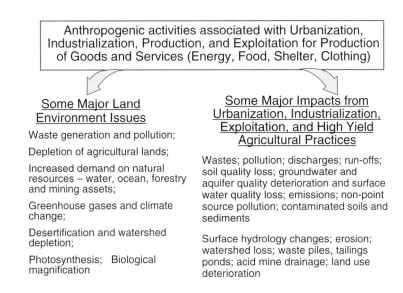

FIGURE 1.4
Some major land environment issues and impacts resulting from activities associated with urbanization, industrialization, production, and exploitation for production of goods and services. (Adapted from Yong, R.N. and Mulligan, C.N., *Natural Attenuation of Contaminants in Soils*, CRC Press, Boca Raton, FL, 2004.)

- The right to development must be fulfilled so as to equitably meet developmental and environmental needs of present and future generations. (Principle 3)
- In order to achieve sustainable development, environmental protection shall constitute an integral part of the development process and cannot be considered in isolation from it. (Principle 4)

There are many who maintain that such declarative statements are designed to show a measure of political will, and are not necessarily practical or supportable from a scientific point of view — given that development at the expense of depletion of scarce resources will never be sustainable. While this book is not the forum for a debate on whether the goals identified by the declarative statements issued by world bodies can be successfully implemented or met, it is nevertheless necessary to take guidance from such statements. By highlighting the need for development in a sustainable manner, in both the WSSD (2002) and the Rio Summit declarative statements of principles, it follows that development of the necessary knowledge and tools to address the goals of sustainability is required. From the geoenvironmental point of view, this means dealing with the impacts to the geoenvironment from developmental activities. Yong and Mulligan (2004) show that to properly address the problems and issues connected with degradation of the land environment, a knowledge of the linkages, interactions, and impacts between the human population and a healthy, robust, and sustainable land environment is required (Figure 1.4). The figure shows the linkages and identifies some of the major issues and land environment impacts. The observations made by Yong and Mulligan (2004) regarding the major land environment or geoenvironmental issues shown in the figure are cited directly as follows:

- Waste generation and pollution: Wastes generated from the various activities associated with resource exploitation, energy production, and industry associated with the production of goods and services will ultimately find their way into one or all of three disposal media; (a) receiving waters, (b) atmosphere, and (c) land. Land disposal of waste products and waste streams appears to be the most popular method for waste containment and management. The various impacts arising from this mode of disposal and containment include degradation of land surface environment and ground contamination by pollutants.

- Depletion of agricultural lands and loss of soil quality: This will arise because of increased urbanization and industrialization pressures, infrastructure development, exploitation of natural resources, and use of intensive agricultural practices. The loss of agricultural lands places greater emphasis and requirement on higher productivity per unit of agricultural land. The end result of this is the development of high-yield agricultural practices. One of the notable effects is soil quality loss. To combat this, there is an inclination to use pesticides, insecticides, fertilizers, more soil amendments and other means to enhance productivity and yield. A resultant land environment impact from such practices is pollution of the ground, groundwater and receiving waters from runoffs and transport of pollutants.

- Increased demand on natural resources and depletion of natural capital: The various issues related with all the generated exploitation activities fall into the categories of: (a) land and surface degradation associated with energy production, mining and forestry activities, and (b) water supply, delivery, and utilization. Surface hydrology changes, erosion, watershed loss, tailings and sludge ponds, acid mine drainage, etc. are some of the many land environment impacts.

- Greenhouse gases, climate change, desertification: To a very large extent, these are consequences of industrialization, urbanization, and production. Their impact on the land environment can be felt for example in acid rain (and snow) interaction with soil and undesirable changes in photosynthesis processes and erosion of coastal areas due to increasing water levels.

- Photosynthesis; biological magnification: The processes associated with photosynthesis are important since they in essence constitute about 20% of the available oxygen in the atmosphere. Desertification, deforestation, and many of the activities associated with mineral and other natural resource exploitations will degrade the capability of the various participants (land and aquatic plants) to engage in these processes. Biological magnification, which concerns itself with the concentration of various toxic elements or pollutants by plants and such biotic receptors as aquatic organisms and animals, is a problem that needs to be addressed in the containment and management of pollutants.

1.4   WEHAB and the Geoenvironment

The Johannesburg World Summit on Sustainable Development (WSSD 2002) identified five priority theme areas that needed to be addressed. These five *thematic areas*, as follows, have the acronym of WEHAB: (1) water and sanitation, (2) energy, (3) health, (4) agriculture, and (5) biodiversity.

The two main sets of forces that, by their interactions, pose potential threats to the geoenvironment are those generated by the activities associated with the following:

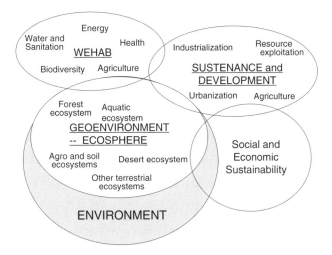

FIGURE 1.5
Links, interactions, and interrelationships between WEHAB, sustainable development, and the geoenvironment.
Note that social and economic sustainability issues cannot be ignored in all determinations and considerations.

1. Sustenance and development, i.e., the four primary components: industrialization, urbanization, resource exploitation (including energy), and agriculture (and food production)
2. Implementation of schemes, projects, and technology to meet the priorities and requirements of WEHAB

Figure 1.5 shows the interactions and interrelationships between development, WEHAB, and the geoenvironment. The net effect or impact of the activities and the stresses on the geoenvironment can be disastrous if measures for protection of the geoenvironment are not put into practice. A reduction or degradation of geoenvironmental resources and the various ecosystems will diminish the capability of the geoenvironment to provide the elements necessary to sustain life. To avoid or minimize the degradations, there needs to be (1) a proper audit of the geoenvironmental impacts on the prominent features that constitute the physical component of the geoenvironment, and also on the ecosystem, and (2) available sensible and logical sets of tools that can be used to execute the principles of sustainability.

1.4.1 Major Geoenvironment Impact Sources

How the various industries (life supporting and manufacturing production) and their associated activities interact with the geoenvironment can be viewed as follows:

1. Resource extraction and processing: The various industries included in this group use the geoenvironment as a resource pool containing materials and substances that can be extracted and processed as value-added products. The common characteristic of the industries in this group is *processing of material extracted from the ground.* Included in this group are (a) the metalliferous mining industries; (b) those industries involved in extraction and processing of other resources from the ground, such as nonmetallic minerals (potash, refractory and clay minerals, phosphates); (c) the industries devoted to extraction of aggregates, sand, and rock for

production of building materials; and (d) the raw energy industries, such as those involved in the extraction of hydrocarbon-associated materials and other fossil fuels (natural gas, oil, tar sands, and coal). Included in this list is the extraction and recovery of uranium for the nuclear power industry.

2. Utilization of land and soil as a resource material in aid of production: Essentially, this group includes the agro- and forest industries, and also the previously mentioned nonmetallic minerals industries.

3. Water, groundwater, and aquifer harvesting: We include the hydroelectric facilities and industries associated with extraction and utilization of groundwater and aquifers.

4. Use of land as a facility: This category considers land as a facility for use, for example, in the land disposal of waste products. Broadly speaking, we can consider the land surface environment here as a resource for treatment and containment of waste products generated by all the industries populating the previous three categories.

Some of the major negative or degradative geoenvironmental impacts resulting from the various activities associated with production technology (e.g., agriculture, forestry, mining, energy, and general production) are shown in Figure 1.6. The nature of the threats to the land environment and the waste streams is shown in Figure 1.7. These affect both soil and water quality. We will discuss the nature of some of the impacts in greater detail in the next few chapters. The diagrams show the nature of the threats originating from the source activities and their immediate physical impact on the land environment. The bottom-most element in Figure 1.7 shows some of the required sets of action for reduction of threats, such as pollution management and toxicity and concentration reduction. Note that problems such as habitat protection and impacts, and air quality are not considered since the attention in this book is focused on the physical land environment itself. In that sense, for the problems and activities shown in Figure 1.6, pollution management and

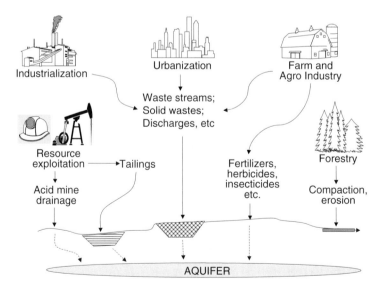

FIGURE 1.6
Nature of geoenvironmental impacts resulting from activities associated with industrialization, urbanization, resource exploitation, farm and agro-industries, and forest harvesting

Industrialization, Urbanization, Resource Exploitation	Agricultural Activities
Waste streams, Waste containment systems, Emissions; Discharges; Tailings ponds; Dams, Landfills; Barrier systems; Liners	Farm wastes, Soil erosion, Compaction, Organic matter loss, Nitrification, Fertilizers, Insecticides; Pesticides, Non-point source pollution

Soil and Water Quality, and Threat Management

Point and non-point source pollution; Aquifer, Groundwater, Surface Water, Watershed, Receiving Waters e.g. lakes, ponds, rivers, streams, etc.

Site Contamination, Management, and Remediation

Soil and sediment contamination; Pollution management and control; Toxicity reduction; Concentration reduction; Site remediation and technology; Land suitability; restoration and rehabilitation; Threat reduction and curtailment

FIGURE 1.7
Threats and waste streams impacting on soil and water quality.

control will have to be exercised to minimize damage to the geoenvironment, which in most cases refers to the surface environment and the receiving surface water and groundwater. Since many of the sources of the impacts cannot be completely reduced to zero, impact management will have to be practiced.

1.4.2 Geoenvironmental Impacts and Land Management

Good land management practice (1) minimizes and mitigates deleterious impacts to the land environment, (2) seeks optimal land use and benefit from the land, and (3) preserves and minimizes depletion of natural capital. The obvious threat to human health linked to detrimental geoenvironmental impacts comes from waste discharge and impoundment, as shown in Figure 1.6 and Figure 1.7. Most of the activities associated with *development* will generate waste in one form or another. Table 1.1 shows some typical waste streams from a representative group of industries.

TABLE 1.1

Typical Composition of Waste Streams from Some Representative Industries

Industry	Waste Streams
Laboratories	Acids, bases, heavy metals, inorganics, ignitable wastes, solvents
Printing, etc.	Acids, bases, heavy metals, inorganic wastes, solvents, ink sludges, spent plating
Pesticide user and services	Metals, inorganics, pesticides, solvents
Construction	Acids, bases, ignitable wastes, solvents
Metal manufacture	Acids, bases, cyanide wastes, reactives, heavy metals, ignitable wastes, solvents, spent plating wastes
Formulators	Acids, bases, ignitable wastes, heavy metals, inorganics, pesticides, reactives, solvents
Chemical manufacture	Same as metal manufacture, except no plating wastes
Laundry/dry cleaning	Dry clean filtration residue, solvents

In terms of geoenvironmental impacts, and in respect to WEHAB, the aims and objectives identified in the priority theme of *water* and *health* are most seriously threatened by (1) wastewater and solid waste discharge, and spills, leaks, and other forms of discharges to the land environment, (2) containment of wastes and other hazardous materials in the land environment, and (3) use of chemical aids in pest control and other agricultural activities. Because of the use of fertilizers, herbicides, insecticides, pesticides, and fungicides in the agro-industry, several conflicting issues can arise if the aims of the WEHAB priority theme area of *agriculture (food production)* are to be fulfilled. The most obvious threat to the geoenvironment from the agro-industry comes from (1) application of pesticides and their like, resulting in pollution of the receiving waters, and (2) use of intensive agricultural practices, resulting in excessive nitrogen and phosphate loading of the soil and a consequent decrease in soil quality. Many of these issues will be discussed in detail in Chapter 6.

1.4.3   Impact on Water and Water Resources

It has been suggested by many that in the future, conflicts among various groups, factions, and nations will arise over drinking water and its availability. We need only consider the availability and distribution of drinking water in the world to see that this suggestion has substance, as shown in Figure 1.8. Less than 5% of the global water is nonsaline water. Of this less than 5% nonsaline water, it is estimated that about 0.2% of the nonsaline water is contained in lakes and rivers, with the remaining proportion existing as snow, ice, wetlands, and groundwater (adapted from Yong, 2001). Values reported much earlier by Leopold (1974) give numbers such as 2.7% of total volume of water (i.e., global water) as freshwater, and of that freshwater, it was estimated by Leopold that about 0.36% was "easily accessible."

Some of the more common and significant impacts to the quality of groundwater and receiving waters have been shown in Figure 1.6 and Figure 1.7. Deterioration of the quality of these waters will not only limit their usefulness, but will also cause distress to the

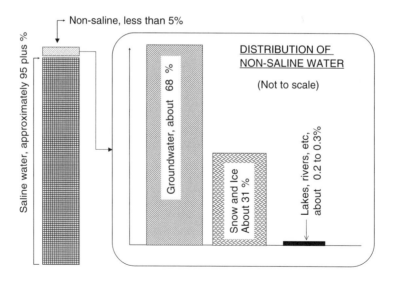

FIGURE 1.8
Distribution of global water. Note that of the less than 5% nonsaline water, about 68% exists in the ground as groundwater.

TABLE 1.2

Chronic Effect of Some Hazardous Wastes on Human Health

Waste Type	Carcinogenic	Mutagenic	Teratogenic	Reproductive System Damage
Halogenated organic pesticide	A	A	A	H
Methyl bromide			H	
Halogenated organic phenoxy herbicide	A	A	A	A
2,4-D[a]				
Organophosphorous pesticide	A	A	A	
Organonitrogen herbicide	A	A	A	
Polychlorinated byphenyls	A		A	
Cyanide wastes[a]				
Halogenated organics	H	H		
Nonhalogenated volatile organics	A	A		
Zn, Cu, Se, Cr, Ni	H			
Hg		H	H	
Cd	H			

Note: H, A = statistically verifiable effects on humans and animals, respectively.

[a] No reportable information available.

Source: Adapted from Governor's Office of Appropriate Technology, Toxic Waste Assessment Group, California, 1981.

animal and plant species that live in these waters. Considering that at least one half or more of the world's plant and animal species live in water, it is clear that any deterioration or decrease of water quality and water availability will have severe consequences on these species. Protection of both surface water and groundwater must be a priority. Chapter 3 discusses these and other issues in greater detail. Water usage by industry, for example, can produce liquid waste streams that are highly toxic by virtue of the chemicals contained in the waste streams, or by virtue of concentration of noxious substances. Before the liquid phase of any waste stream can be returned to the environment, it has to be treated and rendered harmless as a health threat to biotic receptors. As indicated previously, the source of these pollutants can be traced to waste streams and discharges from industrial plants, households, resource exploitation facilities, and farms. Table 1.2 shows some chronic effects from some of these waste products on human health.

Farming and agricultural activities contribute agro-additives to the receiving waters and groundwater through surface runoff and through transport in the ground (Figure 1.6 and Figure 1.7). All the other discharges and waste streams shown in the two figures are most likely contained in storage dumps, landfills, holding ponds, tailings ponds, or other similar systems. All of these containment systems have the potential to deliver pollutants to the receiving waters (ground and surface waters) because of eventual and inevitable leaks, discharges, and failures. Many of these phenomena will be discussed in greater detail in the later chapters of this book.

We highlight the importance of groundwater resources because it is a major water resource (Figure 1.8), and as a rule, groundwater is more accessible than surface water. Furthermore, it is not uncommon for many rural communities to rely heavily on these groundwater resources as a source without proper treatment prior to use. We should note that contamination of receiving waters such as ponds, lakes, and rivers occurs also through leachate transport through the soil and, quite obviously, from surface runoff from point and nonpoint pollutant sources. From the perspective of the geoenvironment, protection of both surface water and groundwater quality requires one to practice impact mitigation and management, shown, for example, in Figure 1.9 for management of liquid waste discharge into the environment to avoid affecting the receiving waters. The decision points

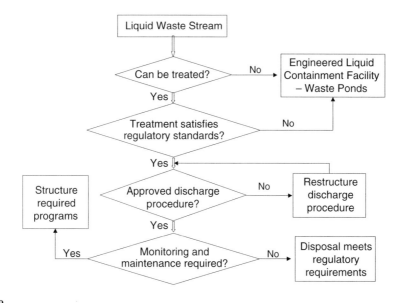

FIGURE 1.9
Program for management of liquid waste stream discharge into receiving waters and land surface areas.

shown in the protocol diagram include criteria, procedures, tests, etc., that need to be conducted to satisfy regulatory requirements.

To provide proper protection of the health of biotic receptors in the geoenvironment, treatment of the liquid waste streams requires detoxification and removal of all toxic and hazardous constituents and suspended solids before discharge. Reuse of the treated waste streams is encouraged. Typical reuse schemes include irrigation (in farming and agriculture activities), process streams (such as resource extraction), and cooling towers. Waste streams that cannot be treated effectively and economically to reach acceptable discharge standards will require impoundment in secure ponds. Procedures have been developed that will reduce the liquid content of these noxious liquid waste streams. To protect the resources in the geoenvironment, the product(s) will most likely need to be incinerated or contained in secure impoundment facilities. Typical containment and impoundment facilities would be landfills. Co-disposal of these kinds of waste products with other types of waste products has been proposed as a means to accommodate these waste products. Figure 1.10 shows two typical barrier systems used to line landfill facilities. The details for these kinds of containment and waste management systems will be discussed in the next chapter.

1.5   Sustainability

In a Gaean world, the earth is a living being where all the living organisms and nonliving entities function and interact independently, but contribute to collectively define and regulate the material conditions necessary for life. When stresses and resultant negative impacts associated with the activities of mankind in the production of *goods and services* arise, the Gaean hypothesis becomes somewhat untenable. This is because living matter in the geoenvironment will be constrained from regulating the material conditions necessary for life. Loss of species diversity is one of the major factors. The paramount terrestrial

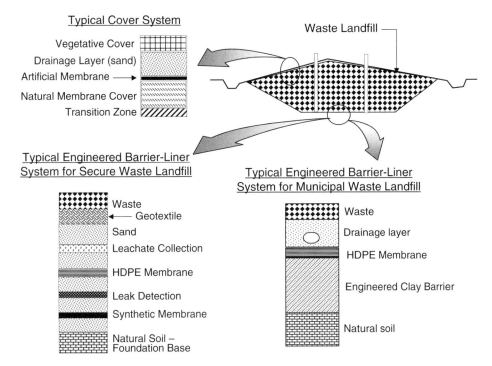

FIGURE 1.10
Waste landfill system showing typical top cover and bottom barrier–liner systems. Two liner systems are shown: the maximum security barrier system (left) and the bottom barrier system generally used for landfills containing municipal solid wastes. (From Yong, R.N. and Mulligan, C.N., *Natural Attenuation of Contaminants in Soils*, CRC Press, Boca Raton, FL, 2004.)

ecosystem imperatives are (1) protection and conservation of the various land environment resources and (2) ensuring that the capability to provide life support is not degraded or diminished. From the viewpoint of the geoenvironment and ecosphere, the pressures from *development stresses* and WEHAB, combined with the processes necessary to satisfy sustainable development objectives, are summed up in Figure 1.11.

1.5.1 Renewable and Nonrenewable Geoenvironmental Natural Resources

Sustainable development in itself may in all probability be a chimera — a nonattainable goal and an illusion. However, this should not deny the fact that proper environmental management and conservation measures are needed if we are to strive to meet the goals and objectives of sustainability. This includes resource conservation and management and preservation of diversity. Failure to do so will result in the diminution of the capability of the geoenvironment to provide the basis for life support. The case of renewable and nonrenewable geoenvironmental natural resources is a good demonstration of this point.

Following the spirit of the systems dynamics model predictions of Meadows et al. (1972), Figure 1.12 speculates on the status of the global population at some future time under conditions identified in the figure caption. The curve identified as A shows the status of population based on the current depletion rate of nonrenewable geoenvironmental natural resources in relation to some future time. The abscissa on the diagram shows years at some future time, and the ordinate gives a qualitative appreciation of the growth or decline of the parameter under consideration. The curve identified as B is the speculative quantity of nonrenewable resources available, assuming that the depletion rate of the nonrenewable

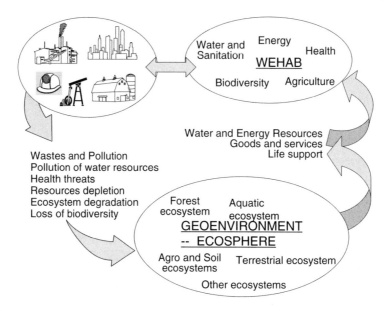

FIGURE 1.11
The continuous cycle of interaction among industry production, WEHAB, and the geoenvironment — from a geoenvironmental perspective.

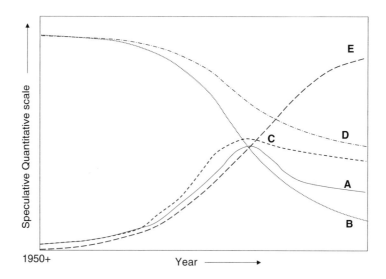

FIGURE 1.12
Speculative chart on status of global population in relation to conservation of the nonrenewable natural geoenvironmental resources — along the lines of reasoning of Meadows et al. (1972). Legend shown in chart: A = status of population based on current usage of resources; B = speculative quantity of nonrenewable resources available; C = status of population based on conservation of resources, use of 4Rs, and use of alternative materials and energy sources; D = available nonrenewable natural geoenvironmental resources using conservation management and 4Rs; E = available alternative materials and energy sources.

geoenvironmental natural resources remains the same, i.e., constant in proportion to the population at hand. We consider the principal nonrenewable geoenvironmental natural resources to consist of fossil fuels, minerals, and geologic building materials (sand, gravel, stone, soil).

Assuming that conservation measures for nonrenewable geoenvironmental natural resources are in place, and that these measures are bolstered by the use of the 4Rs (reduction, reuse, recycle, and recovery), we can further reduce the depletion rate of the nonrenewable resources by using alternative energy sources such as geothermal, wind, solar, etc. Curve C in Figure 1.12 indicates the status of the global population based on conservation of the nonrenewable resources, use of 4Rs, and use of alternative materials and energy sources; curve D indicates available natural geoenvironmental resources using conservation management, the 4Rs, and alternative energy sources. Finally, curve E in Figure 1.12 indicates the resources made available through the use of alternative materials and various other alternative energy sources.

1.5.2 4Rs and Beyond

Although not strictly within the purview of geoenvironmental sustainability consider-ations, the implementation of 4Rs as a means to protect the geoenvironment and also to aid in reduction of nonrenewable resource depletion deserves special attention. Details of many aspects of these will be discussed in subsequent chapters.

By itself, the common understanding of the 4Rs is recovery, reuse, recycle, and reduction. Application of the 4Rs to waste products can reduce the depletion rate of many nonre-newable geoenvironmental resources. A good case in point is the use of coal fly ash as a backfill and liner material. Figure 1.13 shows some of the benefits — as they pertain to the geoenvironment and to the reduction in use of natural geomaterials (Horiuchi et al., 2000; Yamagihara et al., 2000). Traditional disposal of the coal fly ash on land or in the

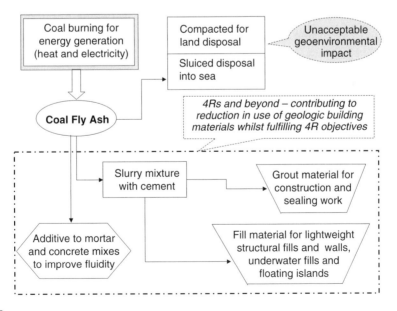

FIGURE 1.13

Example of application of principles of sustainability for waste product obtained in use of a nonrenewable resource for generation of heat and electricity.

ocean is not an acceptable solution. Reuse of the material not only satisfies the aims of the 4Rs, but also contributes to the reduction in use of the nonrenewable geoenvironmental natural resource, i.e., geologic material. Other examples of 4Rs and beyond can be found, for example, in the use of paper sludge ash as backfill slurry (Horiuchi et al., 2002; Asada et al., 2003).

1.6   Concluding Remarks

The impacts of natural and catastrophic events such as earthquakes, hurricanes and typhoons, and associated floods and landslides are not discussed in this book. This does not mean that these are minor events or impacts — in comparison to the impacts on the geoenvironment generated by human activities. It is recognized that these natural cata-strophic events can and do result in considerable loss of life and physical facilities. The problems and impacts generated by these natural disasters on the geoenvironment deserve proper recognition and discussion in a textbook or treatise specifically devoted to such subjects.

Proper management of the geoenvironment is essential if the platform for almost all the life support systems is to be protected for future generations. The principles of *sustainability* require us to recognize a fundamental fact that geoenvironmental natural and cultivated resources are *renewable* and *nonrenewable*. Chapter 11 addresses this issue and the situations where renewable natural resources can become nonrenewable, and hence not sustainable. It is necessary to recognize that renewable geoenvironmental natural resources can be easily threatened and can become ineffective as a resource. A good case in point is *water*. Pollution of receiving waters will render such waters unacceptable for human consumption, therefore rendering this renewable geoenvironmental resource useless. The following items are some of the major issues facing us as we seek to maintain the life support base that provides us with the various goods and services:

- Depletion of nonrenewable resources or natural capital is a reality. Energy production relying on fossil fuels is an example of how nonrenewable resources are continuously depleted.

- Industrial wastes and wastes streams will need to be managed, and it is likely that some of the waste products will find their way into the land environment, resulting thereby in threats to the health and welfare of biotic receptors.

- Loss of soil quality due to various soil degradative forces such as erosion and salinization. In addition to reduction in capability of the soil for crop production, one faces a loss in the capability of the soil to act effectively as a carbon sink.

- Depletion of agricultural lands will occur because of urbanization pressures, thus requiring remaining agricultural lands to be more productive. Implementation of high-yield practices may exacerbate the problem of pollution of both land and water resources.

- Deforestation and inadequate replacement rates, thus contributing to the CO_2 imbalance.

- Pollution of groundwater and surface water resources can reach proportions that render such sources as health threats to biotic receptors.

In the context of the geoenvironmental perspective of environmental management, three particular points need to be stated in regard to the development–environment or sustainable society problematic:

- Soil is a natural resource. In combination with the other geophysical features of the land environment, they constitute at least 90% of the base for sustenance of the human population and production of energy and goods. The depletion rate of the natural capital, represented by all the natural resources, must be minimized.

- Technology and its contribution to environmental management. In addition to the various remediation and impact avoidance tools that technology can develop and contribute to environmental management, perhaps one of the more significant contributions that technology can make would be the development of renewable resources as replacements for the nonrenewable resources that are being depleted.

- Protocols and procedures for management of changes in the environment. It is becoming very evident that changes to the geoenvironment that are presently occurring may reach proportions that require one to develop technology and new social attitudes to manage the change. A particular case in point might be, for example, global warming and the greenhouse effect.

References

Asada, M., Tutsumi, H., Horiuchi, S., and Fukue, M., (2003), Backfill slurry using paper sludge ash, in *Proceedings of the 5th International Conference on the Environmental and Technical Implications of Construction with Alternative Materials*, San Sebastien, Spain, pp. 825–828.

Commoner, B., (1971), The closing circle, in *Nature, Man and Technology*, Alfred A. Knopf, New York, 326 pp.

Glasby, G.P., (2002), Sustainable development: the need for a new paradigm, *J. Environ. Dev. Sustainability*, 4:333–345.

Governor's Office of Appropriate Technology, Toxic Waste Assessment Group, California, 1981.

Horiuchi, S., Asada, M., Tsutsumi, H., and Fukue, M., (2002), Light-weight slurry backfill using paper sludge ash, in *Proceedings of the International Workshop on Lighweight Geo-materials*, Tokyo, pp. 203–206.

Horiuchi, S., Kawaguchi, M., and Yasuhara, K., (2000), Effective use of fly ash slurry as fill material, *J. Hazardous Mat.*, 76:301–337.

Leopold, L.B., (1974), *Water: A Primer*, W.H. Freeman and Co., San Francisco, 172 pp.

Malthus, T., (1798), An Essay on the Principle of Population, as It Affects the Future Improvement of Society with Remarks on the Speculations of Mr. Godwin, M. Condorcet, and Other Writers, London, printed for J. Johnson, in St. Paul's churchyard. (HTML format by Ed Stephan, 10 August 1997.)

Meadows, D.H., Meadows, D.L., and Randers, J., (1992), *Beyond the Limits*, Chelsea Green Publishing Co., White River Junction, VT, 299 pp.

Meadows, D.H., Meadows, D.L., Randers, J., and Behrens, W.W., III, (1972), *The Limits to Growth*, Universe Books, New York, 205 pp.

Naeem, S., Chair, F.S., Chapin, R.C., III, Ehrlich, P.R., Golley, F.B., Hooper, D.U., Lawton, J.H., O'Neill, R.V., Mooney, H.A., Sala, O.E., Symstad, A.J., and Tilman, D., (1999), Biodiversity and Ecosystem Functioning: Maintaining Natural Life Support Processes, *Issues in Ecology*, No. 4, 11 pp.

Trevors, J.T., (2003), Editorial: biodiversity and environmental pollution, *J. Water Air Soil Pollut.*, 150:1–2.

U.S. National Environmental Policy Act, (1969), 42 U.S.C. §§ 4321.

World Commission on Environment and Development, (1987), *Our Common Future*, Oxford University Press, Oxford, 400 pp.

Yamagihara, M., Horiuchi, S., and Kawaguchi, M., (2000), Long-term stability of coal-fly-ash slurry man-made island, in *Proceedings of Coastal Geotechnical Engineering in Practice*, Balkema, the Netherlands, pp. 763–769.

Yong, R.N., (2001), *Geoenvironmental Engineering: Contaminated Soils, Pollutant Fate, and Mitigation*, CRC Press, Boca Raton, FL, 307 pp.

Yong, R.N. and Mulligan, C.N., (2004), *Natural Attenuation of Contaminants in Soils*, CRC Press, Boca Raton, FL, 310 pp.

2

Contamination and Geoenvironmental Land Management

2.1 Introduction

Section 1.4.2 in Chapter 1 points out that good land management practice (1) minimizes and mitigates deleterious impacts to the land environment, (2) seeks optimal land use and benefit from the land, and (3) preserves and minimizes depletion of the geoenvironmental natural capital. Probably the most significant agent responsible for degrading the quality of the land and its ecosystems is contamination of the land environment and its receiving waters by pollutants and hazardous substances. In addition, ground contamination by these same pollutants and hazardous substances poses threats to human health, other biotic receptors, and the environment. In this chapter we will be paying particular attention to the various aspects of ground contamination and land management requirements to meet sustainability goals. We repeat again that we recognize that we live in an unsustainable world, and that depletion of nonrenewable resources and threats to renewable resources render sustainability an impossible goal. The preceding notwithstanding, it is nonetheless necessary and important to undertake measures for protection of the geoenvironment and its natural capital and resources. Failure to do so will exacerbate the conditions that have already led to a compromised geoenvironment. As with all the chapters in this book, the actions discussed and proposed in these chapters are in full recognition that one needs to strive for measures and actions that will relieve the negative pressures and stresses on the geoenvironment.

Contamination of the land environment by hazardous substances, pollutants, and non-pollutants results primarily from man-made activities and events mounted to meet societal and industrial demands. Contamination from natural events can also occur. These include, for example, deposition of ash from volcanic discharge and seepage of sulfuric acid and iron hydroxide when pyrite (FeS_2) is exposed to air and water, according to the following relationship:

$$4\,FeS_2 + 15\,O_2 + 14\,H_2O \leftrightarrow 4\,Fe\,(OH)_3 + 8\,H_2SO_4 \tag{2.1}$$

We can encounter more dramatic discharges of sulfuric acid in the phenomenon commonly referred to as *acid mine drainage* (AMD). In this instance, the contamination is considered to be associated with anthropogenic activities in support of metalliferous mining. This will be discussed in a later section and in more detail in Chapter 5.

By and large, contamination due to anthropogenic activities is by far the greatest contributor to overall contamination of the environment, and especially the geoenvironment. We use the term *contamination* to include contamination by pollutants, toxicants, hazardous

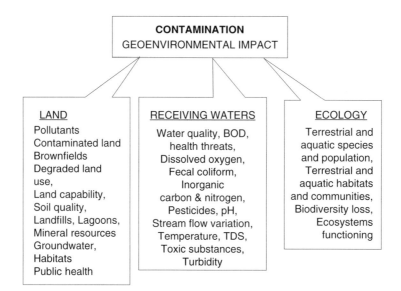

FIGURE 2.1
Simple schematic showing some of the impacts, articulated as concerns and issues, for *land, receiving waters*, and *ecological system* because of contamination of the land environment.

substances, and all substances foreign to the natural state of the particular site and microenvironment. When necessary, the terms *pollutants, toxicants*, and *hazardous substances* will be used to highlight the substance under discussion or to lay emphasis to the problem. Contamination poses the most significant challenge in the maintenance and protection of the many land environment ecosystems and resources needed to support life on earth. Figure 2.1 shows a simple tabular sketch of some of the effects of impacts from contamination of the land environment — illustrated as concerns and issues — for *land, receiving waters*, and *ecological system*.

Because of the far-reaching primary and secondary effects of contamination of the land environment, and because these:

- Have a severe impact on our ability to implement sustainability practices for the geoenvironment

- Directly affect our means to sensibly and responsibly exploit the natural resources in the geoenvironment

- Diminish the quality of the land environment and the very resources needed for mankind to sustain life

- Pose direct health threats to humans and other biotic receptors

- Degrade soil quality and reduce the ability of the soil to function as a resource material

The treatment of the subject of *geoenvironment sustainability* in this book will pay particular attention to the requirements and measures needed to mitigate contamination impacts. Where mitigation is not totally effective, ameliorative procedures and remedial actions required to attain sustainability objectives will be required.

In addition to the very significant problem of contamination of the land environment, there are geoenvironmental impacts that have a considerable effect on the functioning of the land environment itself. These are a group of man-made events that have substantial physical impacts on the geoenvironment landscape features — as will be seen later in Figure 2.3 and discussed in Section 2.2.2. The projects and events included in this group, identified as the primary group, deal directly with the landscape features of the geoenvironment. As such, the effects or impacts arising from the activities associated with this group of events or projects are most often felt as threats to public safety and loss or diminution of natural geoenvironmental resources and natural capital. An example of this is the physical sets of activities associated with mineral resources and hydrocarbon resources recovery. These are discussed in detail in Chapters 5 through 7.

2.1.1   Impacts to the Geoenvironment

We have pointed out in Section 1.1.1 in Chapter 1 that almost anything that happens, i.e., any input or any activity in the ecosphere, will result in the production of some kind of impact on the geoenvironmental landscape and its associated ecosystems, i.e., the land environment. To protect the status and manage the geoenvironment, the nature of these impacts and whether these impacts will add value to the particular ecosystem, or subtract from the functionality of that particular ecosystem in the geoenvironment, need to be determined and better understood. A complete listing of all the impacts on the ecosystems of the geoenvironment is not possible because we are not fully aware of all the kinds of activities and interactions that are included in a functioning ecosystem. As previously defined, a *functioning ecosystem* includes not only the biological and chemical activities in the ecosystem, but also the physicochemical activities and physical interactions characteristic of the type of ecosystem under consideration. With our present stage of knowledge of the geoenvironment and the associated functioning ecosystems, it is difficult (and virtually impossible) to fully catalog all of these activities and interactions. What is possible at this stage is to examine and determine how the known activities and interactions are affected or changed because of the stresses, disturbances, alterations, etc., to the ecosystem of interest.

The major sources of impacts and the resultant nature of the impacts cannot be easily listed without specification of targets of the impacts. Some of these sources may not be immediately evident, and some of the impacts will not be readily perceived or understood. To some extent, this is because the effects of the impacts will not be apparent, and to another extent, this is because the effects or results of the impacts cannot be readily recognized. A simple case in point is the effect of buried toxic substances in the ground on human health — particularly if the impact on the health of those that come in contact with the material is mutagenic or teratogenic.

We have indicated in Section 1.1.1 that geoenvironmental impacts arising from natural causes such as earthquakes, tornadoes, floods, typhoons, and hurricanes do not fall within the purview of this book. The distress and damage to the geoenvironment can be readily perceived because the energy generated in these events in the form of forces and stresses can cause substantial physical damage to the geoenvironment and considerable loss of life. While many of the sources or causes of the impacts are generally obvious, there are many that are not. This is because we do not have any hard and fast rule as to what constitutes an impact and, more importantly, when the impact causes irreversible damage. Geoenvironmental damage of the same scale and magnitude as earthquakes and hurricanes can be obtained as a result of man-made activities, such as landslides and pollution of ground and receiving waters. However, more often than not, the geoenvironmental impacts arising from anthropogenic activities in support of production of goods and services are less dramatic, but nevertheless engender public safety and health-threatening problems and issues.

2.2   Contamination and Geoenvironmental Impacts

Included in Title 1 of the U.S. National Environmental Policy Act (NEPA) of 1969, given in Section 1.3.2 of Chapter 1, are the three specific environmental requirements that we can consider essential in management and control of the contamination impacts to the geoenvironment. These include:

- Environmental inventory: This is essentially an environmental audit, i.e., complete description of the environment as it exists in the area where a particular proposed (or ongoing) action is being considered. The physical, biological, and cultural environments are considered to be integral to the environment under consideration.
- Environmental assessment: The various components included in the assessment package are:
    - Prediction of the anticipated change
    - Determination of the magnitude of change
    - Application of an important or significant factor to the change
- Environmental impact statement (EIS): This is a very crucial document that needs to be written in the format specified by the specific regulatory agency responsible for oversight of the project or event. In respect to the geoenvironment, this document must contain the proper determination of the various geoenvironmental impacts arising from implementation of the project under question or the event being investigated. In respect to the NEPA-type response, this document contains a summary of environmental inventory and findings of environmental assessment (referred to as 102 statements, i.e., section of NEPA relating to requirements for preparation of EIS in NEPA).

To determine the nature of geoenvironmental impacts in general and contaminant-associated impacts to the geoenvironment, it is necessary to:

- Develop a frame of reference. This is essentially a series of targets or receptors that are the victims of the impacts. The reference frame will permit one to examine the effects of the geoenvironmental impacts in relation to the members constituting the reference frame. The following members constitute the essential elements of the reference frame:
    - The separate compartments (terra firma and aquatic) of the land environment. By and large, one determines the impacts of projects and events on the integrity of the landscape (including receiving waters and their boundaries).
    - Health of the human population and other biotic receptors in the geoenvironmental compartment (land and receiving waters). This requires examination of the impacts as threats to human health and the environmental biotic receptors. Generally, this includes a study of waste and waste pollutant streams and other catastrophic phenomena arising from man's activities.
    - Overall health of the environment. Terrestrial and aquatic habitat and community preservation are central to the health of the environment.
- Establish a general or broad-impact identification scheme. By doing so, this allows us to look for the source of the impact. Knowledge of the source of the impact

provides one with a better appreciation of the extent and details of the impact. In looking for the source of the impact, it is necessary to separate the *source* from the *cause*, since it is not always easy to determine the cause itself.

2.2.1 Reference Frame

The geoenvironmental impact reference frame provides or specifies the targets in the total land environment and the receiving waters contained therein. It provides a framework that requires questions to be raised as to whether the actions arising from projects or events will have adverse effects on the various physical and biological elements constituting the geoenvironmental compartment. The specific case of a contaminated ground shown in Figure 2.2 is a good example of the application of the reference frame for determination of geoenvironmental impacts. In this instance, one is concerned with the impacts resulting from the presence of pollutants in the ground. We define a *pollutant* to mean a contaminant that has been identified as a threat to human health or the environment. These are generally toxic elements, chemicals, and compounds and are most often found in priority pollutant lists of regulatory agencies. *Contaminants*, on the other hand, are defined as substances that are not natural to the site or material under consideration. These substances can include hazardous materials or elements, toxic substances, pollutants, and all other substances that are nonthreatening to human health and the environment. In other words, the term *contaminants* is used when we wish to refer to nonindigenous elements, substances, etc., found at a site or in a material under investigation.

In reference to Figure 2.2, the source of the contaminant plume at the site in question is at the ground surface, e.g., dump site, landfill, toxic dump, etc. For this discussion, the reasons for locating the dump site, landfill, toxic dump, etc., are not addressed, even though they result in creating the contaminant source. Our interest is directed toward the impacts generated by the presence of these sources. The terms *sources* and *causes* need to be carefully differentiated. For the particular problem shown in Figure 2.2, the *source of*

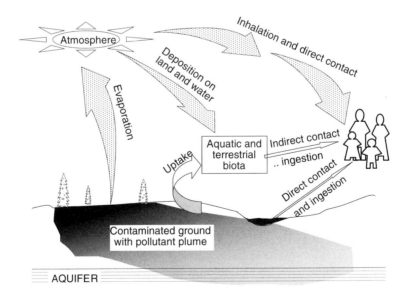

FIGURE 2.2
Schematic diagram showing contaminated ground with a pollutant plume as the source of health and environmental threats. (From Yong, R.N. and Mulligan, C.N., *Natural Attenuation of Contaminants in Soils*, Lewis Publishers, Boca Raton, FL, 2004.)

the contaminant plume is the landfill itself. The *cause* is the fugitive pollutants generated in the leachate stream obtained from the dissolution of the waste material contained in the landfill. Accordingly, one notes that the dissolution of the waste materials (*cause*) in the landfill generates the *source* of the fugitive pollutants in the contaminant plume in the ground. Some of the evident environmental and geoenvironmental impacts resulting from the pollutants in the contaminated ground shown in Figure 2.2 include:

- Pollutants in the atmosphere, carried into the atmosphere by evaporation and volatilization
- Pollutants on the land surface within and outside the contaminated site, resulting from deposition of the airborne pollutants
- Pollutants in the contaminated ground
- Contaminated groundwater and surface water due to progress of the pollutant plume and also to deposition of contaminants in the atmosphere
- Threats to habitats of terrestrial and aquatic biota
- Threats to human health

The impacts shown in the preceding list illustrate the target value of the four members of the reference frame. By using these as targets, one determines not only *what is being impacted*, but also the nature of the impact. There are obviously many more impacts that can be cited for the example shown in Figure 2.2. We can, for example, discuss or speculate on the impact of the contaminants and pollutants in the contaminated ground, on the possible loss of biodiversity in the affected region, and the impact on the quality of the land and how this affects future land use. These impacts can be covered in the impact statements that need to be produced or developed in association with specific projects and events. These will be discussed in greater detail at various times throughout the book when specific projects or events are considered.

2.2.2   Characterization of Geoenvironmental Impacts

In addition to the geoenvironmental impact frame of reference, there needs to be a mechanism or means or criterion for one to determine *what constitutes an impact to the geoenvironment*. A group of questions that need to be addressed include:

1. Whether reversibility (of damage) should be used as one of the decision mechanisms for determination of impacts
2. How man-made improvements, amelioration, mitigation, and remediation procedures to the geoenvironment should be factored into the process of evaluation of geoenvironmental impacts and their effects

There are many ways in which impacts to the geoenvironment can be categorized or classified. A useful and popular method is to categorize the geoenvironmental impacts in relation to natural or man-made causes leading to events that are determined to be responsible for the impact, as shown in Figure 2.3. Described within the two major categories shown in the figure are some typical causes for events leading to geoenvironmental impacts. *Floods* and *landslides* have been singled out as typical examples of both natural and man-made causes for events resulting in geoenvironmental impacts. For example, floods can arise naturally because of hurricanes, as for example shown by the

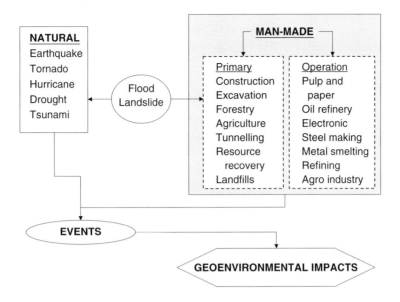

FIGURE 2.3
Categories of some typical causes for events leading to geoenvironmental impacts. Note that floods and land-slides can be both natural and man-made causes for events resulting in geoenvironmental impacts.

massive floods caused in the late summer of 2005 in the gulf region of Central and North America. Floods can also occur naturally because the natural waterways (rivers, brooks, and streams) do not have the capability to carry the excessive water load produced by an undue rainfall occurring over a very short period. It is also possible for floods to occur because of man-made waterway constrictions and shoreline alterations that impair the previously capable performance of the waterways.

Figure 2.3 shows that the man-made events have been divided into two groups. The *primary* group refers to those man-made events that are directly associated with the physical landscape of the geoenvironment. These man-made events are the result of various kinds of anthropogenic activities in support of physical projects, such as excavations and mining, construction of infrastructure and buildings, resource recovery, drilling, tunneling, and waste landfills. The immediate evident geoenvironmental impacts are mostly physical in nature. These generally involve alteration of the surficial and subsurface landscape features. While these landscape alterations are evident as physical impacts of the land surface features, they also have the ability to serve as sources of subsequent impacts. The previously cited example of acid mine drainage (AMD) is a very good example of the subsequent impact generated from physical extraction of metal ores from the ground. In this case, the AMD problem poses a very significant threat to the receiving waters, and ultimately to the biotic receptors. Another less dramatic health threat example is the undercutting of a slope to facilitate the construction of a right-of-way for a highway system. Excessive undercutting without proper analysis of the slope stability could produce a situation where the undercut slope could subsequently trigger a slope failure. When such occurs, one needs to be concerned with the safety of the human and animal populations in the affected region.

The other grouping for man-made events shown in Figure 2.2 is identified as *operations*. The impacts to the geoenvironment arising from this group of events are the direct result of activities associated with agro-industry (outside of physical cultivation of the land), refining, manufacturing, production and process industries, etc. The effects or results of the impacts on the geoenvironment can be physical, chemical, physicochemical, and bio-

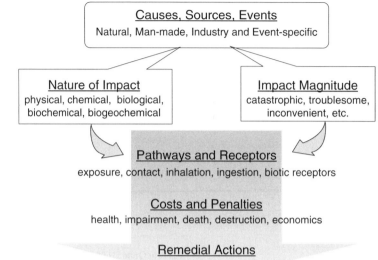

FIGURE 2.4
Useful protocol for geoenvironmental impact scoping exercise. Scoping for impact assessment or determination can omit the last step (remedial actions) if a quick assessment is needed. The last step is needed if one has the ability to influence decision making concerning implementation of the project or event.

geochemical in nature. One needs to factor into the analysis the resultant or potential threats to human health and other biotic receptors. Some examples of these are (1) application of pesticides and fungicides as pest controls in support of agricultural activities, leading to nonpoint source pollution of ground and groundwater; (2) landfilling of hazardous wastes, resulting in the production of pollutants in the fugitive leachate plumes; (3) discharge of waste streams from chemical and electronic industries, resulting in pollution of the receiving waters; and (d) isolation–disposal of high-level nuclear wastes in underground repositories.

2.2.3 Identifying and Assessing for Impact on the Geoenvironment

A useful procedure for performing a geoenvironmental impact scoping exercise is given in Figure 2.4. The various steps shown in the diagram are guidelines and are designed to provide one with specific objectives or targets.

- Causes, sources, events: Knowledge of the cause, source, or event can be helpful in narrowing the field of study or investigation. Take, for example, the case of a chemical plant producing various organic chemicals where inadvertent spills and fugitive discharges during storage are suspected to have occurred. If the spills and discharges are identified as the causes, the source of any contaminant plume or contaminated ground would be the chemicals involved in the spills or discharges. The impacts to the geoenvironment can be readily identified.
- Nature of impacts: It is important to have a proper knowledge of the kind of impact. This allows determination or estimation of the extent of the *damage* done to the geoenvironment. Using the previous chemical plant example, the damage suffered by the geoenvironment is the contaminated site or ground.

Guidelines are necessary to prevent one from rendering judgments on impacts without methodical and proper analyses because the results of the impacts are dramatically visible — as opposed to impacts that do not show visible distress signs. It is not the magnitude of the distress caused by the impact that should determine what constitutes an impact. As we have pointed out in the previous chapter, the results of many impacts do not manifest themselves until many years later. This is especially true for health-related issues. In the most general sense, the guidelines used to determine what constitutes an impact to the geoenvironment should be determined on the basis of whether the geoenvironmental impact will:

- Generate direct or indirect threats and problems relating to public health and the environment. A good case in point is the pollution of receiving waters. These waters serve as habitats for aquatic species and, in many instances, will serve as sources for drinking water. Not only is such an impact a direct threat to the usable water supply for the human population, but it is also a direct threat to the food supply for the same population because of the likely reduction in aquatic food supply for the population.

- Diminish the functioning of the ecosystems in the geoenvironment. An example of this can be found in the degradation of soil quality due to many of the activities associated with high-yield agriculture and mineral resources exploitation. This is particularly important since *soil quality* is a direct measure of the capability of a soil to sustain plant and animal life and their productivity within their particular natural or man-managed ecosystem. The Soil Science Society of America (SSSA) defines soil quality as "the capacity of a specific kind of soil to function, within natural or managed ecosystem boundaries, to sustain plant and animal productivity, maintain or enhance water and air quality, and support human health and habitation" (Karlen et al., 1997). Any diminution of soil quality will have an impact on the capability of the soil to provide the various functions, such as plant and animal life support and forestry and woodland productivity, and will also result in the loss of biodiversity and nutrients. Other prime examples of this can be found in the various activities associated with mineral extraction and energy resource development. To appreciate the impacts resulting from these activities, we can focus on the status of the biological, chemical, and physicochemical activities and physical interactions that define the functioning of the ecosystem of interest. In addition, we can also study the changes in land capabilities or land use options.

2.2.4 Man-Made and Natural Combinations

It is not always easy or simple to distinguish between natural and man-made causes leading to events that impact directly or indirectly on the geoenvironment. This is because many geoenvironmental impacts are the result of a sequence combination of man-made and natural causes. A very good example of this is the previously mentioned acid mine drainage (AMD) problem — a problem that is triggered by the results of mining exposure of pyrite. The presence of pyrite (FeS_2) in rock formations where coal and metalliferous mining occurs will create problems for the environment if the pyrite is exposed to both oxygen and water. Given the favorable geologic and hydrologic conditions, we have the situation where oxidation of the pyrite exposed during mining operations will produce ferrous iron (Fe^{2+}) and sulphate (SO_4^{2-}). For this first chemical reaction step, we can conclude that the trigger for the first sets of reactions is the man-made event, i.e., mining of the rock formation. Subsequent rate-determining reactions, which may or may not be catalyzed by certain bacteria (e.g., *Thiobacillus ferrooxidans*), involve oxidation of the ferrous iron (Fe^{2+}) to ferric iron (Fe^{3+}), to be followed later by hydrolysis of the ferric iron and its ensuing precipitation to ferric hydroxide ($Fe(OH)_3$) if and when the surrounding pH goes above 3.5. Throughout these processes, hydrogen ions are released into the water, thereby reducing the pH of the surrounding medium. The sum total of the reaction products and the reducing pH condition is commonly known as the acid mine drainage (AMD) problem. This problem is a significantly large problem because of the many sets of mining activities conducted all over the world, and particularly because of the presence of pyrite in many of these mines. Acid pollution of groundwater and other receiving waters creates conditions that are adverse to human health and other biotic species. Chapter 4 will discuss the problem and impact of AMD in greater detail — together with procedures for amelioration of AMD. It is recognized that the nature of the activity leading to the generation of AMD will not permit goals of sustainability to be fully achieved. Nevertheless, progress toward fulfillment of the goals is always required.

Another striking example of the arsenic polluted aquifers in West Bengal and Bangladesh demonstrates the point. These aquifers serve a significant portion of the population of these two countries, and unknowing use (initially) of these polluted aquifers has led to what is often referred to as the singular most dramatic case of mass poisoning of the human race. While the direct impact to the geoenvironment from the release of arsenic to the aquifers is arsenic-polluted aquifers, the ingestion of the polluted water has led to the development of arsenicosis in thousands of unfortunate individuals.

Tube wells sunk into the aquifers constitute a major drinking water supply source for the two countries. Investigations on tube well water supply showed concentrations of arsenic far in excess of allowable limits. Tests on these wells at time of installation showed arsenic concentrations, if any, well below threat levels. The present levels of arsenic concentration indicate that arsenic poisoning developed several years after installation of the wells. By switching to groundwater resources to avoid waterborne diseases, the population now faces considerable risk of arsenic poisoning as a result of ingesting arsenic-polluted water obtained from tube wells tapping into the groundwater.

If the aquifers were not polluted by arsenic before extensive harvesting of the aquifer resource, how did the arsenic get there? Is the arsenic pollution due to natural causes? Or indirectly due to a man-made cause? Arsenic occurrence in the hard rock and sedimentary rock aquifers have been reported in Argentina, Nepal, Nigeria, the Czech Republic, and many other countries. While it is not uncommon to find vanishing concentrations of arsenic in the groundwater in many parts of the world, the arsenic concentrations in the samples obtained form the aquifers in Bangladesh and West Bengal, for instance, are too high to be attributable to natural processes for arsenic release from arsenopyrites. In fact, the arsenic concentrations in the Bangladesh and West Bengal aquifers are also too high to be accounted for by direct human activities, such as those associated with metalliferous mining. There appears to be no doubt that the release of arsenic into the aquifers is from source geologic materials. *Why is arsenic being released from the source rocks? Is this a natural process, or is the release of arsenic triggered by some man-made event?*

In the Bangladesh case, two factors appear prominently: (1) presence of arsenopyrites and arseniferrous iron oxyhydroxides in the substrate material and (2) the use of tube wells as a means for abstracting water. The two models proposed to explain the release of arsenic from the arsenic-bearing materials are shown in Figure 2.5. These include:

1. Reduction mechanisms: Reductive dissolution of arseniferrous iron oxyhydroxides releases the arsenic responsible for pollution of the groundwater.

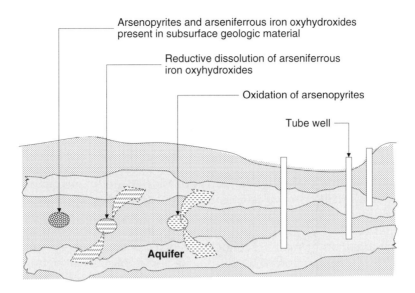

FIGURE 2.5
Schematic of speculative models for release of arsenic from arseniferrous iron oxyhydroxides and arsenopyrites into aquifers.

2. Oxidation processes: Oxygen invades the groundwater because of the lowering of the groundwater from the abstracting tube wells, resulting thereby in the oxidation of the arsenopyrite (FeAsS).

Based on these two possible mechanisms for arsenic release into the aquifer, and based on detailed field studies to determine the presence and distribution of As(III) and As(V) and other reaction products, there are some serious questions as to whether the arsenic-release processes are totally man induced or a case of natural processes hastened and aggravated by man-made events, i.e., abstraction of water from the tube wells. The almost equal proportions of As(III) and As(V) in the aquifer testify to almost equal sets of activity from both arsenic-release mechanisms.

2.3   Wastes, Pollutants, and Threats

The discharge of liquid and solid wastes from industrial and energy-producing plants and facilities, together with inadvertent spills and deliberate dumping of waste materials, combines to introduce contaminants, pollutants, toxicants, and other kinds of hazardous substances into the land environment — including the receiving waters contained therein. These pose significant threats not only to human health and other biotic receptors, but also to the health of the environment and the various ecosystems in the geoenvironment. In the treatment of threats to the environment and public health, there are several ways in which wastes and pollutants can be classified or categorized. One could categorize the wastes and pollutants in terms of source production, i.e., where they come from. Alternatively, one could categorize them in respect to:

- Level of toxicity, e.g., highly toxic, carcinogenic, priority listing, etc.
- Inorganic or organic substances and chemicals, e.g., heavy metals, polycyclic aromatic hydrocarbons (PAHs), perchloroethylene (PCE), styrene, etc.
- Type of industry (source industry or source activity), e.g., pulp and paper, forest, electronic, pharmaceutical, etc.
- Class of pollutants, e.g., pesticides, solvents, etc.
- Nature of impact or threat, e.g., physical, chemical, biological, etc.
- Type of receptor, e.g., land, water, human, other biota, etc.

Experience has shown that although categorization of pollutants according to any single method as described above is not practical or feasible, it is always necessary to obtain a proper identification of the types or species of pollutants. Considerable significance is placed on the potential health threat of pollutants. A popular approach that has gained the attention of many regulatory agencies is the source–pathway–receptor (SPR) method for determination of health threats and impacts created by the presence of pollutants in the geoenvironment and also by events or projects as source pollutants. There have been questions raised as to whether the SPR approach discriminates between levels of treatment or protection from health threat events, depending on the importance of the *receptors*. A school of thought suggests very strongly that risk management should be directly linked to receptor importance and also to certainty of pathways. It is not always clear that pathways to potential receptors are well defined. Accordingly, the same school of thought suggests that the degree of certainty of pathways is integral to the level of risk determination.

2.3.1 Inorganic Pollutants

Evidence of the presence of inorganic pollutants in the geoenvironment (land and water) shows that these are mainly heavy metals, such as Pb, Cr, Cu, etc. Yong (2001) has indicated that while those elements with atomic numbers higher than Sr (atomic number 38) are classified as *heavy metals* (HMs), it is not uncommon to include elements with atomic numbers greater than 20 as heavy metals. The 39 elements commonly considered to be HMs fall into three groups of atomic numbers, as follows:

- From atomic numbers 22 to 34: Ti, V, Cr, Mn, Fe, Co, Ni, Cu, Zn, Ga, Ge, As, and Se
- From 40 to 52: Zr, Nb, Mo, Tc, Ru, Rh, Pd, Ag, Cd, In, Sn, Sb, and Te
- From 72 to 83: Hf, Ta, W, Re, Os, Ir, Pt, Gu, Hg, Tl, Pb, and Bi

The more common heavy metals (HMs) found in the geoenvironment come as a result of anthropogenic activities, such as management and disposal of wastes in landfills, generation and storage of chemical waste leachates and sludges, extraction of metals in metalliferous industries, metal plating works, and even in municipal solid wastes. The more notable HMs include lead (Pb), cadmium (Cd), copper (Cu), chromium, (Cr), nickel (Ni), iron (Fe), mercury (Hg), and zinc (Zn).

2.3.1.1 Arsenic (As)

Strictly speaking, arsenic is a nonmetal, although it is often classified as a metal. It is a metalloid (semimetal) with atomic number 33 and is in Group 5 of the periodic table. Arsenic is found naturally in rocks, most often in iron ores and in sulfide form as magmatic sulfide minerals. The more common ones are arsenopyrite (FeAsS), realgar (AsS), nicolite (NiAsS), and orpiment (As_2S_3). Arsenic is also found naturally in soils in association with hydrous oxides, and sometimes in elemental form in association with silver ores. Arsenic found in the geoenvironment can come directly from weathering of the arsenic-containing rocks and also from industrial sources, which include manufacturing, processing, pharmaceutical, agriculture, and mining. Products such as paints, dyes, preservatives, herbicides, and semiconductors are some of the more common contributors to the arsenic found in the ground and in receiving waters. Extensive use of arsenic-containing (lead arsenate) pesticides, herbicides, and insecticides in agricultural and farm practices can contribute some considerable amounts of arsenic to the subsurface and the receiving waters.

While the existent valence states for arsenic are –3, 0, +3 (arsenite), and +5 (arsenate), arsenite and arsenate are the more common forms of arsenic found in nature. Arsenic is a toxic element, and a regulatory limit of 50 µg/l in groundwater (aquifers) for drinking water has been adopted in many countries and regulatory agencies. In the United States, this was lowered to 10 µg/l for all water systems, to be implemented by 2006. Ingestion of arsenic for a period of time, as for example in the use of arsenic-polluted waters in some regions of the world, can lead to serious health problems, e.g., mortality from hypertensive heart disease traceable to ingestion of arsenic-polluted drinking water (Lewis et al., 1999), and arsenic-associated skin lesions of keratosis and hyperpigmentation (Mazumder et al., 1998). Similarly, inhalation of arsenic dust generated in ore refining processes can also lead to serious health problems, e.g., nasal septal perforation and pulmonary insufficiency (U.S. EPA, 1984).

2.3.1.2　Cadmium (Cd)

Cadmium can be found in nature as greenockite (cadmium sulfide, CdS) or otavite (cadmium carbonate, $CdCO_3$) and is usually associated with zinc, lead, or copper in sulfide form. The two major groups using cadmium include (1) cadmium as a filler, alloy, or active constituent for an industrial product, e.g., nickel–cadmium batteries, enamels, fungicides, phosphatic fertilizers, motor oil, solders, paints, plastics, etc., and (2) cadmium as a coating or plating material, e.g., steel plating, metal coatings. The presence of Cd as a pollutant in the geoenvironment can be traced to:

- Nonpoint sources associated with the use of fungicides and fertilizers
- Deposition of Cd particles in the atmosphere because of mining activities and burning of coal and other Cd-containing wastes
- Specific sources, such as industrial discharges and wastes and municipal wastes where the products manufactured and consumed include Cd as a filler, alloy, or active constituent

From the viewpoint of human health effects and requirements, cadmium is considered to be a nonessential element. The Environmental Protection Agency (EPA) Toxicity Characteristics Leaching Procedure (TCLP) regulatory level for Cd is 1.0 mg/l (roughly equivalent to 1 ppm). The EPA specifies a threshold limit of 5 ppb for drinking water, and the Food and Drug Administration (FDA) specifies a limit of 15 ppm of Cd in food coloring (ATSDR, 1999). Accumulation of Cd in the liver and kidney from oral ingestion can lead to distress to these organs.

2.3.1.3　Chromium (Cr)

Chromium is found naturally as chromite (ferrous chromic oxide, $FeCr_2O_4$) and crocoisite (lead chromate, $PbCrO_4$) minerals. It is an essential element in human nutrition. The three common valence states for chromium are 0, +3, and +6, i.e., chromium (0), chromium (III), and chromium (VI). Chromium (III) is found naturally in the environment, whereas compounds of chromium are generally with chromium (VI). Trivalent chromium (chromium (III)) is stable and is considered to be relatively nontoxic. Cr(III) can form various stable, inert complexes: $Cr(H_2O)_6^{3+}$, $Cr(H_2O)_5(OH)^{2+}$, $Cr(H_2O)_3(OH)_3$, and $Cr(H_2O)(OH)_4^-$. On the other hand, hexavalent chromium (chromium (VI)) is highly toxic and is considered to be a carcinogen. Oxidation of the trivalent chromium to the hexavalent chromium anions chromate (CrO_4^{2-}) and dichromate ($Cr_2O_7^{2-}$) will not only render the previously nontoxic trivalent chromium toxic, but also make it more mobile. The major form of Cr(VI) is CrO_4^{2-} at pH greater than 6.5 and $HCrO_4^-$ at pH less that 6.5. Both ions are very soluble.

Other than the natural sources, chromium found in the geoenvironment can be traced to waste discharges and tailing ponds associated with chromium mining. Principal uses for chromium (Cr) and its compounds include (1) use of chromium as alloys, with iron and nickel, stainless steel, and super alloys as probably the best-known alloys; (2) chromium compounds used in metal plating, tanning of hides, wood preservation, glass, and pottery products; and (3) production of chromic acid. Chromium in the land and aquatic compartments of the geoenvironment can be the result of production and waste discharges associated with the industries, and from the tailing ponds associated with mining activities.

2.3.1.4   Copper (Cu)

Copper is found naturally in sandstones and in other copper-bearing oxidized and sulfide ores. These include such ores as malachite ($Cu_2(CO_3)(OH)_2$), tenorite (CuO), cuprite (Cu_2O), and chalcopyrite ($CuFeS_2$), with chalcopyrite being the most abundant. In addition to its natural occurrence in the land environment, contributions of Cu to the geoenvironment come from (1) deposition of airborne particles from mining of copper and combustion of fossil fuels and wastes, (2) discharges from industrial processes utilizing copper as a metal, and copper compounds (production of electrical products, piping, fixtures, and different alloys), and (3) industrial and domestic discharge of waste water.

Copper deposited on the surface of the land environment from the various sources discussed in the preceding will initially be attached to organic matter and clay minerals — if such are present in the landscape. Degradation of the organic matter through anaerobic or aerobic means will release copper in its monovalent or divalent form, respectively. However, if the subsurface soils contain reactive soil particles, the released copper will be bound to these particles. Environmental mobility of copper in the substratum is not generally a big factor when the soil substratum is composed of fine soil fractions consisting of clay minerals and other soil fractions with reactive particles. The presence of copper in the receiving waters is most often confined to the sediments since that copper will attach itself to the fine particles in water.

In terms of health considerations, copper is considered to be an essential trace element in both human and animal nutrition. The amounts required, however, are extremely small. Threshold limits for human ingestion of copper vary between different countries and jurisdictions, with values of about 1.3 ppm for drinking water and 0.1 mg/m³ for airborne concentrations being reported.

2.3.1.5   Lead (Pb)

Lead is found in nature in sulfide, carbonate, and oxide forms. These are galena (lead sulfide, PbS), anglesite (lead sulfate, $PbSO_4$), cerrusite (lead carbonate, $PbCO_3$), and minium (lead oxide, Pb_3O_4). Although it has three valence states (0, +2, and +4), the most common state is +2. Compounds of Pb(II) have ionic bonds, whereas the higher valence state, Pb(IV)

compounds, has covalent bonds. Lead found in the land compartment of the geoenvironment will most often be bonded to reactive soil particles. It is a nonessential element.

Lead is used to a very large extent in the manufacture of lead–acid batteries and in the
electronics and munitions industries. Other, lesser uses for lead are in production of crystal
glass, lead liner material, weights, insecticides, and in construction. Lead found in the
ground and in the receiving waters can be traced to deposition of airborne lead obtained
from emissions, such as burning of wastes and fuel, and from transport of lead compounds
in the soil. Lead is considered to be a nonessential toxic element, and ingestion or inhalation of lead will result in consequences to the central nervous system and damage to
kidneys and the reproductive system.

2.3.1.6 Nickel (Ni)

Nickel is quite widely found in nature in various soil deposits, e.g., laterite deposits, and
generally in mineral form in combination with oxygen or sulfur as oxides or sulfides.
These include nickel sulfide (NiS), nickel arsenide ($NiAs$), nickel diarsenide ($NiAs_2$), and
nickel thioarsenide ($NiAsS$). Some debate exists concerning whether nickel is an essential
element. It is maintained that small amounts of nickel are essential for maintaining the
good health of animals and, to a lesser extent, humans.

Nickel found in the geoenvironment, other than from natural sources, can be traced
to fugitive atmospheric nickel and waste discharge associated with nickel mining activities, burning of waste, operation of oil-burning and coal-burning power plants, and
discharges from manufacturing industries using nickel alloys and compounds. Nickel
does not precipitate but is sorbed onto clays, oxides of manganese, and iron, and organic
material occurs. Mobility increases with the formation of complexes with organic and
inorganic ligands.

2.3.1.7 Zinc (Zn)

Similar to nickel, zinc is found in soil deposits and does not generally exist as a free
element. Instead, it is found in mineral form in combination with oxides, sulfides, and
carbonates to form zinc compounds. The sulfide form is perhaps the more common form
for zinc found naturally in the environment. Some of the naturally occurring zinc compounds are zincite (zinc oxide, ZnO), hemimorphite (zinc silicate, $2\,ZnO\cdot SiO_2H_2O$), smith-
sonite (zinc carbonate, $ZnCO_3$), and sphalerite (zinc sulfide, ZnS). Natural levels of zinc
in soils are 30 to 150 ppm.

Typical uses and products for zinc in element form as oxide and sulfide compounds
include alloys, batteries, paints, dyes, galvanized metals, pharmaceuticals, cosmetics, plastics, electronics, and ointments. It follows that non-naturally-occurring zinc found in the
geoenvironment would be the zinc compounds associated with the production and use
of industrial products. Deposition of airborne fugitive zinc from mining and extraction of
zinc, together with discharges (spills, wastes, and waste streams) from the processing and
production of products utilizing zinc compounds, account for the major sources of non-
naturally-occurring zinc found in the geoenvironment. Some of these sources are galva-
nizing plant effluents, coal and waste burning, leachates from galvanized structures,
natural ores, and municipal waste treatment plant discharge.

Although not as toxic as cadmium, zinc is quite often associated with this metal. Under
acidic conditions below its precipitation pH, zinc is usually divalent and quite mobile. In
the divalent state, sorption onto the surfaces of reactive soil particles includes ionic bonding and sequestering by organic matter. At high pH, the solubility of its organic and
mineral colloids can render zinc bioavailable. Zinc hydrolyzes at pH 7.0 to 7.5, forming

$Zn(OH)_2$ at pH values higher than 8. Under anoxic conditions, ZnS can form upon precipitation, whereas the unprecipitated zinc can form $ZnOH^+$, $ZnCO_3$, and $ZnCl^+$.

2.3.2   Organic Chemical Pollutants

Organic chemical pollutants found in the geoenvironment have origins from (1) industries producing various chemicals and pharmaceuticals, e.g., refineries, production of specialty chemicals, etc.; (2) waste streams and disposal of chemical products, e.g., sludges and spills; and (3) utilization of various chemical products, e.g., use of petroleum products, pesticides, organic solvents, paints, oils, creosotes, greases, etc. There are at least one million organic chemical compounds registered in the various Chemical Abstracts Services available, with many thousands of these in commercial use. The more common organic chemicals found in the physical landscape of the land environment can be grouped as follows:

- Hydrocarbons, including the petroleum hydrocarbons (PHCs), the various alkanes and alkenes, and aromatic hydrocarbons, such as benzene, multicyclic aromatic hydrocarbons (MAHs) (e.g., naphthalene), and polycyclic aromatic hydrocarbons (PAHs) (e.g., benzo-pyrene)
- Organohalide compounds, of which the chlorinated hydrocarbons are perhaps the best known. These include trichloroethylene (TCE), carbon tetrachloride, vinyl chloride, hexachlorobutadiene, polychlorinated biphenyls (PCBs), and polybrominated biphenyls (PBBs)
- Oxygen-containing and nitrogen-containing organic compounds, such as phenol, methanol, and trinitrotoluene (TNT)

By and large, organic chemicals found in the geoenvironment can be traced to sources and activities associated with humans. Figure 2.6 shows some of the main sources of pollutants (inorganic and organic) found in the land compartment of the geoenvironment. Not all of the organic pollutants are soluble in water. Those that are not are identified as nonaqueous phase organics. Separation of the nonaqueous phase organic compounds into two classes that distinguish between whether they are lighter or denser than water is useful because this tells one about the transport characteristics of the organic compound. These nonaqueous phase organics are called nonaqueous phase liquids (NAPLs), and the distinction between lighter than water and heavier than water is given as LNAPL and DNAPL, respectively; i.e. the LNAPLs are lighter than water and the DNAPLs are heavier than water. Figure 2.7 shows that because the LNAPL is lighter than water, it stays on the surface of or above the water table. Because the DNAPL is denser than water, it will sink through the water table and will come to rest at the impermeable bottom (bedrock). Some typical LNAPLs include gasoline, heating oil, kerosene, and aviation gas. DNAPLs include the organohalide and oxygen-containing organic compounds, such as 1,1,1-trichloroethane, creasote, carbon tetrachloride, pentachlorophenols, dichlorobenzenes, and tetrachloroethylene.

2.3.2.1   *Persistent Organic Chemical Pollutants (POPs)*

Not all the organic chemicals found in the geoenvironment will be biologically and chemically degraded. The characteristic term used to describe organic chemicals that persist in their original form or in altered forms that pose threats to human health is *persistence*. The acronym used to describe persistent organic chemical pollutants is *POPs* (persistent organic pollutants). These are generally defined as organic pollutants that are toxic, per-

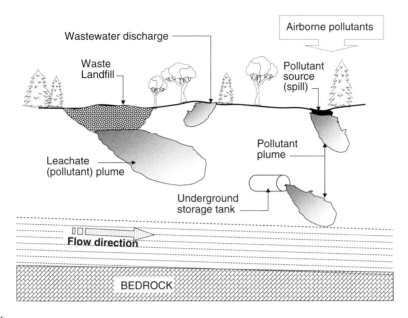

FIGURE 2.6
Schematic showing (a) leachate plume with pollutants emanating from a waste landfill, (b) pollutant plumes from a leaking underground storage tank and wastewater discharge, (c) deposition of airborne pollutants onto land surface, and (d) a surface pollutant source (spill). (Adapted from Yong, R.N. and Mulligan, C.N., *Natural Attenuation of Contaminants in Soils*, Lewis Publishers, Boca Raton, FL, 2004.)

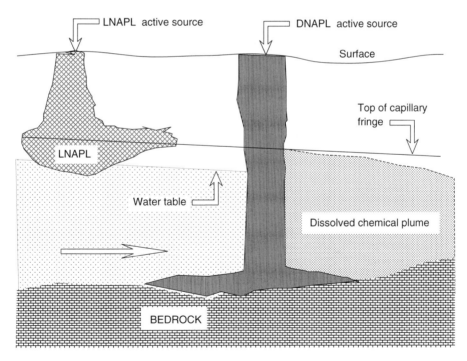

FIGURE 2.7
Schematic diagram showing LNAPL and DNAPL in the subsurface. Note that the LNAPL stays above the water table, whereas the DNAPL penetrates into substratum and rests on the impermeable rock bottom. (From Yong, R.N., *Geoenvironmental Engineering: Contaminated Soils, Pollutant Fate and Mitigation*, CRC Press, Boca Raton, FL, 2001.)

sistent, and bioaccumulative. Included in the POPs are dioxins, furans, the pesticides and insecticides (aldrin, chlordane, DDT, etc.), and a whole host of industrial chemicals grouped as polycyclic aromatic hydrocarbons (PAHs) and halogenated hydrocarbons (including the chlorinated organics). The top 12 POPs that have been identified by the United Nations Environmental Program as POPs for reduction and elimination are dioxins, furans, PCBs, hexachlorabenzene, aldrin, dieldrin, endrin, chlordane, DDT, heptachlor, mirex, and toxaphene. The majority of the top 12 POPs are pesticides. Because of their heavy use in agriculture, golf courses, and even at the household level for control of insects and other pests, it is not difficult to see how these find their way into the geoenvironment.

2.4   Surface and Subsurface Soils

Surface and subsurface soils constitute the uppermost portion of the mantle of the land environment, i.e., the unconsolidated material in the upper layer of the lithosphere. This upper layer is an integral part of the terrestrial ecosystem. When combined with the flora and fauna, this upper mantle constitutes the habitat for terrestrial living organisms. Soil, as a material, can be considered a natural capital of the geoenvironment. There are many functions served by surface and subsurface soils. They provide the physical, chemical, and biological habitat for animals and soil microorganisms. In addition, they support growth of plants and trees and are the vital medium for agricultural production — the virtual host for food production. Soil materials in the subsurface are very useful in the mitigation of the impacts of liquid wastes discharged on (and in) the land surface — because of their inherent chemical and physical buffering capabilities. Soil is a renewable resource that is in danger of becoming a nonrenewable resource (see Chapter 10).

2.4.1   Soil as a Resource Material

In addition, as noted also in Chapter 1, soils are essential as a resource material. Food production, forestry, and extraction of minerals are some of the life support activities that depend on soil. Surface and subsurface soils constitute the primary host or recipient of pollutants and contaminants. From the schematic shown in Figure 2.7, it is evident that the transport and fate of pollutants that find their way into this land compartment of the geoenvironment will be a function of (1) the properties of the soil, (2) the properties of the pollutants themselves, (3) the geological and hydrogeological settings, and (4) the microenvironment (regional controls). It is not the intent of this section or this book to deal in detail with the properties and characteristics of soils; nor is it intended to develop the basic fundamental principles of contaminant–soil interactions. These can be found in Yong and Warkentin (1975), Yong (2001), and Yong and Mulligan (2004). Instead, we will focus on the aspects of soils and their interactions that help us develop a better understanding of soils as they relate to the control of pollutant fate and transport.

2.4.2   Nature of Soils

Soils are three-phased systems: solids, aqueous, and gaseous. All three phases coexist together to form a soil mass. The composition of the solids, the chemistry of the aqueous phase, and the proportions of each phase at any location are functions of how the soil was

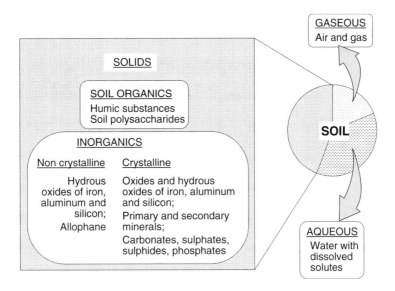

FIGURE 2.8
Soil constituents. Proportioning of the three phases (gaseous, aqueous, and solids) is approximately representative of a partly saturated soil.

formed and how it got there. Soils are formed by natural processes associated with the weathering of rock and the decomposition of organic matter. Weathering or disintegration of the parent rock material can be either physical or chemical. In both cases, rock disintegration will produce smaller fragments, and ultimately, soil material is formed. Unless the soil material is transported to other regions after formation, as for example by wind forces, by water movement, or even by glacial action, the in-place and resident soil material will reflect the primary compositional features of the parent rock. Further weathering of the soil material will produce the compositional features seen in the soil in place. Weathering is at its highest intensity in the upper soil zone in temperate humid climate regions and deeper soil zones in humid tropics. The transformations occur principally in the regolith, the region between solid rock and the topsoil.

Figure 2.8 shows the three phases (gaseous, aqueous, and solids) of soils and the various kinds of constituents that make up a soil. The various constituents in the solid phases are generally identified as *soil fractions* since each type of constituent is a fraction of the soil solids that comprise the total soil itself. The various soil fractions combine to form the natural soils that one sees at any one site. The soils may have been transported and deposited as sedimentary deposits of alluvial, fluvial, or marine action. The soils derived from these actions are appropriately called fluvial soils, alluvium, and marine soils. Figure 2.9 shows an idealized schematic of the various soil fractions grouped into a soil unit. Not all soils have all the various fractions shown in Figure 2.8. How the proportions and distributions of the different soil fractions occur will depend on not only the geologic origin of the soil, but also the regional controls and weathering processes existent at the soil location. At least five factors and four different processes are involved in the production of individual soil fractions. The five factors are:

- Parent rock material: Composition and texture are important. The influence of these features depends on where weathering occurs. In extreme humid conditions and temperatures, the influence of composition and texture are short-lived. However, in arctic and arid regions, the influence of composition and texture of the

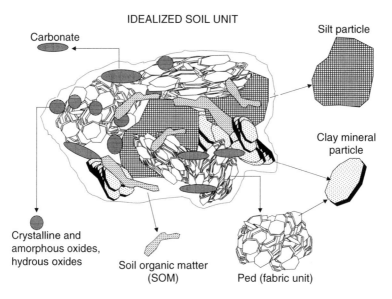

IDEALIZED SOIL UNIT

Carbonate

Silt particle

Clay mineral
particle

Crystalline and
amorphous oxides,
hydrous oxides

Soil organic matter
(SOM)

Ped (fabric unit)

FIGURE 2.9
An idealized typical soil unit in a soil mass consisting of various soil fractions. The positions of the various fractions and the configuration define the structure of the soil. (From Yong, R.N. and Mulligan, C.N., *Natural Attenuation of Contaminants in Soils*, Lewis Publishers, Boca Raton, FL, 2004.)

parent rock material are long-lived, and can even remain indefinitely. Alkali and alkaline earth cations are important factors in determining the weathering products. Thus, for example, rocks containing no alkali can only produce kaolinite or lateritic soils as weathering products. On the other hand, weathering of igneous rocks, shales, slates, schists, and argillaceous carbonates will produce a large variety of weathering products because of the presence of alkalis, alkaline earth cations, alumina, silica, etc.

- Climate: Temperature and rainfall are important climatic factors. Warm and humid climates encourage rapid weathering of the minerals of the parent rock material. Decaying vegetative products and organic acids contribute significantly to the weathering process.

- Topography: This affects how water infiltrates into the ground. The greater the residence time of water, as for example found at low-lying areas that impound water, the greater are the reactions between the solutes in water and the soil material.

- Vegetation: Decaying vegetation is a significant factor since this reacts with the parent silicate minerals.

- Time: This is an important factor in situations where reaction rates are slow.

The four processes that are influential in the weathering sequences include:

- Hydrolysis: This is the reaction between the H^+ and OH^- ions of water and other mineral ions, particularly for the rock-forming silicates.

- Hydration: This is important for the formation of hydrous compounds with the minerals in the rocks such as the silicates, oxides of iron and aluminum, and the sulfates.

- Oxidation: Since most rocks carry iron in the form of sulfides or oxides, oxidation of the Fe to FeS or FeS_2 could easily occur in the presence of moisture since this promotes the process of oxidation.
- Carbonation: The interaction or reaction of carbonic acid with bases will yield carbonates. The process of carbonation in silicates is accompanied by the liberation of silica. The silica may remain as quartz or may be removed as colloidal silica.

2.4.3 Soil Composition

Figure 2.8 and Figure 2.9 show that the solid phase of soils contains both inorganic and organic constituents, and that the inorganic components can be minerals as well as other quasi-crystalline and noncrystalline materials.

2.4.3.1 *Primary Minerals*

We define *primary minerals* as those minerals derived in unaltered form from parent rock material — generally through physical weathering processes. The more common ones found in soils are quartz, feldspar, micas, amphiboles, and pyroxenes. By and large, primary minerals are generally found as sands and silts, with a small portion of clay-size fractions qualifying as primary minerals. We classify particles less than 2 microns in effective diameter as clay-size. This classification is made because it is necessary to distinguish between *clay-size* particles and *clay minerals*.

2.4.3.2 *Secondary Minerals*

Secondary minerals are derived as altered products of physical, chemical, or biological weathering processes. These minerals are layer silicates, commonly identified as phyllosilicates, and they constitute the major portion of the clay-size fraction of soil materials in clays. Because of the possibility for confusion in usage of terms and names, it is important to distinguish between the terms *clays*, *clay soils*, *clay-size*, and *clay minerals*.

Clays and *clay soils* refer to soils that have particle sizes less than 2 micron effective diameters. *Clay-size* refers to soil particles with effective diameters less than 2 microns. No specific reference to the kind or species of particles is required, since attention is directed toward the size of the particles. *Clay minerals* refer specifically to the layer silicates. These are secondary minerals. They consist of oxides of aluminum and silicon with small amounts of metal ions substituted within the crystal structure of the minerals. Because of their size and their structure, secondary minerals have large specific surface areas and significant surface charges. The major groups of clay minerals include kaolinites, smectites (montmorillonites, beidellites, and nontronites), illites, chlorites, and vermiculites.

2.4.3.3 *Soil Organic Matter*

Soil organic matter (SOM) can exist in soils in proportions as low as 0.5 to 5%. Although their proportions may be small, their influence on the bonding of soil particles and aggregate groups, together with their ability to attenuate contaminants, cannot be overstated. SOM originates from vegetation and animal sources and is generally categorized in accord with its state of degradation into humic and nonhumic material. Humic materials or substances are those organics that result from the chemical and biological degradation of nonhumic material. Nonhumic material or compounds, on the other hand, are organics that remain undecomposed or are partly degraded. Humic substances are classified into humic acids, fulvic acids, and humins, with the distinction being made on the basis of their solubility to acid and base.

2.4.3.4 Oxides and Hydrous Oxides

The general list of oxide and hydrous oxide minerals includes the oxides, hydroxides, and oxyhydroxides of iron, aluminum, manganese, titanium, and silicon. The common crystalline form of these minerals includes hematite, goethite, gibbsite, boehmite, anatase, and quartz. They differ from layer silicate minerals (secondary minerals) in that their surfaces essentially consist of broken bonds. In an aqueous environment, these broken bonds are satisfied by the OH groups of disassociated water molecules. The surfaces exhibit pH-dependent charges; i.e., the surfaces have variable charged properties.

2.4.3.5 Carbonates and Sulfates

The most common carbonate mineral found in soils is calcite ($CaCO_3$). Some of the other less common ones are magnesite ($MgCO_3$) and dolomite ($CaMg(CO_3)_2$). Gypsum ($CaSO_4 \cdot 2 H_2O$) is the most common of sulfate minerals found in soils.

2.4.4 Soil Properties Pertinent to Pollutant Transport and Fate

The reactions between pollutants and soil during the time when the pollutant is in contact with a soil will determine its transport through the soil, and also its fate. Figure 2.10 shows a simplified schematic of an influent pollutant leachate entering a soil unit. Interaction between the various pollutants in the leachate stream with the exposed surfaces of the soil particles of the various soil fractions will ultimately determine the transport characteristics and fate of the pollutants in the leachate stream. The controlling factors involved, other than the properties and functional groups (chemically reactive groups) of the pol-

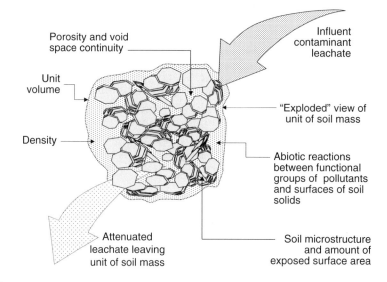

FIGURE 2.10
Schematic diagram showing the major physical soil factors involved in controlling contaminant transport through the unit soil mass.

lutants, are listed in the diagram. The two types of interactions involved in the pollutant transport processes are:

- Physical interactions: These interactions involve fluid and pollutant movement through the soil fabric and structure. As such, the nature of the void spaces and the amount of surface area presented by the soil particles to the liquid waste and leachate stream are important factors. The principal factors include:
  - Void spaces: The size distribution of voids and the continuity of voids are important factors in determining the rate of transport of the leachate stream through the soil. The nature of these voids, their continuity, and their distribution are all dependent on the density and structure of the soil. These in turn are functions of soil composition and manner in which the soil was formed *in situ*.
  - Surface area and microstructure: The interactions of the various pollutants in the leachate stream with the surfaces of the soil particles will be dictated by the amount of surfaces exposed to the pollutants. Because of the existence of soil microstructural units, i.e., packets of soil particles grouped together to form aggregate groups or peds or clusters of particles, not every single particle will have its total surface area exposed to the leachate stream. The sizes and types of microstructural units that comprise the soil in question will determine the amount of surface areas exposed to the influent leachate. A detailed discussion on soil microstructure and hydraulic conductivity can be found in Sections 10.2 and 10.3 in Chapter 10, in the discussion on mitigation of impacts from pollutant transport.
- Chemical interactions: This grouping of interactions includes all the types of chemical reactions that occur when two chemically reactive participants interact with each other. The surface properties, especially the surface chemistry of the soil solids, are very important factors. The surface chemically reactive groups for the soil solids and for the pollutants are identified as functional groups. These will be discussed in a later section in this chapter.

From the schematics shown in Figure 2.8 and Figure 2.9, we obtain an appreciation of the constituents of a typical soil unit and have learned that the different soil solids are known as soil fractions. Although the types of soil fractions can range from sands to clay minerals to soil organics, it is the soil fractions with reactive surfaces that are of interest in the study of transport and fate of pollutants. *Reactive surfaces*, in the present soil unit context, are defined as those surfaces that can react chemically with dissolved solutes in the pore water of the soil. We should also note that pollutants in the ground will also have reactive surfaces, and in this case, these reactive surfaces can react chemically with other dissolved solutes in the pore water and also with the soil solids.

2.4.4.1 Specific Surface Area (SSA) and Cation Exchange Capacity (CEC)

The soil fractions that have more particles with significant reactive surfaces are the clay minerals, oxides and hydrous oxides, soil organics, and carbonates. Table 2.1 gives the surface charge characteristics, SSA, and CEC for some clay minerals.

We define the *specific surface area* (SSA) as the total surface area of all the soil solids or particles per unit volume. Since theoretical calculations for specific surface area can become both complex and tedious, because of the irregular shapes and sizes of the soil particles, laboratory techniques are often used. A popular procedure is to determine the amount of

TABLE 2.1

Charge Characteristics, SSA, and CEC for Some Clay Minerals

Soil Fraction	Cation Exchange Capacity (CEC), meq/100 g	Surface Area, m²/g	Range of Charge, meq/100 g	Reciprocal of Charge Density, nm²/charge	Isomorphous Substitution	Source of Charges
Kaolinite	5–15	10–15	5–15	0.25	Dioctahedral: 2/3 of positions filled with Al	Surface silanol and edge silanol and aluminol groups (ionization of hydroxyls and broken bonds)
Clay micas and chlorite	10–40	70–90	20–40	0.5	Dioctahedral: Al for Si Trioctahedral or mixed Al for Mg	Silanol groups, plus isomorphous substitution and some broken bonds at edges
Illite	20–30	80–120	20–40	0.5	Usually octahedral substitution Al for Si	Isomorphous substitution, silanol groups and some edge contribution
Montmorillonite[a]	80–100	800	80–100	1.0	Dioctahedral: Mg for Al	Primarily from isomorphous substitution, with very little edge contribution
Vermiculite[b]	100–150	700	100–150	1.0	Usually trioctahedral substitution Al for Si	Primarily from isomorphous substitution, with very little edge contribution

Note: Ratios of external:internal surface areas are highly approximate since surface area measurements are operationally defined; i.e., they depend on the technique used to determine the measurement.

[a] Surface area includes both external and intralayer surfaces. Ratio of external particle surface area to internal (intralayer) surface area is approximately 5:80.

[b] Surface area includes both external and internal surfaces. Ratio of external to internal surface area is approximately 1:120.

Source: Adapted from Yong, R.N., *Geoenvironmental Engineering: Contaminated Soils, Pollutant Fate and Mitigation*, CRC Press, Boca Raton, FL, 2001.

gas or liquid (adsorbate) that forms a monolayer coating on the surface of the particles. The choice of the adsorbate and the availability of soil particles in a totally dispersed state are important factors in production of the final sets of data. Because of the dependence on techniques used, we consider laboratory measurement of the specific surface area (SSA) of a soil sample to be an operationally defined property, i.e., dependent on technique, adsorbate used, and degree to which the soil has been properly dispersed.

To explain the *reciprocal of charge density* shown in Table 2.1, we need to explain what surface charge density means. The *surface charge density* is the total number of electrostatic charges on the clay particles' surfaces divided by the total surface area of the particles. The common procedure is to express this surface charge density in terms of its reciprocal, as shown in Table 2.1. We have omitted the values for the hydrous oxides such as goethite ($-FeOOH$) and gibbsite ($-Al(OH)_3$) from the table because the range of values for these types of soil fractions are dependent upon (1) their structure, (2) the specifically adsorbed potential-determining ions, and (3) the pH of the pore water.

The *cation exchange capacity* (CEC) is defined as the quantity of exchangeable ions held by a soil and is generally equal to the amount of negative charge in the soil. This is usually expressed in terms of milliequivalents per 100 g of soil (meq/100 g soil). Exchangeable cations are associated with clay minerals, amorphous materials, and natural soil organics. Many of the surface functional groups of these soil fractions are direct participants in cation exchange, e.g., the oxygen-containing functional groups of soil organic matter (SOM), such as the carboxyl and phenolic functional groups. While not reported in Table 2.1, we see measured values for CEC ranging from 15 to 24 meq/100 g soil for Fe oxides, from 10 to 18 meq/100 g soil for Al oxides, and from 20 to 30 meq/100 g for allophanes. CEC values of up to 100 meq/100 g soil for goethites and hematites, and from 150 to 400 meq/100 g soil for organic matter at a pH of 8 have been reported by Appelo and Postma (1993). Their empirical relationship for the CEC of a soil is given in terms of the percentage of clay less than 2 μm and the organic carbon as follows:

$$CEC \ (meq/100 \ g \ soil) = 0.7 \ Clay\% + 3.5 \ OC\%$$

where Clay% refers to the percentage of clay less than 2 μm and OC% refers to the percentage of organic carbon in the soil.

By combining the density of charges with the amount of surface areas available and the cation exchange capacity of the specific clay mineral, we will obtain some appreciation of the degree of reactivity of the clay mineral in question. This should not be construed as a quantitative estimate since actual field soils will not have all particles and their surfaces available for exposure to pollutants. Aggregate groups of particles such as flocs, domains, peds, and clusters will diminish the total calculated surface area obtained from the single-particle theory.

2.4.5 Surface Properties

The surface properties of soils are important because it is these properties, together with those surface properties of pollutants themselves, and the geometry and continuity of the pore spaces that will control the transport processes of the pollutants. We have previously defined *reactive surfaces* to mean those surfaces that, by virtue of their properties, are capable of reacting physically and chemically with solutes and other dissolved matter in the pore water. The chemically reactive groups, which are molecular units, are found on the surfaces of the various soil fractions and are defined as *surface functional groups*. These surface functional groups give the surfaces their reactive properties.

The soil fractions that possess significant reactive surfaces include layer silicates (clay minerals), soil organics, hydrous oxides, carbonates, and sulfates. The surface hydroxyls (OH group) are the most common surface functional group in inorganic soil fractions (soil solids), such as clay minerals with disrupted layers (e.g., broken crystallites), hydrous oxides, and amorphous silicate minerals. The common functional groups for soil organic matter (SOM) include the hydroxyls, carboxyls, phenolic groups, and amines. More detailed explanations concerning the nature of these functional groups and their manner of interaction with the functional groups associated with pollutants can be found in soil science and geoenvironmental engineering textbooks (e.g., Sposito, 1984; Greenland and Hayes, 1981; Huang et al., 1995a, 1995b; Knox et al., 1993; Yong, 2001; Yong and Mulligan, 2004).

2.5   Pollutant Transport and Land Contamination

The transport of pollutants (or contaminants) in soils refers to the movement of pollutants through the pore spaces in soils. Liquid pollutants such as organic chemical compounds and inorganic/organic pollutants carried in waste streams and leachate streams that pass through the soil pore spaces will interact with the exposed surfaces of the soil fractions. The reactions arising from the interactions between pollutants and soil fractions will dictate the nature of the transport of pollutants and, indirectly or directly, the fate of the pollutants. It is useful to remember that except for liquid chemicals, water is the primary carrier or transport agent for pollutants. The liquid phase of a soil–water system, i.e., the pore water, consists of water and dissolved substances such as free salts, solutes, colloidal material, and organic solutes. All dissolved ions, and probably all dissolved molecules, are to some extent surrounded by water molecules.

The questions relating to *what happens to pollutants in the ground* are perhaps the most critical concerns at hand. Will they eventually disappear? How long will they likely stay in the ground? Will they move to other locations, i.e., be transported? Will they be harmful to human health and the environment? To address the questions, it is necessary to obtain an understanding of the partitioning mechanisms, i.e., the chemical mass transfer of pollutants from the pore water to the surfaces of the soil solids, and how these are related to the soil and pollutant properties. A detailed consideration of these will be found in Sections 9.4 and 9.5 in Chapter 9 and in Section 10.3 in Chapter 10. For this section, we will highlight some of the main elements of the interactions as they relate to the mass transfer of dissolved solutes in the pore water (partitioning of dissolved solutes).

2.5.1   Mechanisms of Interaction of Heavy Metal Pollutants in Soil

The processes of transfer of metal cations from the soil pore water can be grouped as follows:

- Sorption: This includes physical adsorption (physisorption), occurring principally as a result of ion exchange reactions and van der Waals forces, and chemical adsorption (chemisorption), which involves short-range chemical valence bonds. The general term *sorption* is used to indicate the process in which the solutes (ions, molecules, and compounds) are partitioned between the liquid phase and the soil

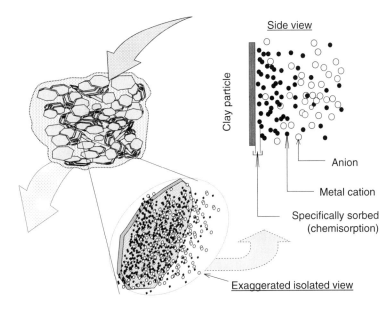

Side view

Clay particle

Anion

Metal cation

Specifically sorbed
(chemisorption)

Exaggerated isolated view

FIGURE 2.11
Exaggerated isolated view of interaction of metal cations (in the pore water) with soil particles. The soil particle shown is a clay mineral particle.

particle interface. When it is difficult to fully distinguish between the mechanisms of physical adsorption, chemical adsorption, and precipitation, the term *sorption* is used to indicate the general transfer of material to the interfaces.

- *Physical adsorption* occurs when the pollutants or contaminants in the soil solution (aqueous phase, pore water) are attracted to the surfaces of the soil solids because of the unsatisfied charges of the soil particles. In the case of the heavy metals (metal cations), for example, they are attracted to the negative charges exhibited by the surfaces of the soil solids (Figure 2.11). When the cations are held primarily by electrostatic forces, this is called *nonspecific* cation adsorption. *Specific cation adsorption* refers to the situation where the ions penetrate the coordination shell of the structural atom and are bonded by covalent bonds via O and OH groups to the structural cations. The valence forces are of the type that binds atoms to form chemical compounds of definite shapes and energies. This type of adsorption is also referred to as *chemisorption.*

- Complexation with various ligands: *Complexation* occurs when a metallic cation reacts with an anion that functions as an inorganic ligand. Metallic ions that can be complexed by inorganic ligands include the transitional metals and alkaline earth metals. The inorganic ligands that will complex with the metallic ions include most of the common anions, e.g., OH^-, Cl^-, SO_4^{2-}, CO_3^{2-}, PO_3^{3-}, etc. Complexes formed between metal ions and inorganic ligands are much weaker than those complexes formed with organic ligands.

- Precipitation: Accumulation of material (solutes, substances) on the interface of the soil solids to form new (insoluble) bulk solid phases. Precipitation occurs when the transfer of solutes from the aqueous phase to the interface results in accumulation of a new substance in the form a new soluble solid phase. The Gibbs' phase rule restricts the number of solid phases that can be formed. Precipitation can occur on the surfaces of the soil solids or in the pore water.

2.5.2 Chemically Reactive Groups of Organic Chemical Pollutants

When organic chemical compounds come in contact with soil, the nature of the chemically reactive groups in the organic molecules, their shape, size, configuration, polarity, polarizability, and water solubility are important factors in determining the adsorption of these chemicals by the soil solids. These chemically reactive groups, which are also known as functional groups, populate both the surfaces of pollutants and soil solids. The chemical properties of the functional groups of the organic chemicals will influence the surface acidity of the soil particles. This is important in the adsorption of ionizable organic molecules by the soil solids (clays).

The mechanisms of interaction between organic chemicals and soil fractions include (1) London–van der Waals forces, (2) hydrophobic bonding, (3) charge transfer, (4) ligand and ion exchange, and (5) chemisorption. Sorption of organic chemicals is enhanced when there is no hydration layer (of water) on the surfaces of soil particles. Further sorption of other organic chemicals occurs through van der Waals type forces and hydrogen bond formation between functional groups, such as the hydroxyl (OH) group on the soil particles and the carboxyl (COOH) group on the organic chemicals.

The *hydroxyl group* (OH) consists of a hydrogen atom and an oxygen atom bonded together. This group is by far the most common reactive surface functional group for soil fractions such as clay minerals, amorphous silicate minerals, metal oxides, and the other oxides (oxyhydroxides and hydroxides). The hydroxyl group is also present in two groups of organic chemicals:

1. Alcohols: Methyl, ethyl, isopropyl, and n-butyl. Alcohols can be considered hydroxyl alkylcompounds (R–OH) and are neutral in reaction since the OH group does not ionize.
2. Phenols: Monohydric (aerosols) and polyhydric (obtained by oxidation of acclimated activated sludge [pyrocatechol, trihydroxybenzene]).

The two other kinds of functional groups associated with organic chemical compounds (Figure 2.12), in addition to the hydroxyl (OH) group, are:

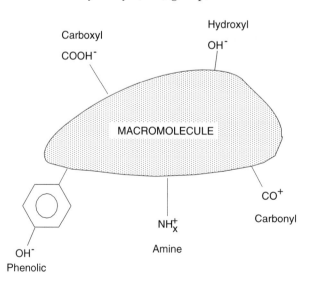

FIGURE 2.12
Some typical functional groups associated with organic chemicals. The macromolecule shown in the diagram is an organic chemical.

1. Functional groups having a C–O bond: These include the carboxyl, carbonyl, methoxyl, and ester groups. Compounds possessing the *carbonyl group*, called carbonyl compounds, include aldehydes, ketones, and carboxylic acids. The carboxyl group, which combines the carbonyl and hydroxyl groups into a single unit to form a new functional group, is the characteristic functional group of carboxylic acids, e.g., benzoic and acetic acids.

2. Nitrogen-bonding functional groups such as the amine and nitrile groups: The *amino group* NH_2 is found in primary amines. The amines may be aliphatic, aromatic, or mixed, depending on the nature of the functional groups, and are classified as:

   • Primary: For example, methylamine (primary aliphatic), aniline (primary aromatic).

   • Secondary: For example, dimethylamine (secondary aliphatic), diphenalamine (secondary aromatic).

   • Tertiary: For example, trimethylamine (tertiary aliphatic).

Surface acidity is very important in the adsorption of ionizable organic molecules of clays. The chemical properties of the functional groups of the soil fractions contribute appreciably to the acidity of the soil particles. Surface acidity is an important factor in clay adsorption of amines, s-triazines, amides, and substituted urea. This is due to protonation on the carbonyl group — as demonstrated by the hydroxyl groups in organic chemical compounds. As shown at the beginning of this section, there are two broad classes of these compounds: (1) alcohols — ethyl, methyl, isopropyl, etc., and (2) phenols — monohydric and polyhydric. In addition, there are two types of compound functional groups, i.e., those having a C–O bond (carboxyl, carbonyl, methoxyl, etc.) and the nitrogen-bonding group (amine and nitrile). Amine, alcohol, and other organic chemicals that possess dominant carbonyl groups that are positively charged by protonation can be readily sorbed by clays. In amines, for example, the NH_2 functional group of amines can protonate in soil, thereby replacing inorganic cations from the clay complex by ion exchange. The extent of sorption of these kinds of organic molecules depends on (1) the CEC of the clay minerals, (2) the composition of the clay soil (soil organics and amorphous materials present in the soil), (3) the amount of reactive surfaces, and (4) the molecular weight of the organic cations. Because they are longer and have higher molecular weights, large organic cations are adsorbed more strongly than inorganic cations. Polymeric hydroxyl cations are adsorbed in preference to monomeric species because of the lower hydration energies and higher positive charges and stronger interactive electrostatic forces.

The unsymmetrically shared electrons in the double bond endow carbonyl compounds with dipole moments, thus allowing for hydrogen bonding between the OH group of the adsorbent (soil particles) and the carbonyl group of the ketone or through a water bridge. Sorption onto soil particles, especially clays, for the carbonyl group or organic acids (e.g., benzoic and acetic acids) occurs directly with the interlayer of cation or by formation of hydrogen bonds with the water molecules (water bridging) coordinated to the exchangeable cation of the clay complex.

2.5.3 Partitioning of Pollutants and Partition Coefficients

Section 9.5 in Chapter 9 provides a more detailed discussion of pollutant partitioning in transport through a soil. The introduction to this subject in this section provides an overall appreciation of what happens when waste leachate streams enter the subsurface

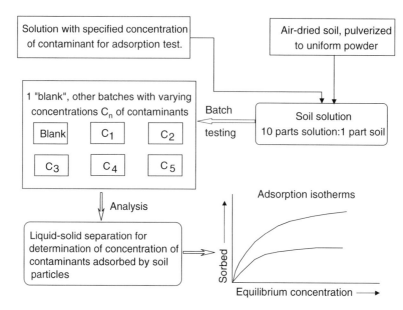

FIGURE 2.13

Batch equilibrium procedure for determination of adsorption isotherms. (From Yong, R.N. and Mulligan, C.N., *Natural Attenuation of Contaminants in Soils*, Lewis Publishers, Boca Raton, FL, 2004.)

soil that constitutes the land environment. The *partitioning* of pollutants refers to the transfer of pollutants from the pore water in the soil to the soil solids by processes that include all of those described in the previous subsections. It is important to determine the partitioning of target pollutants because this will tell us something about its distribution of the pollutants — pollutants sorbed by the soil particles and pollutants remaining in the pore water. This can be interpreted in terms of the quantity or proportion of pollutants likely to move from one location to another. Determination of partitioning of inorganic contaminants and pollutants is generally conducted using batch equilibrium tests. The conventional procedure shown in Figure 2.13 uses soil solutions. The candidate soil is used with an aqueous solution consisting of the contaminant or pollutant of interest to form a soil solution. Figure 2.13 shows the various steps and analyses required. Results obtained from the tests are called adsorption isotherms. Graphical representation of these results show sorbed concentration (of contaminants or pollutants) on the ordinate and equilibrium concentration of contaminants (pollutants) on the abscissa, as seen at the bottom right-hand part of Figure 2.13. The three common types of adsorption isotherms (Freundlich, Langmuir, and constant) found in reported literature on sorption characteristics of inorganic contaminants are shown in Figure 2.14. The parameter k_n in the equations shown with the various curves denotes the slope of the curves.

In the case of organic chemicals, partitioning is denoted by an *equilibrium partition coefficient* k_{ow}, i.e., coefficient describing the ratio of the concentration of a specific organic pollutant in other solvents to that in water. This coefficient k_{ow}, which relates the water solubility of an organic chemical with its *n*-octanol solubility, is more correctly referred to as the *n*-octanol–water partition coefficient. It has also been found to be sufficiently correlated to soil sorption coefficients. The relationship for the *n*-octanol–water partition coefficient k_{ow} has been given in terms of the solubility S by Chiou et al. (1982) as

$$\log k_{ow} = 4.5 - 0.75 \log S \text{ (ppm)}$$

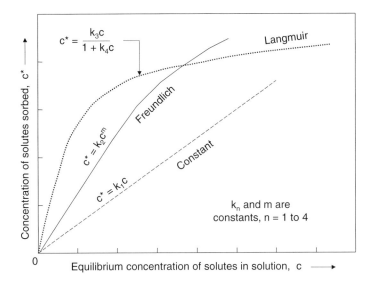

FIGURE 2.14

Freundlich, Langmuir, and constant types of adsorption isotherms obtained from batch equilibrium tests. c = concentration of solutes or contaminants; c^* = concentration of solutes or contaminants sorbed by soil fractions.

Organic chemicals with k_{ow} < 10 are generally considered to be relatively hydrophillic. They tend to have high water solubilities and small soil adsorption coefficients. Organic chemicals with k_{ow} > 10^4 have low water solubilities and are considered to be very hydrophobic. The greater the hydrophobicity, the greater is its bioaccumulation potential (Section 9.5 in Chapter 9).

2.5.4 Predicting Pollutant Transport

There are many models that purport to describe the movement of contaminants, solutes, pollutants, etc., in soil. For convenience in discussion, we will use the term *solutes* to mean contaminants and pollutants and all other kinds of solutes found in the pore water. Most of the models deal with movement of solutes in saturated soils and are best applied to inorganic contaminants and pollutants such as heavy metals. Application of the models for determination of transport of organic chemicals has been attempted by some researchers — with varied degrees of success. So long as the movement of the pollutants is governed by Fick's law, some success in prediction of transport can be obtained. The ability to properly model partitioning of the pollutants remains one of the central issues in determination of success or failure of predictions. The problem is complicated by the fact that a reasonably complete knowledge of the initial and boundary conditions is not always available. Additionally, the presence of multicomponent pollutants and their individual and collective reactions with the soil fractions will make partitioning determinations difficult. Detailed discussions of the modeling problems and the physicochemical interactions and partitioning of pollutants in soils can be found in textbooks dedicated to the study of pollutant fate and transport in soils (e.g., Knox et al., 1993; Fetter, 1993; Huang et al., 1995a, 1995b; Yong, 2001; Yong and Mulligan, 2004). The most common and widely used transport model has the following relationship:

$$\frac{\partial c}{\partial t} = D_L \frac{\partial^2 c}{\partial x^2} - v \frac{\partial c}{\partial x} - \frac{\rho}{n \rho_w} \frac{\partial c^*}{\partial t} \tag{2.2}$$

where c = concentration of solutes or contaminants, t = time, D_L = longitudinal dispersion coefficient, v = advective velocity, x = spatial coordinate, ρ = bulk density of soil, ρ_w = density of water, n = porosity of soil, and c^* = concentration of solutes or contaminants adsorbed by soil fractions (Figure 2.14). If we assume a slope constant $k_d = k_1$, as shown in Figure 2.14 for the constant adsorption isotherm, the concentration of solutes sorbed by the soil fractions c^* can be written as $c^* = k_d c$. The slope constant k_d is defined as the distribution coefficient and is meant to indicate the manner of distribution of the solutes being transported in the pore water of a soil–water system. Equation 2.2 can be written in a more compact form to take into account the distribution coefficient as follows:

$$R \frac{\partial c}{\partial t} = D_L \frac{\partial^2 c}{\partial x^2} - v \frac{\partial c}{\partial x} \qquad (2.3)$$

where R = retardation factor = $\left[1 + \dfrac{\rho}{n\rho_w} k_d \right]$.

A more detailed discussion of the transport and fate of pollutants will be found in Chapter 9.

2.6   Geoenvironmental Land Management

Since the various discussions presented in this and subsequent chapters in this book concern *geoenvironmental land management* for the purpose of meeting the basic requirements leading to *geoenvironmental sustainability*, we will direct our attention here to some of these basic elements. As stated at the beginning of this chapter, the focus is on the geoenvironment portion of land management because of the very significant problems associated with contamination of the various constituents in the land environment.

The major geosphere and hydrosphere features that constitute the geoenvironment components for land management attention are shown in Figure 2.15. Land management, as the term implies, is the utilization of management practices to a land environment to meet a set of *land use* objectives. Because of the use of several common terms in the various geosciences and geoenvironmental communities, we will define the following terms as they relate to their use in *geoenvironmental land management*:

- *Land use*: Utilization of a land to fit a particular set of objectives or purposes. A simple example of such can be found in such usages as eco-parks, playing fields, forests, wetlands, etc. It is important to appreciate that land use should be considered within the bounds of proper environment and natural resource protection.

- *Land capability*: The performance capability of a piece of land, i.e., what a land is capable of *doing*. Land capability requires one to determine the natural capital of the land and to determine how these assets can be utilized. In the context of geoenvironmental sustainability, it follows that utilization of these assets must be consistent with the principles of sustainability.

- *Natural capital and assets*: This includes the physical attributes, such as landforms and natural resources (including biodiversity). This grouping of properties is not only the most important, but also perhaps the most difficult to fully delineate.

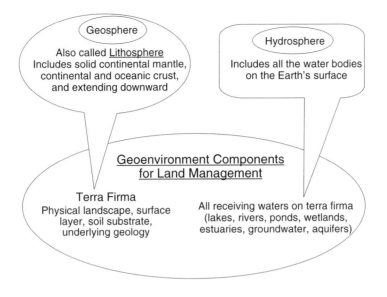

FIGURE 2.15
Geosphere and hydrosphere features included as geoenvironment components for land management.

Other than the obvious assets of a piece of land, e.g., mineral and hydrocarbon resources, what constitutes an asset or a natural capital is to some extent dependent on the sets of criteria and guidelines developed for the particular problem at hand.

The manner in which a land is exploited is dictated by several factors and forces, not the least of which is *land capability*. Figure 2.16 gives an illustration of the procedure that might be used in land use planning and implementation to satisfy geoenvironmental land management concerns. As with any project, the impacts to the geoenvironment must be evaluated, and procedures for avoidance and mitigation of these impacts need to be

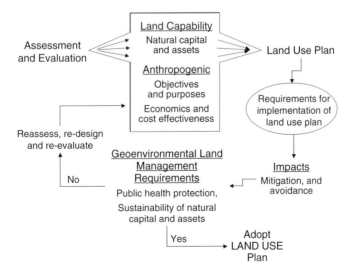

FIGURE 2.16
Procedure to be used to satisfy geoenvironmental land management issues in adoption and implementation of land use plans.

established. The end result of all of these must satisfy geoenvironmental land management requirements; i.e., they must ensure that there are no threats to public health and that the natural capital and assets of the site are maintained. Reassessment and redesign are necessary if initial land use plans do not satisfy geoenvironmental impact concerns. The key elements that must be satisfied are preservation of natural capital and assets.

2.7 Concluding Remarks

We have focused on contamination of the land environment as one of the key issues in the need to protect the natural capital and assets of the land environment.

- Geoenvironmental land management requires one to practice the principles of sustainability.
- As far as the land environment is concerned, environmental impact associated with anthropogenic activities can be in the form of changes in the quantity or quality of the various features that constitute the land environment.
- Not all anthropogenic activities will result in adverse impacts on the environment.
- The degree of environmental impact due to pollutants in a contaminated ground site is dependent on (1) the nature and distribution of the pollutants; (2) the various physical, geological, and environmental features of the site; and (3) existent land use.
- Each type of land use imposes different demands and requirements from the land. The ideal situation in land utilization matches land suitability with land development consistent with environmental sensitivity and sustainability requirements.
- Groundwater is an integral part of land use considerations.
- Causes and sources of groundwater contamination include wastewater discharges, injection wells, leachates from landfills and surface stockpiles, open dumps and illegal dumping, underground storage tanks, pipelines, irrigation practices, production wells, use of pesticides and herbicides, urban runoff, mining activities, etc.
- To evaluate and determine the nature of geoenvironmental impacts created by the presence of pollutants in the geoenvironment, one needs to have an appreciation of the nature of the pollutants and associated events that are responsible for the impacts. The various activities associated with the production of goods and services generate waste streams and products. In most instances, these waste streams and products find their way into the land environment — either inadvertently, as in the case of runoffs and spills, or by design, i.e., constructed safe land disposal facilities.

References

Agency for Toxic Substances and Disease Registry (ATSDR), (1999), Toxicological Profile for Cadmium, U.S. Department of Health and Human Services, Public Health Service, Atlanta, GA.

Appelo, C.A.J. and Postma, D., (1993), *Geochemistry, Groundwater and Pollution*, Balkema, Rotterdam, 536 pp.

Chiou, G.T., Schmedding, D.W., and Manes, M., (1982), Partition of organic compounds on octanol-water system, *Environ. Sci. Technol.*, 16:4–10.

Fetter, C.W., (1993), *Contaminant Hydrogeology*, Macmillan Publishing Co., New York, 458 pp.

Greenland, D.G. and Hayes, M.H.B. (Eds.), (1981), *The Chemistry of Soil Processes*, John Wiley & Sons, New York, 714 pp.

Huang, P.M., Berthelin, J., Bollag, J.-M., McGill, W.B., and Page, A.L. (Eds.), (1995a), *Environmental Impact of Soil Component Interactions: Natural and Anthropogenic Organics*, CRC Press, Boca Raton, FL, 450 pp.

Huang, P.M., Berthelin, J., Bollag, J.-M., McGill, W.B., and Page, A.L. (Eds.), (1995b), *Environmental Impact of Soil Component Interactions: Metals, Other Inorganics, and Microbial Activities*, CRC Press, Boca Raton, FL, 263 pp.

Karlen, D.L., Mausbach, M.J., Doran, J.W., Cline, R.G., Harris, R.F., and Schuman, G.E., (1997), Soil quality: a concept, definition, and framework for evaluation, *Soil Sci. Soc. Am. J.*, 61:4–10.

Knox, R.C., Sabatini, D.A., and Canter, L.W., (1993), *Subsurface Transport and Fate Processes*, Lewis Publishers, Boca Raton, FL, 430 pp.

Lewis, D.R., Southwick, J.W., Ouellet-Hellstrom, R., Rench, J., and Calderon, R.L., (1999), Drinking water arsenic in Utah: a cohort mortality study, *Environ. Health Perspect.*, 107:359–365.

Mazumder, D.N.G., Haque, R., Ghosh, N., De, B.K., Santra, A., Chakraborty, D., and Smith, A.H., (1998), Arsenic levels in drinking water and the prevalence of skin lesions in West Bengal, India, *Int. J. Epidemiol.*, 27:871–877.

Sposito, G., (1984), *The Surface Chemistry of Soils*, Oxford University Press, New York, 234 pp.

U.S. EPA, (1984), Health Assessment Document for Inorganic Arsenic, EPA-600/8-83-012F, Final Report, U.S. EPA Office of Research and Development, Office of Health and Environmental Assessment, Environmental Criteria and Assessment, Research Triangle Park, NC, pp. 5–20.

Yong, R.N., (2001), *Geoenvironmental Engineering: Contaminated Soils, Pollutant Fate and Mitigation*, CRC Press, Boca Raton, FL, 307 pp.

Yong, R.N. and Mulligan, C.N., (2004), *Natural Attenuation of Contaminants in Soils*, Lewis Publishers, Boca Raton, FL, 319 pp.

Yong, R.N. and Warkentin, B.P., (1975), *Soil Properties and Behaviour*, Elsevier Scientific Publishing Co., Amsterdam, 449 pp.

3

Sustainable Water Management

3.1 Introduction

As we have seen from Chapter 1, the water component of the geoenvironment includes all rivers, lakes, ponds, inlets, wetlands, estuaries, coastal water, groundwaters, and aquifers. These contribute as inputs to the oceans that make up 70% of the earth's surface water. Water is required for many needs, such as drinking, agriculture, cooking, domestic and industrial uses, transportation, recreation, electrical power production, and support for aquatic life. Among the many reasons for increasing water shortages are (1) demand in excess of supply, (2) depletion of aquifers, (3) lack of rain and other forms of precipitation, (4) watershed mismanagement, and (5) diversion of rivers. It is noted that irrigation requirements for agriculture are increasing. Recent estimates indicate that more than 70% of the world's freshwater is utilized for agricultural purposes (Postel, 1999).

The importance of water has been highlighted at the Johannesburg World Summit on Sustainable Development (WSSD 2002). Water and sanitation were identified as one of the five thematic areas at WSSD 2002. The quality of the water can be, and is, significantly affected by the other four thematic areas defined by the WSSD 2002: sustenance and development, industrialization, urbanization, and resource exploitation and agriculture. Adequate quantities of good quality are also essential for health, agriculture, energy, and biodiversity. Lack of water and poverty are intimately linked. The discussion in this chapter will focus on the uses (and misuse) of water in the geoenvironment and will examine some of the main elements required to address, contain, and manage the impacts to water quality — as a step toward water management for sustainability of water resources.

3.2 Uses of Water and Its Importance

Safe and adequate amounts of water are essential. The first Dublin–Rio principle has emphasized the need for sustainable water resource practices. It states that "fresh water is a finite and vulnerable resource, essential to sustain *life, development and the environment.*" In other words, water is essential for these three categories. Note that the use of the term *environment* is meant to include the life-supporting functions of ecosystems, as discussed in Chapter 1.

3.2.1 Hydrologic Cycle

Although it is convenient to think in terms of essentially two primary sources of drinking water, i.e., surface water and groundwater (aquifer and soil pore water), it is more useful to bear in mind the total hydrologic cycle when one wishes to consider (1) the need for water for life support systems and (2) the impact of mankind on the sources of water. The hydrologic cycle reflects the constant or continuous movement of water within the earth, on the earth, and in the atmosphere. It is a cycle because of the various processes that are part of the hydrologic cycle. If we visualize the ocean and other surface water bodies, together with open land surfaces as the beginning source point, we can identify these processes as (1) evaporation and transpiration, (2) condensation, (3) precipitation, (4) infiltration (percolation), and (5) runoff. It is necessary to note that the terms *infiltration* and *percolation* are quite often used to mean the same event, i.e., entry of water into the ground surface from water on the surface, either from precipitation or from ponded water, or other similar sources. Direct anthropogenic interference in the natural hydrologic cycle occurs most often in processes such as infiltration and runoff, as for example in the quality of the water that infiltrates into the ground and also the quality of the runoff that enters the receiving waters.

3.2.1.1 *Human Interference on Infiltration and Runoff*

Human interference in the hydrologic cycle is most significant in the processes of infiltration and runoff. The forms of some of the major interferences on infiltration and their impacts are as follows:

- Development, production, and construction of impermeable surface areas: These impermeable surfaces include housing and similar structures, roads and pavements, runways and aprons, parking lots, and other generally paved surfaces constructed in urban areas. The effect of these impermeable surfaces is to deny infiltration into the ground, hence denying recharge of any underlying aquifer. Runoffs obtained on the impermeable surfaces are generally fed to storm drains or other similar drainage systems.

- Compacted surface layers: These are surfaces of natural soil compacted by agricultural and construction machinery and other similar devices. The effect of compacted surface layers is to reduce the infiltration rate and infiltration capability in general. Surface runoff occurs when the rate of precipitation onto the surface is greater than the infiltration rate. Unlike the paved impermeable surfaces, these runoffs are not generally fed to storm drains or other catchment facilities.

- Soil pore water and aquifer pollution by infiltration: The term *soil pore water* is used to mean the water in the pores of the soil matrix. Normally, in compact clay soils, this pore water is not easily or readily extractable. However, in more granular materials such as silts and sands, this pore water can be harvested. Aquifers in general are seams or layers of primarily granular materials that are full of water. We use the term *groundwater* to mean both soil pore water and water in the aquifers, but mainly to denote water in the ground that is normally considered a water resource. When it is necessary to talk about soil pore water, the term *pore water* is generally used. Contaminants and pollutants on the land surfaces will be transported into the subsoil via infiltration. Pesticides, herbicides, insecticides, organic and animal wastes, ground spills, other deliberate and inadvertent discharges of hazardous and noxious substances on surfaces, etc., serve as candidates for transport into the subsoil. Pollution of groundwater will occur when communication between the pollutants and these water bodies is established, i.e., when the contaminated infiltration plume reaches the groundwater. The end result is impairment of groundwater quality. Ingestion of polluted groundwater can be detrimental to human health.

As with infiltration processes, surface runoffs can occur on natural ground surfaces devoid of much human contact. This occurs when the rainfall rate exceeds the infiltration rate, and when the natural surface cover (vegetation and plants, etc.) is so dense that it acts as a shield or umbrella. As we have seen from the preceding, natural infiltration properties of soils can be severely compromised by human activities, resulting in runoffs. There are at least two types of runoffs: (1) managed runoffs, where the runoffs are channeled into drains and sewers, and (2) unmanaged runoffs, where the runoffs take directions controlled by surface topography and permeability properties of the surface cover material. In both cases, the runoffs will end up in receiving waters — lakes, rivers, oceans, etc. In the case of managed runoffs, there are at least two options for the runoffs before they meet the receiving waters: (1) treatment (full or partial) of the runoff water and (2) no treatment before discharge into receiving waters. The result of untreated runoffs into receiving waters is obvious — degradation of the quality of the receiving waters. When such occurs, the receiving waters will require treatment to reach drinking water quality standards. A point can be reached where the accumulated pollutants will be at a level where treatment of the degraded receiving water will not be effective — neither from an economic standpoint nor from a regulatory point of view.

The combination of infiltration and runoffs as potential carriers of pollutants into the ground and receiving waters (including groundwater) is a prospect that should alert one to the need for proper management of the sources of pollution of water resources. Treatment of water to achieve levels of quality dictated by drinking water standards is only one means for water resource management. The other has to be directed toward eliminating or mitigating the sources of pollution of water resources.

3.2.2   Harvesting of Groundwater

We have seen in Chapter 1 and in Figure 1.8 that only about 5% of the water in the world is freshwater — with the rest being seawater. This tells us that our drinkable freshwater resource is severely limited. The graphics in Figure 1.8 inform one that about two thirds of the available freshwater is groundwater. Infiltration and runoffs carrying pollutants into the ground will result in contamination of groundwater. Groundwater abstraction for drinking water purposes will require treatment aids. Since reliance on groundwater as a drinking water source is considerable, it is clear that maintaining acceptable groundwater quality is a high priority not only because of the need for treatment after abstraction, but also because once the aquifer is contaminated, cleanup of the aquifer to acceptable standards is almost impossible. When aquifers become contaminated, one key element for aquifer sustainability is lost. In effect, sustainability of that aquifer as a water resource is lost.

Depletion of aquifers is also another sustainability loss that requires attention. Depletion occurs when the rainfall is insufficient and recharge is nonexistent or exceedingly slow. Water use from that aquifer is thus not sustainable. This is occurring with more frequency as water use has increased by a factor of 6 since the beginning of the twentieth century. Water depletion has been estimated at 160 billion m^3 per year (Postel et al., 1996). The Ogallala aquifer in the central United States is used at 140% above its recharge rate (Gleick, 1993). Freshwater is depleting rapidly in countries such as India and China, and even in the United States, rivers are drying up and water table levels are decreasing. In China, there is a lack of water in more than 300 cities (WRI, 1994). It was estimated in the *Global Environment Outlook 3* report by the United Nations Environmental Program (UNEP, 2002) that about half of the world's population will not have sufficient water by the year 2032. This would be the result of the currently unsustainable practices that are using about 50% of the earth's freshwater supply (UN-FPA, 2001). What will happen to the ecosphere and to human life when the population reaches 9 billion by 2100 (UNDP, 2001)? Even now, dehydration is occurring in many areas to the extent that this will affect biodiversity and increase the requirements for agricultural irrigation.

Per hectare of land, 10 million liters of water each season is required for production of 8000 kg of corn (Pimentel et al., 1997). Water use for irrigation has decreased due to improved agricultural practices. Treatment of groundwater as a scarce resource may be required. Water markets have been proposed (Dorf, 2001) as a means of valuing water as a resource. Conservation by farmers and urban users will increase as the price of water increases.

3.2.2.1 Excessive Groundwater Abstraction and Land Subsidence

In regions where the underlying geology consists of interlayering of soft aquitards and aquifers, excessive groundwater abstraction from the interlayered aquifers can cause subsidence of the ground surface. A good case in point is the Quarternary sediments that underlie many coastal cities, such as Shanghai, Bangkok, and Jakarta. In Bangkok, the capital of Thailand, for example, the city is situated on a low-lying, flat deltaic plain known as the Lower Central Plain, or as the Lower Chao Phraya Basin, about 30 km north of the Gulf of Thailand. The basement bedrock gently inclines southward toward the Gulf of Thailand, and the strata overlying the basement bedrock consist of a complex mix of unconsolidated and semiconsolidated sediments of the Tertiary to Quarternary geologic age. The thickness of the strata ranges from about 400 m in the north to more than 1800 m in the south, with a stratigraphic profile that shows five discernable separate aquifer layers overlain by a stiff surface clay layer. Excessive long-term groundwater abstraction has resulted in subsidence in the region. This causes severe flooding of the region. Yong et al. (1991) reported that with the drainage system existent at that time, extensive flooding lasting for periods of 6 to 24 h occurred with rainfall exceeding 60 mm.

For coastal cities in similar situations, land subsidence due to prolonged excessive groundwater abstraction can reach the stage where the land surface will reach levels below sea level. When such occurs, unless containment dikes are built, seawater intrusion causing local flooding and contamination of the aquifers can occur. The impacts to human health in respect to contaminated water, in addition to physical problems associated with flooding, can be severe. Building and other structures have been known to suffer considerable structural distress due to uneven settlement of footings and foundations. By some accounts, delivery of potable water from a central source located in a safe region is required when flood waters compromise the existent drinking water sources.

3.2.2.2 Uses of Water

By all indications, if sustainable water usage is to be obtained, water treatment and reuse will need to increase and demand will need to decrease. Decreased water quality through pollution decreases the amount of available water. In the industrialized world, although water quality is in general good, water use is increasing. Many cities lose up to 40% of their water through leaking sewer and distribution systems (United Nations Economic Commission for Europe, 1998). In developing countries, water is both scarce and unhealthy due to rapid urbanization. Water management practices must be improved to reduce scarcities and impacts on ecosystems.

Water can also be used as geothermal energy. The sites are mainly in volcanic and seismic regions. Countries such as Iceland, Spain, France, Hungary, Japan, Mexico, Russia, and the United States (in the states of California and Hawaii) are exploiting this form of energy (Chamley, 2003). Iceland uses this form of energy for domestic heating, greenhouse cultivation, and electricity production. Exploitation can be from the steam of geysers or very hot liquids of 100°C or more. Cooled water can then be reinjected for reheating and reuse. Geothermal systems can also be utilized where groundwater at a depth of 100 m is about 15°C. This is warm enough to heat buildings in the winter using a heat pump principle. There are some hurdles that need to be overcome in exploiting geothermal energy. These include (1) transportation of warm water over long distances without major loss of heat, and (2) earthquakes may be induced. The advantages in use of geothermal energy include (1) negligible amounts of pollution produced, (2) little or negligible production of greenhouse gases, and (3) almost no waste production.

As water serves multipurposes, these demands can be highly competitive. Agriculture is an intensive user, and thus water use must be more efficient per crop grown. Since ecosystems can be depleted for human water use, public awareness of these links is necessary if conservation and sustainability goals are to be achieved. In parallel with WSSD (2002), the WaterDome, which was organized by the African Water Task Force (AWTF), highlighted the importance of water and sustainable development — an imperative that was not really addressed in the summit in Rio in 1992. They addressed the link of poverty with lack of water and sanitation, and the importance of hydroelectric dams as a means of electricity generation.

3.3 Characterization of Water Quality, Management, and Monitoring

3.3.1 Classes of Contaminants

Organic contaminants in water will deplete oxygen required for fish and other water organisms. These contaminants generally originate from the discharge of domestic and industrial wastewaters into water bodies. Increases in population density near these water bodies generally result in corresponding increases in the levels of organic matter in the water. High levels of biological oxygen demand (BOD) or chemical oxygen demand (COD) deplete the oxygen in the water through microbial degradation of the organic matter. This depletion of oxygen can lead to severe effects on aquatic biota. Color, taste, and odor of the drinking water may also be affected.

Farming and agricultural activities have added pollutant sources such as insecticides, pesticides, fungicides, and fertilizers. The impact from these pollutants is clearly demonstrated in the United States, where more than 147 million ha of land is known to be affected by groundwater pollution (Gleick, 1993). Herbicides and pesticides (1) are persistent in the environment, (2) are highly mobile, and (3) can accumulate in the tissues of animals, producing a variety of ill effects. Data on their concentrations in groundwater are quite limited. So as not to appear prejudicial, all of these pollutant sources will be identified as a group under the name of agro-additives. These agro-additives find their way into the receiving waters and groundwater through surface runoff and through transport in the ground. Nutrients such as nitrates from animal wastes from poultry or pigs are other sources of pollutants into the groundwater (Figure 3.1). Agricultural pollutants are typi-

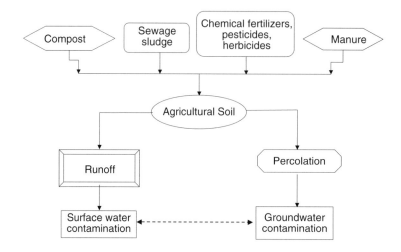

FIGURE 3.1
Mechanisms for surface and groundwater contamination from agricultural soil. Dashed line between ground-water and surface water contamination indicates potential contamination of groundwater from surface water or surface water contamination from groundwater.

cally difficult to monitor and estimate due to their dispersion and seepage through soils into the groundwater.

High levels of nutrients (nitrogen and phosphorus) from fertilizers, detergents, and other sources can reach surface waters and lead to eutrophication and excessive algal growth. Nitrogen, in particular, and, to a lesser degree, phosphorus demand in fertilizers has grown significantly in the past 50 years (USGS, 1998). Eutrophication causes oxygen depletion in the water and production of bad odors, tastes, and colors. Nitrogen, in particular, originates from fertilizer and manure addition to the soil. Although nitrates have an important role within the nitrogen cycle, overapplication of manure and fertilizers has a negative impacts negatively on plant and microorganism biodiversity. Although ammonium ions adsorb onto clay, nitrate compounds are easily transported. Movement of nitrates in the groundwater depends on aquifer hydraulic conductivity, soil type and moisture, temperature, vegetation, and amount of precipitation. Shallow unconfined aquifers have been found to be highly susceptible to nitrate contamination from agricultural sources (Burkart and Stoner, 2002).

Another major class of contaminants is microorganisms from agricultural runoff and septic and sewage systems. Microorganisms contribute turbidity, odors, and elevated levels of oxygen demand. Drinking contaminated water can lead to severe gastrointestinal illnesses and even death. The case of Walkerton, Ontario, a small town 200 km northwest of Toronto, is an example of this. In May 2000, heavy rain washed manure into swampy land, which subsequently contaminated the drinking well. More than 2300 people became ill and 7 died due to drinking this contaminated water supply. This is discussed further in Chapter 11.

Numerous organic compounds are potentially mutagenic and carcinogenic to humans, animals, and plants. We have previously seen from Figure 1.3 how industrial, agricultural, and urban discharges can contaminate aquifers and surface waters. As indicated previously, the sources of these pollutants can be traced to waste streams and discharges from industrial plants, households, resource exploitation facilities, and farms. Oil spills are a major cause of devastation to marine and land ecosystems. Some examples of contaminant groups include pesticides and herbicides such as dichlorodiphenyltrichloroethane (DDT), aldrin, chlordane, diazonin, and partinon; volatile organic compounds (VOCs) such as vinyl chloride, carbon tetrachloride, and trichloroethylene (TCE); and heavy metals (e.g., chromium, cobalt, copper, iron, mercury, molybdenum, strontium, vanadium, and zinc). VOCs enter the water systems as industrial and municipal discharges. Due to their higher volatility, they are less persistent than herbicides and pesticides. Metals originate from industrial processing, runoff from mining operations, and atmospheric disposition from incinerator emissions and other processes.

All the other sources of pollutants shown, i.e., discharges and waste streams, are most likely contained in storage dumps, landfills, holding ponds, tailings ponds, or other similar systems. All of these containment systems have the potential to deliver pollutants to the receiving waters (surface and groundwaters) because of eventual leaks, discharges, and failures. Some of these will be discussed in greater detail in a later section dealing with containment systems.

In urban regions, leakage of sewers and other wastewater sources can significantly contribute to recharge and pollution of aquifers. It was estimated in 1997 that more than 950 million m^3 per year of wastewater is lost due to broken sewers in the United States (Pedley and Howard, 1997). Aquifers under cities can be highly polluted, making them unsuitable for drinking water. This is particularly significant in regions where (1) wastewater is untreated, (2) source pollutants such as nitrates, ammonia, fecal coliforms, and dissolved organic carbon abound, and (3) urbanization is rapid and essentially uncontrolled.

Pollution of the surface waters and groundwater can occur as a result of industrial or municipal discharges or runoff from agricultural land, mining operations, or construction. Industrial pollutants in the groundwater, such as benzene, toluene, xylene, and petroleum products, originate from (1) leakage of underground storage tanks, (2) chemical spills, and (3) discharges of organic chemicals and heavy metals such as cadmium, zinc, mercury, and chromium. The numbers of affected sites in the United States have been reported to be at least five or more orders of magnitude (Gleick, 1993). Runoff and seepage from mining operations can contribute significant levels of heavy metals — as for example in the illustration shown in Figure 3.2 of runoff of iron from a coal mine.

Natural sources of pollutants can also contribute to the contamination of groundwater. Salt water intrusion from coastal aquifers can also degrade groundwater quality by bringing salt into groundwater with freshwater (Melloul and Goldenberg, 1997). The arsenic polluted aquifers in West Bengal and Bangladesh, discussed previously in Section 2.2.4 in Chapter 2, provide good examples of a combined man and nature impact on the geoenvironment. Since these aquifers provide potable water for the majority of the population of Bangladesh and some significant proportion of the population in West Bengal, it has been estimated that from 35 to 50 million people are at risk of arsenic poisoning. Often referred to as the singular most dramatic case of mass poisoning of the human race, the arsenic polluted aquifers serving the tube wells in the two countries contain arsenic concentrations far in excess of the prescribed limits of the World Health Organization (WHO).

In the United States, a survey showed that more than 36% of the surface water does not meet water quality objectives (U.S. EPA, 1996). Increasing population and economic growth are contributing to this problem. Major sources of contaminants are the agricultural industry and urban runoff from storm sewers. Leading contaminants are summarized in Figure 3.3. Although problematic in some locations, the presence of pesticides and fertilizers is currently not a huge concern. There is not a large amount of data concerning groundwater quality in the United States.

3.3.2 Monitoring of Water Quality

In industrialized countries, concern over the quality of rivers has resulted in a considerable amount of public funds being invested in water quality monitoring during the past decade.

FIGURE 3.2
(See color insert following page 142.) Iron runoff from a coal mine into canal area.

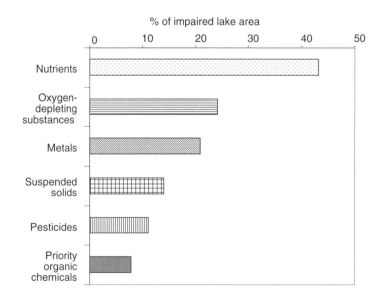

FIGURE 3.3
Percentage of lakes contaminated by various components. (Data from U.S. EPA, *Environmental Indicators of Water Quality in the U.S.*, EPA-841-R-95-005, U.S. Environmental Protection Agency, Office of Water, Washington, DC, 1996.)

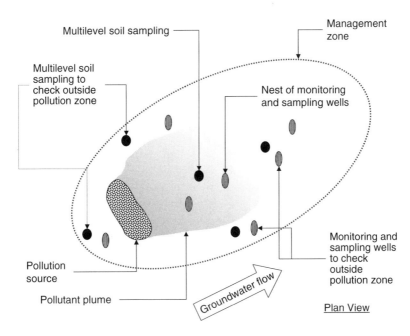

Multilevel soil sampling

Management zone

Multilevel soil sampling to check outside pollution zone

Nest of monitoring and sampling wells

Monitoring and sampling wells to check outside pollution zone

Pollution source

Pollutant plume

Groundwater flow

Plan View

FIGURE 3.4

Plan view of distribution of monitoring wells and soil sampling boreholes for verification monitoring and long-term conformance monitoring. (From Yong, R.N. and Mulligan, C.N., *Natural Attenuation of Contaminants in Soils*, CRC Press, Boca Raton, FL, 2004.)

Accordingly, monitoring of chemical pollutants in the environmental matrices has entered a new phase. Modifications in instrumentation, sampling, and sample preparation techniques have become essential in keeping pace with the requirements for (1) achieving low detection levels, (2) high-speed analysis capability, and (3) convenience and cost efficiency.

Environmental indicators such as water quality can be used as indicators of sustainability. The term *monitoring* is used in many different ways. In the context of monitoring of a particular site to determine whether the events expected to occur in the site have indeed transpired, it is necessary to gather all pertinent pieces of information providing evidence that those events had occurred. We interpret, from the definition of monitoring in the previous chapter, the term to mean a program of sampling, testing, and evaluation of status of the situation being monitored. In the situation being monitored, a *management zone* needs to be established — as shown, for example, in Figure 3.4. To determine whether attenuation of pollutants in a contaminated site has been effective, it is necessary to obtain information pertaining to the nature, concentrations, toxicity, characteristics, and properties of the pollutants in the attenuation zone. The pollutants have residence in both the pore water (or groundwater) and on the surfaces of the soil solids. Residence associated with the soil solids can take the form of sorbates and co-precipitates. In turn, the sorbates can be complexed with the soil solids and will remain totally fixed within the structure of the soil solids. However, the sorbates can also be held by ionic forces, which can be easily disrupted, thus releasing the sorbates.

What the preceding discussion of residence status of the pollutants tells us is that we need to monitor and sample not only the pore water or groundwater, but also the soil fractions in the contaminant attenuation zone. Two types of monitoring sampling systems are needed. For pore water or groundwater, monitoring wells are generally used. These wells are necessary to provide access to groundwater at various locations (vertically and spatially) in a chosen location. The choice of type of monitoring wells and distribution or

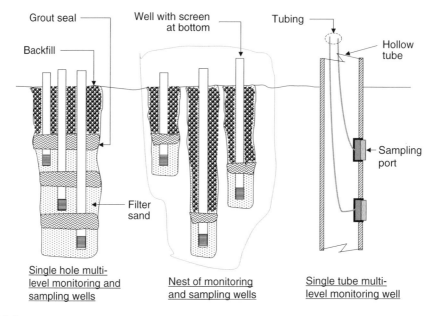

Grout seal — Well with screen at bottom — Tubing —

Backfill —

Hollow tube

Sampling port

Filter sand

Single hole multi-level monitoring and sampling wells

Nest of monitoring and sampling wells

Single tube multi-level monitoring well

FIGURE 3.5

Some typical groundwater monitoring and sampling wells. (From Yong, R.N. and Mulligan, C.N., *Natural Attenuation of Contaminants in Soils*, CRC Press, Boca Raton, FL, 2004.)

location of wells will depend on the purpose for the wells. In respect to determination of whether natural attenuation can be used as a treatment process, there are at least three separate and distinct monitoring schemes that need to be considered. These range from the initial site characterization studies to verification monitoring and long-term conformance monitoring. The term *monitoring scheme* is used deliberately to indicate the use of monitoring and sampling devices to obtain both soil and water samples. Figure 3.5 shows some typical devices used as monitoring wells to permit monitoring groundwater at various levels. In the left-hand group, we see individual monitoring wells with sampling ports located at different depths but grouped together in a shared borehole. This is generally identified as a single-borehole multilevel monitoring well system. The middle drawing shows a nest of single monitoring wells in their own separate boreholes, and the right-hand drawing shows a single tube system with monitoring portholes located at the desired depths. With present technological capabilities, monitoring wells and the manner of operation have reached levels of sophistication where downhole sample analysis of groundwater can be performed without the need for recovery of water samples.

Site characterization monitoring is necessary to provide information on the hydrogeology of the site. It is necessary to properly characterize subsurface flow to fully delineate or anticipate the transport direction and extent of the pollutant plume. Determination of the direction and magnitude of groundwater flow is most important. Obviously this means a judicious distribution of monitoring wells upgradient and downgradient. A proper siting of the monitoring wells and analysis of the results should provide one with knowledge of the source of the pollutants and the characteristics of the pollutant plume.

Verification monitoring requires placement of monitoring wells and soil sampling devices within the heart of the pollutant plume and also at positions beyond the plume. Figure 3.4 and Figure 3.6 show the vertical and plane views of how the wells and sampling stations might be distributed. It is a truism to state that the more monitoring and sampling devices there are, the better one is able to properly characterize the nature of the pollutant plume — assuming that the monitoring wells and sampling devices are properly located.

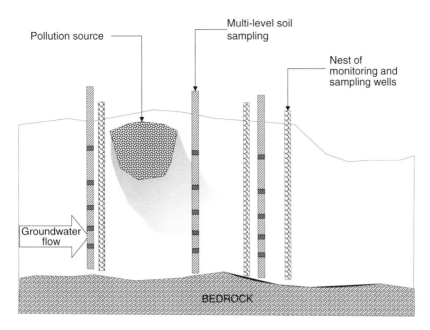

FIGURE 3.6
Simple monitoring and sampling scheme for evaluating groundwater quality. (From Yong, R.N. and Mulligan, C.N., *Natural Attenuation of Contaminants in Soils*, CRC Press, Boca Raton, FL, 2004.)

The monitoring wells and sampling devices placed outside the pollutant plume, shown in Figure 3.4, will also serve as monitoring wells and sampling devices for long-term conformance assessment.

The tests required of samples retrieved from monitoring wells are designed to determine the nature of the pollutants in the pore or groundwater. These will inform one about the concentration, composition, and toxicity of the target pollutant. For prediction of further or continued attenuation of the target pollutant, the partition coefficients and solubilities of the various contaminants are needed as input to transport and fate models. If biotransformation of the target pollutants has occurred, supporting laboratory research would be needed to determine the likely fate of the transformed or intermediate products.

3.3.2.1 Remote Sensing

Most of the existing technologies for monitoring algae and cyanobacteria rely on microscopic techniques that are laborious and highly variable. Recently, algae and algal blooms have been detected by satellite imaging systems and spectrofluorimetry (Gitelson and Yacobi, 1995). Subsequently, Millie et al. (2002) have demonstrated that absorbance and fluorescence spectra can be used to discriminate microalgae. This method can potentially be applied as part of an *in situ* monitoring program.

Remote sensing by Landsat satellites has been evaluated by Zhu et al. (2002) for the Pearl River in South China, Hong Kong, and the Macau region. Sea color was monitored and the images were analyzed by an algorithm of gradient transition. *In situ* optical, chemical, and biochemical measurements for chlorophyll-a and phytoplankton and other parameters correlated well with ocean color.

Water quality in urban runoff can also be monitored by remote sensing. Ha et al. (2003) combined Landsat, the Korea Multi-Purpose Satellite (KOMPSAT), with two types of neural networks. Runoff, peak time, and pollutant data could be obtained to determine water quality in an urban area. In the Netherlands, a remote sensing system was tested to construct maps of water quality (Hakvoort et al., 2002). A hyperspectral scanner was set up on an airplane to retrieve information on subsurface reflectance spectra, specific inherent optical properties (SIOPs) of the water components, and the Gordon reflectance model using a matrix inversion technique. Data for both total suspended matter and chlorophyll could be obtained when circumstances were good, and reliable data mainly only for total suspended matter under less ideal conditions. Information on dissolved organic matter levels could not be obtained. The matrix inversion technique for airborne images was particularly promising.

Significant advances have been made in the past decade concerning the levels of detection and information about contaminants, including speciation and the speed of monitoring of contaminants in water, as regulatory requirements have become more demanding. For example, the U.S. Groundwater Rule is being established to provide a risk-based regulatory strategy for groundwater systems that can be sensitive to fecal contamination.

Another example is the Arsenic Rule, where the rule agreed that 10 µg/l of arsenic would be the standard from 2002 for drinking water (www.epa.gov/safewater/arsenic.html). Drinking water systems must comply by 2006. Analytical methods for determining the different arsenic species include solid-phase microextraction (SPME) and solid-phase extraction (SPE) with gas chromatography/mass spectrometry (GS/MS), liquid chromatography (LC)/electrospray ionization mass spectrometry (ESI-MS), LC/inductively coupled plasma mass spectrometry (ICP-MS), and ion chromatography (IC)/ICP-MS. Field kits have also been used, particularly in Bangladesh and West Bengal. However, there have been problems with their accuracy (Erickson, 2003).

Methyl tert-butyl ether (MTBE) has recently been of concern because of leaking underground storage tanks and other fuel discharges. Monitoring of this compound is now required (www.epa.gov/OGDS/mtbe.html). In a study by Williams (2001) in California from 1995 to 2000, 1.3% of all drinking water samples contained detectable levels of MTBE. More than 75% were below 13 µg/l. In Germany, Achten et al. (2002) found average levels of 88 ng/l in well water and filtered riverbank samples. Lacorte et al. (2002) found that 7 of 21 groundwater samples in Catalonia, Spain had levels exceeding 20 to 40 µg/l. Methods to measure MTBE have been reviewed by Richardson (2003), including headspace, purge and trap, or SPME combined with GC or GC/MS analysis. Detection levels are in the order of the nanograms per liter range.

Following water pollution cases such as Walkerton in Ontario, where *Escherichia coli* gastroenteritis occurred in 2000, the importance of analytical methods for microorganism detection has increased. Due to the deficiencies in the standard methods for detection of viruses, bacteria, and protozoa, particularly length of time, various new methods have been developed. These include (1) immunofluorescent antibody techniques, (2) fluorescent *in situ* hybridization, (3) magnetic bead cell sorting, (4) electrochemiluminescence, (5) amperometric sensors, (6) various polymerase chain reactions (PCRs) and reverse transcriptase (RT)-PCR, and (7) real-time PCR methods (Straub and Chandler, 2003). Richardson (2003) has also indicated that several reviews have been published using mass spectrometric techniques for characterization of microorganisms.

Analysis of herbicides and pesticides is also of interest because of their effects on human health and the environment. A field kit has been evaluated by Ballesteros et al. (2001) for analysis of triazine herbicides in water samples. Detection levels by this enzyme-linked immunosorbent assay (ELISA)-based technique were 0.1 µg/l for atrazine and 0.5 µg/l for triazine. Other methods have now been developed to analyze enantiomeric pesticides. Studies were performed by enantioselective GC/MS from 1997 to 2001 on a lake that received agricultural runoff in Switzerland (Poiger et al., 2002). Pre-1998, samples showed the dominance of the racemic metalochlor, whereas samples collected in 2000 to 2001 showed a clear dominance of the S-isomer. This coincided with the commercial switch from the racemic to the S form for agricultural use. Pharmaceutically active compounds and other personal care products have also been identified in surface runoff from fields irrigated with wastewater treatment effluents (Pedersen et al., 2002).

Methods such as GC, HPLC, and atomic absorption spectrometry (AA) are accurate, and limits of detections are continually decreasing due to combined techniques (e.g., GC/MS, LC/MS, CE (capillary electrophoresis)/MS, CE/ICP (inductively coupled plasma)-MS, and ICP-MS). Polar compounds, prior to LC/MS, were extremely difficult to detect. Improved methods of sample preparation have reduced solvent usage and are more environmentally friendly and rapid. Although efforts have also been made to automate and simplify many of the technologies, they must continue to enable more widespread use of the technology. Recoveries from solid-phase extraction will need to improve to enable one to obtain more reliable data from this technique. More extensive testing needs to be done with real environmental samples — not just standards — to more fully understand interferences within the samples. This will enable one to (1) monitor water quality, (2) determine the origin or source of biological and chemical contaminants, and (3) determine the transport and fate of the contaminants in the environment.

3.3.2.2   *Biomonitoring*

Biomonitoring is used to indicate the effect and extent of contaminants in the water. It includes determining changes in species diversity, composition in a community, and the mortality rates of a species. Buildup of pollutants in the tissues of individuals can also be evaluated, in addition to physiological, behavioral, and morphological changes in individuals. The effect of specific contaminants is difficult to determine.

Biomonitoring involves the determination of the numbers, health, and presence of various species of algae, fish, plants, benthic macroinvertebrates, insects, or other organisms as a way of determining water quality (U.S. EPA, 2000). Knowledge of background information is essential. Attached algae (known as periphyton) are good indicators of water quality since they grow on rocks and other plants in the water. The advantages for using these as indicators are (1) high numbers of species are available, (2) their responses to changes in the environment are well known, (3) they respond quickly to exposures, and (4) they are easy to sample. An assessment could include (1) determination of the biomass by chlorophyll or on an ash-free dry basis, (2) species, (3) distribution of species, and (4) condition of the attached algae assemblages. As yet, their use has not been widely incorporated in monitoring programs.

Benthic macroinvertebrates have numerous advantages as bioindicators (protocols). They do not move very far and thus can be used for upstream–downstream studies. Their life span is about a year, enabling their use for short-term environmental changes. Sampling is easy. They are numerous, and experienced biologists can easily detect changes in macroinvertebrate assemblages. In addition, different species respond differently to various pollutants. They are also food sources for fish and other commercial species. Many states in the United States have more information on the relationship between invertebrates and pollutants than for fish (Southerland and Stribling, 1995).

Fish are good indicators of water quality since (1) they always live in water, (2) they live for long periods (2 to over 10 years) (Karr et al., 1986), (3) they are easily identifiable and easy to collect, and (4) they can quickly recover from natural disturbances. They are also consumed by humans and are of importance to sport and commercial fisherman. Fish make up almost 50% of the endangered vertebrate species in the U.S. (Warren et al., 2000).

Aquatic plants (macrophytes) grow near or in water, and many of them can serve as indicators of water quality. A lack of macrophytes can indicate quality problems caused by turbidity, excessive salinity, or the presence of herbicides (Crowder and Painter, 1991).

Excessive numbers can be caused by high nutrient levels. They are good indicators since they respond to light, turbidity, contaminants such as metals and herbicides, and salt. No laboratory analysis is required, and sampling can be performed through aerial photography.

Biosurveys are useful in identifying if a problem exists. Chemical and toxicity tests would then be required to determine the exact cause and source (U.S. EPA, 1991). Routine biomonitoring can be less expensive than chemical tests over the short term, but more expensive over the long term. Field bioassessment experts are required to obtain and interpret data. However, there are no established protocols. More knowledge is required to determine the effects of contaminants on populations of organisms and better coordination of background data before site contamination. Recently, data (Zhang et al., 2000) on toxicity and chemicals have been combined to evaluate the sustainability of reaction pathways. Risk indices were developed for aquatic life or human health as part of the environmental index determination.

3.4   Sustainable Water Treatment and Management

To enable the adequacy of water resources for future generations, management practices must control the sources of pollution and limit water use. This requires sufficiency in recharge of aquifers and prevention of pollution of surface water and groundwater. Remediation of polluted water is required, but as is well known, effective and complete remediation of aquifers is not easily accomplished. As discussed previously, the quality of both surface water and groundwater needs to be protected by mitigation and management procedures. Reuse of treated waste streams, in particular, needs to be practiced in farming and agricultural activities by irrigation.

Real-time monitoring and remote sensing and graphical information systems (GIS) are essential for water management. Calera Belmonte et al. (1999) examined the use of GIS tools to manage water resources in an aquifer system using remote sensing from a satellite to determine the spatial distribution of irrigated crops and water pumping estimates. The information obtained enables the GIS to be used as a tool for monitoring and control of water exploitation for agricultural uses.

3.4.1   Techniques for Soil and Groundwater Treatment

Several techniques are available to manage and control contaminant and pollution plumes to minimize environmental and health impacts. These include (1) construction of impermeable barriers and liner systems for containment facilities, (2) remediation techniques designed to remove or reduce (attenuate) the pollutants in the ground, such as soil flushing, and (3) passive procedures relying on the properties of the ground to reduce contaminant concentrations in leachate streams and pollution plumes.

3.4.1.1   *Isolation and Containment*

Contaminants can be isolated and contained to (1) prevent further movement, (2) reduce the permeability to less than 1×10^{-7} m/s, and (3) increase the strength (U.S. EPA, 1994). Physical barriers made of steel, concrete, bentonite, and grout walls can be used for capping and vertical and horizontal containment. Liners and membranes are mainly used for protection of groundwater systems, particularly from landfill leachates. A variety of materials are used, including polyethylene, polyvinyl chlorides, asphalt materials, and soil–bentonite or cement mixtures. Monitoring is a key requirement to ensure that the contaminants are not mobilized.

Most *in situ* remediation techniques are potentially less expensive and disruptive than *ex situ* ones, particularly for large contaminated areas. Natural or synthetic additives can be utilized to enhance precipitation, ion exchange, sorption, and redox reactions (Mench et al., 2000). The sustainability of reducing and maintaining reduced solubility conditions is key to the long-term success of the treatment. *Ex situ* techniques are expensive and can disrupt the ecosystem and the landscape. For shallow contamination, remediation costs, worker exposure, and environmental disruption can be reduced by using *in situ* remediation techniques.

Solidification/stabilization techniques are common for contaminated soils since they are designed to incorporate the contaminants in a solid matrix. Some metals such as arsenic, mercury, and chromium (VI) are less suitable for these techniques. As mentioned previously, monitoring is required to ensure that the process is stable and the contaminants are not mobilized. For inorganic contaminants, the two groups of stabilizing agents used include (1) cement, fly ash, kiln dust, clays, zeolites, and pozzolanic materials, and (2) bitumen products, epoxy, polyethylene, and resins. Strict requirements for weathering (leachability, etc.) and durability of solidified and stabilized products have been specified by many regulatory agencies. Performance assessment of such products — as with most treatment procedures — is a standard requirement.

Vitrification, a high-temperature solidification process, leads to the formation of a glassy solid and is especially applicable for treatment of arsenic-contaminated soils since arsenic possesses low volatility. The melting ability of the contaminated soil depends on the soil's silica content. The maximum allowable oxide content in a soil containing arsenic as a contaminant is 5% (Smith et al., 1995). It is the best demonstrated available technology (BDAT) for Resource Conservation and Recovery Act (RCRA) wastes.

As an example, information regarding the sustainability of arsenic immobilization could be obtained for sequential extraction techniques to evaluate the adsorption, sequestration, and bioavailability of arsenic in soils (Yong and Mulligan, 2004) before and after treatment with additives such as iron, aluminum, calcium and manganese, cement, lime, and pozzolanic materials, as was previously shown for phosphate treatment.

3.4.1.2 *Extraction Treatment Techniques*

To remove NAPLs from the groundwater, extraction of the groundwater can be performed by extraction pumping of the contaminated dissolved phase or free-phase NAPL zone. Drinking water standards can be achieved with the method of treatment. However, substantial periods of time can be required before this occurs. To enhance the removal rates of the contaminants, extraction solutions can be infiltrated into the soil using surface flooding, sprinklers, leach fields, and horizontal or vertical drains. Water with or without additives is employed to solubilize and extract the contaminants, as shown in Figure 3.7. Chemical additives include organic or inorganic acids, bases, water-soluble solvents, complexing agents, and surfactants. Removal efficiencies are related to, and affected by, soil pH, soil type, cation exchange capacity, particle size, permeability, and the type of contaminants. High soil permeabilities (greater than 1×10^{-5} m/s) are considered to be beneficial for such procedures.

As an alternative to groundwater pumping, soil vapor extraction (SVE) (Figure 3.8) may be utilized for the vaporization of volatile and semivolatile components in the unsaturated zone (Yong, 1998; Rathfelder et al., 1991). Soil can be decontaminated by applying a vacuum to pull the volatile emissions through the soil pore spaces. The air may then be

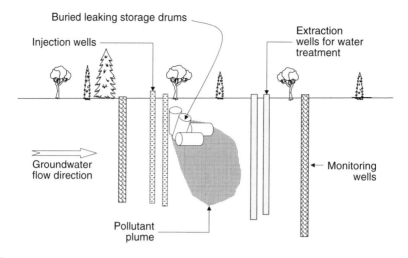

FIGURE 3.7
Schematic diagram of a soil flushing process for removal of contaminants.

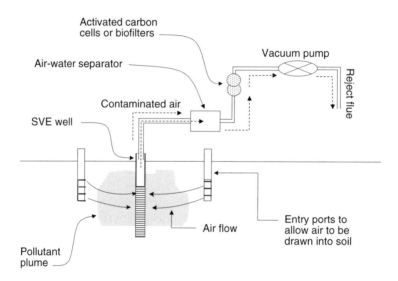

FIGURE 3.8
Schematic of a soil vapor extraction (SVE) process. Note that a series of SVE wells can be introduced into the ground — connected in series or in parallel. Obviously the number of SVE wells that can be introduced will depend on the capacity of the vacuum pump system. (Adapted from Yong, R.N., *Compatible Technology for Treatment and Rehabilitation of Contaminated Sites*, NNGI Report 5, Nikken Sekkei Geotechnical Institute, Japan, 1998.)

treated at the surface with activated carbon filters or biofilters. This technique is applicable to highly permeable soils and volatile contaminants such as gasoline or solvents. Other parameters, such as the octanol–water coefficient, Henry's law constant and solubility of the contaminant, and moisture and organic contents of the soil, also affect the removal efficiencies. Volatile components can also be removed by air sparging. In this technique, bubbles of air are injected into the groundwater to strip NAPLs and to add oxygen for *in situ* bioremediation. It has been successfully used for dissolved hydrocarbon plumes (Bass et al., 2000). Reduction of the NAPL zone may then allow natural attenuation processes to proceed. Bioventing is a variation of this technique, where lower aeration rates are used to promote aerobic biodegradation instead of volatilization.

Phytoremediation is the use of plants to remove, contain, or render harmless environmental contaminants. The various mechanisms involved in phytoremediation include (1) phytoextraction, (2) uptake of contaminants through the roots and subsequent accumulation in the plants, (3) phytodegradation, (4) metabolism of contaminants in the leaves, shoots, and roots, (5) release of enzymes and other components for stimulation of bacterial activity or biochemical conversion and rhizodegradation, and (6) mineralization of contaminants in the soil by microbial activity in the rhizosphere. Phytoremediation is a low-cost *in situ* technology that causes minimal disturbance. It is acceptable to the public and generates low amounts of waste. This technology is presently being developed for a treatment of a wide variety of organic and inorganic contaminants.

3.4.1.3   Electrokinetic Applications

Electrokinetics involves the use of electrodes and electrical current to mobilize inorganic contaminants. It is more effective for treatment of silty soils than for clay soils, where energy requirements can be substantial. Energy levels must be higher than the energy that binds the contaminants to the soil. The processes involved have been described in detail by Yong (2001). Electro-osmosis and electrophoretic phenomena are the principal mechanisms in the treatment process. Conditioning fluids are required to enhance contaminant ion movement, and electrode fouling is a substantial problem.

3.4.1.4   Natural Attenuation

The attenuation of contaminants or pollutants due to the assimilative processes of soils refers to the reduction of concentrations and toxicity of contaminants and pollutants during transport in soils. This process is discussed in detail in Section 10.2 in Chapter 10, where the use of soils as a waste management tool is addressed. For the present, we will examine some of the phenomena pertinent to the present context of water and groundwater controls. Reduction in concentrations and toxicity of pollutants in the groundwater can be accomplished by (1) dilution because of mixing with uncontaminated groundwater, (2) interactions and reactions between contaminants and soil solids that can lead to partitioning of the pollutants between the soil solids and pore water, and (3) transformations that reduce the toxicity threat posed by the original pollutants. Short of overwhelming dilution with groundwater, it is generally acknowledged that partitioning is by far the more significant factor in attenuation of contaminants or pollutants.

Natural attenuation refers to the situation when attenuation of contaminants results because of the processes that contribute to the natural assimilative capacity of soil. This means that contaminant attenuation occurs as a result of the natural processes occurring in the soil during contaminant–soil interaction. Broadly speaking, therefore, natural attenuation refers to natural processes occurring in the soil that serve to reduce the toxicity of the contaminants or the concentration of the contaminants. The term *contaminants* is used instead of *pollutants* because the attenuation processes apply to the broad range of contaminants and not solely to pollutants. These natural processes of contaminant attenuation include dilution, partitioning of contaminants, and transformations. They involve a range of physical actions, chemical and biologically mediated reactions, and combinations of all of these.

According to the U.S. National Research Council (NRC), the sustainability of natural attenuation is dependent on the sustainability of the mechanisms for immobilizing or destroying contaminants while the contaminants are released into the groundwater. A mass balance analysis can be used to estimate the long-term destruction or immobilization rates (NRC, 2000). For hydrocarbons, the availability of electron acceptors or donors may be evaluated to determine the sustainability of remediation techniques such as natural attenuation for hydrocarbons. However, in the case of metals and metalloids such as arsenic, this approach is only applicable if the attenuation is biologically driven.

Monitored natural attenuation (MNA), because of its adherence to remedy by natural processes, necessitates a proper understanding of the many principles involved in the natural processes that contribute to the end result. Monitoring of the pollutant plume at various positions away from the source is a key element of the use of MNA. Remembering that MNA is a pollutant and soil-specific phenomenon, one generally tracks a very limited number of pollutants, specifically the ones considered to be the most noxious. Historically, more attention has been paid to documenting the properties and characteristics of the pollutants. By and large, the pollutants tracked have primarily been the organic chemicals, including, for example, chlorinated solvents (PCE, TCE, and DCE) and hydrocarbons such as benzene, toluene, ethyl benzene, and xylene (BTEX).

When active controls or agents are introduced into the soil to render attenuation more effective, this is called *enhanced natural attenuation*. This is to be distinguished from *engineered natural attenuation* (EngNA), which is probably best illustrated by the permeable reactive barrier shown in Figure 3.9 and the barrier–liner system shown in Figure 3.10. Enhanced natural attenuation (ENA) refers to the situation where, for example, nutrient packages are added to the soil system to permit enhanced biodegradation to occur, or where catalysts are added to the soil to permit chemical reactions to occur more effectively. ENA could include biostimulation or bioaugmentation. These subjects are discussed in greater detail in Chapter 10.

3.4.1.5   Biostimulation

Probably the simplest procedure for improving the intrinsic bioremediation capability of a soil is to provide a stimulus to the microorganisms that already exist in the site. This procedure is called *biostimulation*, i.e., adding nutrients and other growth substrates, together with electron donors and acceptors. The intent of biostimulation is to promote increased microbial activity with the set of stimuli to better degrade the organic chemical pollutants in the soil. With the addition of nitrates, Fe(III) oxides, Mn(IV) oxides, sulfates, and CO_2, for example, anaerobic degradation can proceed. This technique is used for sites contaminated with organic chemical pollutants and is perhaps one of the least intrusive of the methods of enhancement of natural attenuation. The other method of enhancement that falls in the same class of less intrusive enhancement procedures is bioaugmentation.

3.4.1.6   Bioaugmentation

If the native or indigenous microbial population is not capable of degrading the organic chemicals in the soil — for whatever reason, e.g., concentrations, inappropriate consortia, etc. — other microorganisms can be introduced into the soil. These are called exogenous microorganisms. Their function is to augment the indigenous microbial population such that effective degradative capability can be obtained. If need be, biostimulation can also be added to the bioaugmentation to further increase the likelihood of effective degradative capability. We need to be conscious of the risks that arise when unknown results are obtained from interactions between the genetically engineered microorganisms and the various chemicals in the contaminated ground. The use of microorganisms grown in uncharacterized consortia, which include bacteria, fungi, and viruses, can produce toxic metabolites (Strauss, 1991). In addition, the interaction of chemicals with microorganisms may result in mutations in the microorganisms themselves or microbial adaptations.

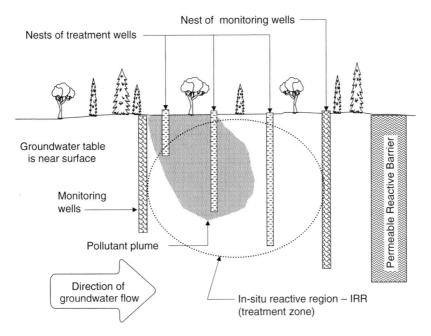

FIGURE 3.9

Enhancement of natural attenuation using treatment wells. Treatments for enhancement can be any or all of the following: geochemical intervention, biostimulation, and bioaugmentation. Treatment occurs in the pollutant plume and downgradient from the plume. (From Yong, R.N. and Mulligan, C.N., *Natural Attenuation of Contaminants in Soils*, CRC Press, Boca Raton, FL, 2004.)

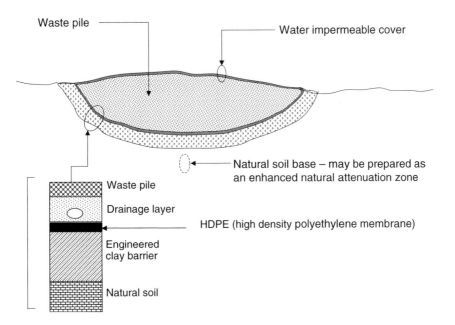

FIGURE 3.10

Pollutant attenuation layer constructed as part of an engineered barrier system. The dimensions of the attenuation layer and the specification of the various elements that constitute the filter, membrane, and leachate collection system are generally determined by regulations or by performance criteria. (From Yong, R.N. and Mulligan, C.N., *Natural Attenuation of Contaminants in Soils*, CRC Press, Boca Raton, FL, 2004.)

3.4.1.7 *Enhanced Natural Attenuation*

We show in Figure 3.9 a direct application of enhanced natural attenuation (ENA) as an *in situ* remediation process. Enhanced treatment of a region (spatial and vertical) of the site downgradient from the contaminated site permits the ENA to function as planned. The treated region is called the in situ *reactive region* (IRR) or *treatment zone*, and can be used in conjunction with other treatment procedures. Figure 3.9 illustrates the use of the IRR as a treatment procedure for the pollutant plume in the region in front of the permeable reactive barrier. Treatment procedures using treatment wells or boreholes and associated technology include:

- Geochemical procedures such as pH and Eh manipulation
- Soil improvement techniques such as introduction of inorganic and organic ligands; introduction of electron acceptors and donors
- Various other biostimulation procedures and bioaugmentation

The choice of any of these, or a combination of these methods of augmentation, will depend on the type, distribution, and concentration of pollutants in the contaminated site, and also on the results obtained from microcosm and treatability studies.

3.4.1.8 **In Situ *Reactive Regions: Treatment Zones***

We have seen from Figure 3.9 an example of the use of a treatment zone, known also as an *in situ* reactive region (IRR), i.e., the region immediately in front of the permeable reactive barrier. The purpose of an IRR is to provide not only pretreatment or preconditioning in support of another treatment procedure, but also as a post-treatment process for sites previously remediated by other technological procedures. In Figure 3.9, we show the IRR used in support of the permeable reactive barrier (PRB) treatment procedure. Other treatment procedures can also be used in place of the PRB. The presence of heavy metals in combination with organic chemicals in the pollutant plume is not an uncommon occurrence. One could, for example, envisage using IRR as a treatment procedure in combination with a subsequent procedure designed to fix or remove the metals.

In application of IRR as a posttreatment process, one is looking toward the IRR as the *final cap* for some kind of design or technological process for remediation of a contaminated site. This is generally part of a multiple-treatment process — as opposed to the use of IRR in a pretreatment or preconditioning process. A good example of this is the use of pump-and-treat as the first phase of the remediation program, followed by the IRR as a post-treatment process where the treated pollutant plume will receive its final cleanup. The efficiency of cleanup using pump-and-treat methods rapidly decreases as greater pollutant extraction from the groundwater or pore water is required. It is not unusual to remove some large proportion of the pollutants from the groundwater or pore water, and to leave the remaining proportion to be removed via natural attenuation processes in an IRR.

3.4.1.9 *Permeable Reactive Barriers*

The intent of a permeable reactive barrier (PRB) is to provide treatment as a remediation procedure to a pollutant plume as it is transported through the PRB, so that the plume no longer poses a threat to biotic receptors when it exits the PRB. Figure 3.11 shows a funnel-and-gate arrangement of a PRB application, where the pollutant plume is channeled to the PRB gate by the impermeable walls. Transport of the pollutant plume through the PRB

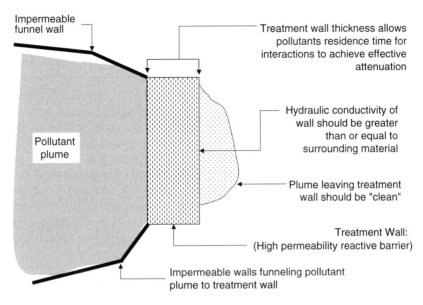

Plan View of Funnel and Treatment Gate

Impermeable funnel wall

Treatment wall thickness allows pollutants residence time for interactions to achieve effective attenuation

Hydraulic conductivity of wall should be greater than or equal to surrounding material

Pollutant plume

Plume leaving treatment wall should be "clean"

Treatment Wall: (High permeability reactive barrier)

Impermeable walls funneling pollutant plume to treatment wall

FIGURE 3.11
Funnel-and-gate arrangement of PRB treatment of pollutant plume. Funnel effect is provided by the impermeable walls that channel pollutant plume transport to the PRB gate. (From Yong, R.N. and Mulligan, C.N., *Natural Attenuation of Contaminants in Soils*, CRC Press, Boca Raton, FL, 2004.)

allows the various assimilative and biodegradative mechanisms to attenuate the pollutants. The PRB needs to be strategically located downgradient to intercept the pollutant.

PRBs are also known as treatment walls. The soil materials in these walls or barriers can include a range of oxidants and reductants, chelating agents, catalysts, microorganisms, zero-valent metals, zeolite, reactive clays, ferrous hydroxides, carbonates and sulfates, ferric oxides and oxyhydroxides, activated carbon and alumina, nutrients, phosphates, and soil organic materials. The choice of any of these treatment materials is made on the basis of site-specific knowledge of the interaction processes between the target pollutants and material in the PRB. Laboratory tests and treatability studies are essential elements of the design procedure for the treatment walls (PRBs). When designed properly, a PRB provides the capability for assimilation of the pollutants in the pollutant plume as it migrates through the barrier. In that sense, PRBs function in much the same manner as IRRs, except that the region is a constructed barrier. Some of the assimilative processes in the PRB include the following:

- Inorganic pollutants: Sorption, precipitation, substitution, transformation, complexation, oxidation, and reduction.
- Organic pollutants: Sorption, abiotic transformation, biotransformation, abiotic degradation, biodegradation.

Use of natural attenuation for management of pollutant transport and transmission in soil: We have at least three ways in which natural attenuation can be used to manage or control the transport of pollutants in soil. These include monitored natural attenuation (MNA), enhanced natural attenuation (ENA), and engineered natural attenuation (EngNA). These have various benefits and are used as effective tools in the control and

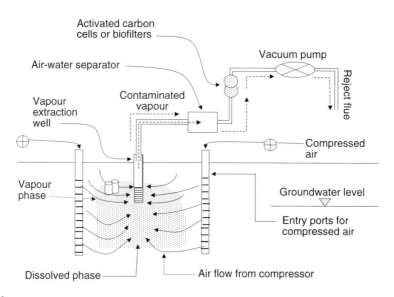

FIGURE 3.12

Schematic of a biosparging process in combination with SVE. Note that a series of compressed air wells and SVE wells can be introduced into the ground — connected in series or in parallel. Obviously the number of wells that can be introduced will depend on the capacity of the compressor and vacuum pump systems.

management of contaminant and pollutant leachate plumes, especially in specific bioremediation schemes.

Other techniques, such as air sparging, are employed to volatilize the VOCs from the groundwater. Biological techniques can also be combined with the extraction techniques in biosparging (Figure 3.12) and bioslurping processes. Essentially, both techniques add another component to the bioventing technique, as for example shown as the solvent extraction procedure in Figure 3.8. In the bioslurping technique, another dimension to the SVE process is added by using vacuum-enhanced pumping to recover free product (NAPLs).

3.4.1.10   Ex Situ *Processes*

For extracted groundwater, treatment is required before discharge or utilization of the abstracted groundwater as drinking water. These techniques are usually quite extensive, involving extraction of substantial groundwater. Standard wastewater treatment techniques, physical, chemical, or biological, are utilized. Physical-chemical techniques include physical and chemical procedures for removal of the pollutants, including precipitation, air-stripping, ion exchange, reverse osmosis, electrochemical oxidation, etc.

Techniques for groundwater treatment for arsenic and waste considerations are highlighted in Table 3.1. Treatment methods need to minimize the wastes produced to ensure that these processes are sustainable. To evaluate the sustainability of these methods, several factors, including materials, energy, transportation, and waste management requirements for the treatment process, need to be taken into consideration. One of the principal methods is ion exchange. However, in some cases, simple ion exchange techniques are insufficient. An example of this is As(III). Oxidation of this form to As(V) must be required and performed with a preoxidation filter. Although this method is highly efficient, disposal of a toxic arsenic waste from the regeneration of these filters and the ion exchange resins as a result of the water treatment procedures affects the sustainability

TABLE 3.1

Comparison of Technologies for the Remediation of Arsenic-Contaminated Groundwater

Technology	Waste Stream	Treatment of Waste	Disposal Options
Coagulation/filtration	Ferric sludge, redox sensitive, 97% water content	Dewatering and drying	Landfill after dewatering, brick manufacture (Rouf and Hossain, 2003)
Activated alumina with regeneration	Alkaline and acidic liquids	Neutralization and precipitation with ferric salts	Sewer, residual into landfill
Iron oxide filters	Exhausted adsorbent, redox sensitive, <50% solids, passes TCLP test	No treatment	Landfill, immobilization, brick manufacture (Rouf and Hossain, 2003)
Ion exchange	Liquid saline brine	Precipitation with ferric salts	Sewer, brine discharge, landfill for residual, possible recycling of brines
Membrane techniques such as reverse osmosis or nanofiltration	Concentrated liquids	None performed	Sewer or brine discharge

Note: TCLP = Toxicity Characteristics Leaching Procedure; see Figure 4.4 and discussion in Section 4.3.5 in Chapter 4.

Source: Adapted from Driehaus, W., in *Natural Arsenic in Groundwater: Occurrences, Remediation and Management,* Bundschuh, J. et al., Eds., Taylor & Francis, London, 2005, pp. 189–203.

of these processes. These purification activities generate significant wastes that can severely affect the environment, causing more harm than good. Due to the problems of arsenic in the groundwater, economic solutions need to be found to ensure the safety of the drinking water.

Several common treatment technologies are used for removal of inorganic contaminants, including arsenic, from drinking water supplies. Large-scale treatment facilities often use conventional coagulation with alum or iron salts, followed by filtration to remove arsenic. Lime softening and iron removal also are common conventional treatment processes that can potentially remove arsenic from source waters. Treatment options identified by EPA include ion exchange, reverse osmosis, activated alumina, nanofiltration, electrodialysis reversal, coagulation/filtration, lime softening, greensand filtration and other iron/manganese removal processes, and emerging technologies not yet identified (U.S. EPA, 2003). Treatment facilities or alternative water sources are required. We can see that this treatment method, along with the others, does not eliminate the arsenic source before delivery, and that other methods must be developed to accomplish this. Other groundwater methods, such as oxidation, can lead to the formation of toxic by-products and sludge. These will require disposal.

Methods for treatment of MTBE include bioremediation, granular activated carbon (GAC), air-stripping, ozonation, ozone/hydrogen peroxide, or phytoremediation (Richardson, 2003). Thermal techniques include pyrolysis and supercritical water oxidation. For a detailed discussion of standard chemical and biological treatment processes, refer to various textbooks, such as Metcalf and Eddy (2003).

Previously, effluent quality was the only basis for evaluating treatment capabilities of water treatment processes. Capital, energy, nutrient, and other requirements need to be included to determine if the process under consideration is sustainable for future generations. Recycling of resources needs to be practiced as much as possible. Mulder (2003) compared the sustainability of nitrogen removal systems that included (1) conventional activated sludge systems, (2) an activated sludge system that relied on autotrophic nitro-

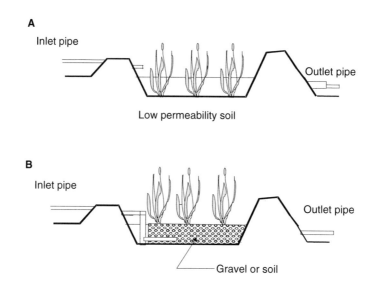

FIGURE 3.13
Overview of (A) surface and (B) subsurface flow-constructed wetlands. (Adapted from Mulligan, C.N., *Environmental Biotreatment*, Government Institutes, Rockville, MD, 2002.)

gen removal, (3) algal or duckweed ponds, and (4) constructed wetlands (Figure 3.13). The author used six sustainability indicators: production of sludge, energy consumption, resource recovery, space and area requirements, and N_2O emissions. They determined that the system that combines nitrification and anaerobic ammonia oxidation (autotrophic nitrogen removal) is the most sustainable because (1) organic matter is not required, (2) sludge production is low, and (3) nitrogen removal is high.

3.4.2   Groundwater and Water Management

An effective groundwater management policy must first involve an evaluation of present practices — beginning with a determination of the basic needs for the water. Laws and regulations then need to be established to ensure water quality and quantity. As shown in Figure 3.14, one must first determine if adequate quantities of groundwater are available to meet the required needs. If not, one needs to manage the system to allow for recharge before depleting this resource. If water budget analyses show that the quantities are sufficient, the quality of the groundwater will then need to be determined. It may be adequate for industrial or irrigation purposes without treatment, or perhaps with *in situ* treatments — as indicated in the previous sections. Drinking water quality may require further treatment. Most often this is accomplished by extraction pumping and treatment with a suitable sustainable method — to avoid harming aspects of the environment to protect others. Evaluation of the most sustainable water treatment processes can be performed through procedures similar to that previously described.

Twelve principles of green engineering have been suggested to engineers as a way to improve the sustainability of industrial processes (Anastas and Zimmerman, 2003). The second principle is particularly relevant for the prevention of water pollution. It says that "it is better to prevent waste than to treat or clean up waste after it is formed." In other words, processes should be designed to reduce water use and the amount of pollutants that reach the water, so that the water will not have to be treated later on. In the past, many models have been developed for prediction of the impact of certain chemicals in the environment, such as the transport of contaminants from point and nonpoint sources

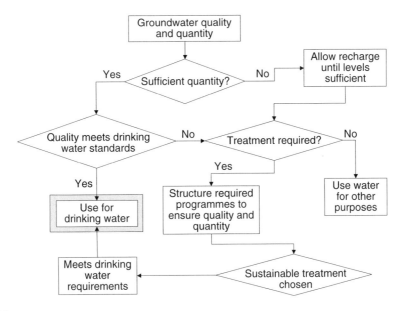

FIGURE 3.14
Flow diagram for groundwater management for drinking water purposes.

(Mihelcic et al., 2003). However, they have not focused on how to reduce or prevent the pollution. End-of-pipe solutions were the main waste management strategy until recently, when green engineering has become more prominent.

Upon determination of impaired water quality, strategies would need to be developed to prevent the introduction of the pollutants. Groundwater and surface water monitoring and GIS will enable the development of the management strategies. The GIS would incorporate all aspects of land use, including the types of ecosystems, landscapes, and water use. Monitoring will include determination of the quantities of water, the quality in terms of nutrient and contaminant contents, and biological monitoring, as described previously. Models would need to be developed to predict water discharge and recharges. All the information could then be combined to determine the water management strategy for avoidance of water pollution and optimal water use (Figure 3.15). All of these cannot be achieved, however, if the society is not educated concerning water usage and its importance. Legal guidelines must also be issued and followed to protect the quantity of resource water.

3.4.2.1 *Evaluation of the Sustainability of Remediation Alternatives*

Attention is focused on the problem of arsenic-polluted groundwater because pollution of groundwater from arsenic is a major threat to human health, and because this is both a man-induced and a naturally occurring phenomenon. In choosing the remediation technologies to treat this problem, it is necessary to factor in the targets, exposure routes, future land use, acceptable risks, legislation, and resultant emissions. A schematic illustration of the criteria and tools for evaluating technologies and protocols for environmental management of contaminated soils and groundwater is shown in Figure 3.16. Specific comments are included in Table 3.2 for the various technologies. Other factors that need to be considered to evaluate site remediation technologies include (1) disturbance to the environment, (2) energy use and consumption, (3) solid wastes generated, (4) emissions of contaminants and greenhouse gases into the air, and (5) water and materials used.

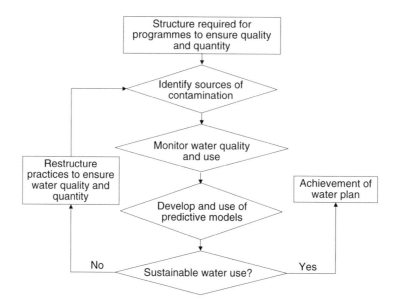

FIGURE 3.15

Flowchart demonstrating development of a program to ensure sustainable water quantity and quality.

TABLE 3.2

Comparison of Remediation Technologies in Soil and Groundwater

Technology	Cost[a]	Long-Term Effectiveness	Reduction of Toxicity	Reduction in Mobility
Capping	Good	Low	Low	Good
Solidification (*in situ*)	Avg.	Avg.	Low	Good
Solidification (*ex situ*)	Good	Avg.	Low	Good
Vitrification	High	Good	Low	Good
Biological treatment	Good	Low	Good	Good
Soil washing	Avg.	Good	Low	Low
Pyrometallurgical extraction	High	Good	Low	Low
Electrokinetics	Avg.	Good	Low	Low

[a] High is in the range of $300 to 900/tonne; average (Avg.), from $100 to 300/tonne; and low, up to $100/tonne.

Source: Adapted from Evanko, C.R. and Dzombak, D.A., *Remediation of Metals-Contaminated Soils and Groundwater Technology Evaluation Report*, TE 97-01, GWRTAC, Pittsburgh, PA, October 1997.

3.5   Concluding Remarks

The W (water) in WEHAB is of utmost importance because (1) without water, one would perish, and (2) the wide and varied requirements for water usage. The demand and use of water can often produce situations that result in conflicts between humankind and the environment. Degradation or impairment of water quality results from various usages associated with agriculture, industrialization, and urbanization. Management and education in sustainable water usage are required, and sources of pollution must be eliminated to maintain water quality and supply for future generations. Water must be conserved and managed properly for preservation of biodiversity. Failure to do so will result in the

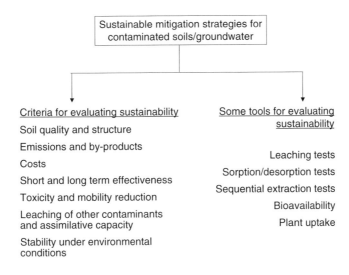

FIGURE 3.16

Criteria and tools for evaluating technologies and protocols for environmental management of contaminated soils and groundwater.

diminution of the capability of the geoenvironment to provide the basis for life support. Various remediation tools have been developed to treat the water once it has been contaminated. These techniques each need to be compared on the basis of resource depletion, energy requirements, and emissions to enable more informed choices regarding sustainable treatment processes.

The record shows that outside of the global distribution of water resources, the two great threats to the availability and quality of water resources (surface water and groundwater) are (1) overuse of the available water resources and (2) pollution of these same water resources. The overuse problem arises from poor management practices and lack of knowledge or ignorance of the nature of the various water budget items and how they contribute to the health of the water budget. Control and mitigation of the impact of pollution of available water resources from the many pollutant sources are measures that must be undertaken as critical procedures in structuring water sustainability protocols and requirements.

While implementation of remediation technology to improve compromised or impaired water quality is an admission that management and controls on water resource pollution have failed, it is nevertheless a remedy that needs more attention and research. Development of more capable technology to clean polluted water resources is necessary — to meet the critical demand for clean water by communities. Since aquifers or groundwater in general serve as a primary water resource for many developed and remote communities, it is imperative that protection of groundwater quality be mandated as the first priority by regulatory agencies. This in essence requires (1) attention to the many forces that individually and collectively produce the contaminants and pollutants that find their way into the groundwater resource, and (2) control, treatment, and remediation technology to manage these forces.

References

Achten, C., Kolb, A., and Putmann, W., (2002), Occurrence of methyl tert-butyl ether (MTBE) in riverbank filtered water and drinking water produced by riverbank filtration: 2, *Environ. Sci. Technol.*, 36:3662–3670.

Anastas, P. and Zimmerman, J., (2003), Design through the 12 principles of green engineering, *Environ. Sci. Technol.*, 37:94A–101A.

Ballesteros, B., Barcelo, D., Dankwardt, A., Schneider, P., and Marco, M.P., (2001), Evaluation of a field-test kit for triazine herbicides (SensioScreen(R) TR500) as a fast assay to detect pesticide contamination in water samples, *Anal. Chim. Acta*, 475:105–116.

Bass, D.H., Hastings, N.A., and Brown, R.A., (2000), Performance of air sparging systems: a review of case studies, *J. Hazardous Mat.*, 72:101–120.

Burkart, M.R. and Stoner, J.D., (2002), Nitrate in aquifers beneath agricultural systems, *Water Sci. Technol.*, 45:19–29.

Calera Belmonte, A., Medrano Gonzalez, J., Vela Mayorga, A., and Castano Fernandez, S., (1999), GIS tools applied to the sustainable management of water resources: application to the aquifer system 08-29, *Agric. Water Manage.*, 40:207–220.

Chamley, H., (2003), *Geosciences, Environment and Management*, Elsevier, Amsterdam, 450 pp.

Crowder, A. and Painter, D.S., (1991), Submerged macrophytes in Lake Ontario: current knowledge, importance, threats to stability and needed studies, *Can. J. Fisheries Aquatic Stud.*, 48:1539–1545.

Dorf, R.C., (2001), *Technology, Humans and Society: Toward a Sustainable World*, Academic Press, San Diego, 500 pp.

Driehaus, W., (2005), Technologies for arsenic removal from potable water, in *Natural Arsenic in Groundwater: Occurrences, Remediation and Management*, Bundschuh, J., Bhattacharaya, P., and Chandrasekharam, D. (Eds.), Taylor & Francis, London, pp. 189–203.

Erickson, B.E., (2003), Field kits fail to provide accurate measure of arsenic in groundwater, *Environ. Sci. Technol.*, 37:35A–38A.

Evanko, C.R. and Dzombak, D.A., (1997), *Remediation of Metals-Contaminated Soils and Groundwater Technology Evaluation Report*, TE 97-01, GWRTAC, Pittsburgh, PA, October.

Gitelson, A. and Yacobi, Y., (1995), *Spectral Features of Reflectance and Algorithm Development for Remote Sensing of Chlorophyll in Lake Kinneret, Air Toxics and Water Monitoring*, Europto, SPIE, Vol. 2503, 21 June.

Gleick, P.H., (1993), *Water in Crisis*, Oxford University Press, New York, 504 pp.

Ha, S.R., Park, S.Y., and Park, D.H., (2003), Estimation of urban runoff and water quality using remote sensing and artificial intelligence, *Water Sci. Technol.*, 47:319–325.

Hakvoort, H., de Haan, J., Jordans, R., Vos, R., Peters, S., and Rijkeboer, M., (2002), Towards airborne remote sensing of water quality in the Netherlands: validation and error analysis, *ISPRS J. Photogram. Remote Sens.*, 57:171–183.

Karr, J.R., Fausch, K.D., Angermeier, P.L., Yant, P.R., and Schlosser, I.J., (1986), *Assessing Biological Integrity in Running Waters: A Method and Its Rationale*, Special Publication 5, Illinois Natural History Survey.

LaCorte, S., Olivella, L., Rosell, M., Figueras, M., Ginebreda, A., Barcello, D. (2002), Gas chromatography-cross-validation of methods used for analysis of MTBE and other gasoline components in groundwater, *Chromatographia* 56:739–744.

Melloul, A.J. and Goldenberg, L.C., (1997), Monitoring of seawater intrusion in coastal aquifers: basic and local concerns, *J. Environ. Manage.*, 51:73–86.

Mench, M., Vangronsveld, J., Clijsters, H., Lepp, N.W., and Edwards, R., (2000), *In situ* metal immobilization and phytostabilization of contaminated soils, in *Phytoremediation of Contaminated Soil and Water*, Terry, N. and Bauelos, G. (Eds.), Lewis Publishers, Boca Raton, FL, pp. 323–358.

Metcalf and Eddy, Inc., (2003), *Wastewater Engineering: Treatment and Reuse*, McGraw-Hill, Dubuque, IA, 1858 pp.

Mihelcic, J.R., Crittenden, J.C., Small, M.J., Shonnard, D.R., Hokanson, D.R., Zhang, Q., Chen, H., Sorby, S.A., James, V.U., Sutherland, J.W., and Schnoor, J.L., (2003), Sustainability science and engineering: the emergence of a new metadiscipline, *Environ. Sci. Technol.*, 37:5314–5324.

Millie, D.F., Schofield, O.M., Kikpatrick, G.J., Johnson, G., and Evens, T.J., (2002), Using absorbance and fluorescence spectra to discriminate microalgae, *Eur. J. Phycol.*, 27:313–322.

Mulder, A., (2003), The quest for sustainable nitrogen removal technologies, *Water Sci Technol.*, 48:67–75.

Mulligan, C.N., (2002), *Environmental Biotreatment*, Government Institutes, Rockville, MD, 395 pp.

National Research Council (NRC), (2000), *Natural Attenuation for Groundwater Remediation*, National Academy Press, Washington, DC, 292 pp.

Pedersen, J.A., Yeager, M.A., and (Mel)Suffet, I.H., (2002), Characterization and mass load estimates of organic compounds in agricultural irrigation runoff, *Water Sci.Technol.*, 45:103–110.

Pedley, S. and Howard, G., (1997), The public health implications of microbiological contamination of groundwater, *Q. J. Eng. Geol.*, 30:179–188.

Pimentel, D., Wilson, C., McCullum, C., Huang, R., Dwen, P., Flack, J., Tran, Q., Saltman, T., and Cliff, B., (1997), Economic and environmental benefits of biodiversity, *BioScience*, 47:747–757.

Poiger, T., Muller, M.D., and Buser, H.R., (2002), Verifying the chiral switch of the pesticide metolachlor on the basis of the enantiomer composition of environmental residues, *Chimia*, 56:300–303.

Postel, S., (1999), When Will the World's Wells Run Dry?, *World Watch*, September, pp. 30–38.

Postel, S.L., Daily, G.C., and Ehrlich, P.R., (1996), Human appropriation of renewable fresh water, *Science*, 271:785–788.

Rathfelder, K., Yeh, W.W.G., and Mackay, D., (1991) Mathematical simulation of soil vapour extraction systems: model development and numerical examples, *J. Contam. Hydrol.*, 8:263–297.

Richardson, S.D., (2003), Water analysis: emerging contaminants and current issues, *Anal. Chem.*, 75:2831–2857.

Rouf, Md.A. and Hossain, Md.D., (2003), Effects of using arsenic-iron sludge in brick making, in *Fate of Arsenic in the Environment, Proceedings of the BUET-UNU International Symposium*, Dhaka, Bangladesh, 5–6 February 2003.

Smith, L.A., Means, J.L., Chen, A., Alleman, B., Chapman, C.C., Tixier, J.S., Jr., Brauning, S.E., Gavaskar, A.R., and Royer, M.D., (1995), *Remedial Options for Metals-Contaminated Sites*, Lewis Publishers, Boca Raton, FL, 221 pp.

Southerland, M.T. and Stribling, J.B., (1995), Status of biological criteria development and implementation, in *Biological Assessment and Criteria: Tools for Water Resource Planning and Decision Making*, Davis, W.D. and Simon, T.P. (Eds.), Lewis Publishers, Boca Raton, FL, pp. 81–96.

Straub, Ti.M. and Chandler, D.P., (2003), Towards a unified system for detecting waterborne pathogens, *J. Microbiol. Methods*, 53:185–198.

Strauss, H., (1991), *Final Report: An Overview of Potential Health Concerns of Bioremediation*, Environmental Health Directorate, Health Canada, Ottawa, 54 pp.

UNDP (United Nations Development Program), (2001), *Making New Technologies Work for Human Development: The Human Development Report 2001*, Oxford University Press, Oxford.

UNEP (United Nations Environmental Program), (2002), *GEO: Global Environment Outlook*, Earthscan Publisher, London.

UN-FPA (United Nations Populations Fund), (2001), *The State of World Population 2001 Footprints and Milestones: Population and Environmental Change*, Phoenix-Trykkkeriet AS, Denmark.

United Nations Economic Commission for Europe, Environments and Human Settlements Division, (1998), *Burst Water Main Floods Central Manhattan*, ECE/ENV/98/1, Geneva, 1 January.

U.S. EPA, (1991), *Technical Support Document for Water Quality Based Toxics Control*, EPA 505-2-90-001, U.S. Environmental Protection Agency, Office of Water, Washington, DC.

U.S. EPA, (1994), *Selection of Control Technologies for Remediation of Soil with Arsenic, Cadmium, Chromium, Lead or Mercury*, revised draft engineering bulletin, 31 January.

U.S. EPA, (1996), *Environmental Indicators of Water Quality in the U.S.*, EPA-841-R-95-005, U.S. Environmental Protection Agency, Office of Water, Washington, DC.

U.S. EPA, (2000), *Methods for the Determination of Organic and Inorganic Compounds in Drinking Water*, Vol. I, EPA 815-R-00-014, U.S. Environmental Protection Agency, Office of Water, Washington, DC.

U.S. EPA, (2003), *Arsenic Treatment Technology Handbook*, EPA 816-R-03-014, Office of Water (4606M), July.

USGS, (1998), Materials Flow and Sustainability, USGC Fact Sheet FS-068-98, U.S. Geological Survey, U.S. Department of the Interior, Washington, DC.

Warren, M.L., Jr., Burr, B.M., Walsh, S.J., Bart, H.L., Jr., Cashner, R.C., Etnier, D.A., Freeman, B.J., Kuhajda, B.R., Mayden, R.L., Robison, H.W., Ross, S.T., and Starnes, W.C., (2000), Diversity, distribution, and conservation status of the native freshwater fishes of the southern United States: a comprehensive review of the diversity, distribution, and conservation status of native freshwater fishes of the southern United States reveals formidable challenges for conservation management, *Fisheries Am. Fisheries Soc.*, 25:7–31.

Williams, P.R.D., (2001), MTBE in California drinking water: an analysis of patterns and trends II, *Environ. Forensics*, 2:75–86.

World Resources Institute (WRI), (1994), *World Resources 1994–95*, Oxford University Press, New York, 217 pp.

Yong, R.N., (1998), *Compatible Technology for Treatment and Rehabilitation of Contaminated Sites*, NNGI Report 5, Nikken Sekkei Geotechnical Institute, Japan, pp. 1–33.

Yong, R.N., (2001), *Geoenvironmental Engineering, Contaminated Soils, Pollutant Fate and Mitigation*, CRC Press, Boca Raton, FL, 307 pp.

Yong, R.N. and Mulligan, C.N., (2004), *Natural Attenuation of Contaminants in Soils*, CRC Press, Boca Raton, FL, 310 pp.

Yong, R.N., Nutalaya, P., Mohamed, A.M.O., and Xu, D.M., (1991), Land subsidence and flooding in Bangkok, in *Proceedings of the Fourth International Symposium on Land Subsidence*, IAHS Publication 200, pp. 407–416.

Zhang, Q., Crittenden, J.C., and Mihelcic, J.R., (2000), Does simplifying transport and exposure yield reliable results? An analysis of four risk assessment methods, *Environ. Sci. Technol.*, 35:1282–1288.

Zhu, X., He, Z., and Deng, M., (2002), Remote sensing monitoring of ocean colour in Pearl River estuary, *Int. J. Remote Sens.*, 23:4487–4497.

4

Urbanization and the Geoenvironment

4.1 Introduction

Almost 50% of the world's population lives in cities, compared to 34.2% 40 years ago (Chamley, 2003). More than 60% of the population will live in urban areas by the year 2020, making this an increasingly significant component in the global environment. Urban centers, together with their suburbs, constitute what is now called the *built environment*. This built environment includes (1) the various physical structures that serve the community; (2) the resultant products and discharges associated with the various industrial, municipal, and domestic activities, such as waste piles, dumps, aeration ponds, gravel pits, etc.; (3) the infrastructure, such as pipelines, transmission towers, roads, runways, bridges, etc.; (4) the various utilities necessary to service the community, such as power plants, gas plants, wastewater treatment plants, reservoirs, etc.; and (5) the other kinds of resources associated with and necessary to sustain the urban population and the welfare of the community (e.g., parks, lakes, forests, recreational and sporting facilities, etc.). By its very nature, the man-made environment that defines the built environment is often in conflict with the natural environment, and in particular with the goals of sustainability of the land environment and its natural resources. The general perception is that urban centers consume significant resources and pollute the air, land, and water. Populations within the cities require clean air, clean water, sewage and waste management systems, adequate food supply, housing, and transportation. It is estimated that more than 200 million people live in cities that do not have access to clean drinking water, and that more than 400 million people live in cities that do not have access to solid waste collection services and facilities. It is often argued that these demands are currently not well met, and that the demand deficit will continue to escalate with time. Some typical types of urban problems are summarized in the illustration shown in Figure 4.1.

A recent definition of urban sustainability offered at the Sustainable City Conference in Rio (2000) stated that

> The concept of sustainability as applied to a city is the ability of the urban area and its region to function at levels of quality of life desired by the community, without restricting the option available to the present and future generations and without causing adverse impacts inside and outside the urban boundary.

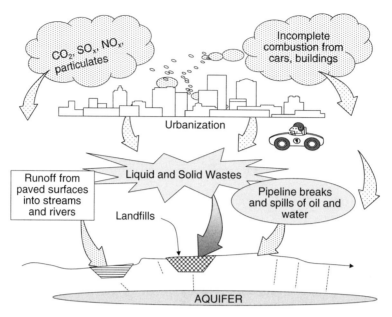

FIGURE 4.1
Urban sources of contamination and their effect on the geoenvironment.

4.2   Land Uses by Urbanization

Urban development is a major consumer of land. Natural landscape areas around the cities are converted into housing estates, industrial parks, and other kinds of facilities designed to serve the community. Land is typically used for housing, businesses, industry, surface and subsurface infrastructures such as roads, wastewater supply, sewers, and power lines, and recreational purposes such as parks and playgrounds. In the U.S., it is estimated that about 100 m^2 of land/sec is lost to urban uses; Germany loses about 14 m^2/sec, and Switzerland, a more environmentally aware country, loses about 1 m^2/sec (RSU, 2000). Abandoned industrial lands are not included in the estimates.

   In the urban context, surface and subsurface soils may be contaminated and degraded. They serve a variety of functions, as for example (1) foundation base for buildings, (2) medium for plant growth, (3) open spaces, (4) park space, (5) urban gardens, (6) bases for roads, ponds, and reservoirs, and (7) sources and sinks of pollutants. Soils in urban areas tend to be more diverse because of the introduction of additives such as buried waste, debris, fuel ash, and other residues. Assessment of soil impairment requires specification of intended use of the soil. Particle size distribution, porosity, erodability, structural stability, hydraulic conductivity, and rootability are some properties that need to be assessed to determine the degree of impairment of the soil. For example, if playing fields, under repeated use in wet conditions, are no longer usable for playing, the situation will need to be remediated to return the land to its intended land use (Mullins, 1990). Signs of deterioration could include ponding, runoff, soil erosion, or poor grass growth. Soil properties relevant for use, soil type, and sensitivity of the properties to changes in soil use are important factors in mitigating soil damage. Some of the properties required for various uses are summarized in Table 4.1.

TABLE 4.1

Soil Physical Properties for Various Urban Uses

Requirement	Soil Properties	Application
Drainage	Hydrology, hydraulic conductivity (soil structure), porosity	Playing fields, effluent disposal
Load-bearing capacity	Bulk density, compactability, water content/potential, penetration resistance, shear strength, compressibility, consolidation	Playing fields, foot paths, load-bearing support for surface infrastructures, and spread footings for light structures
Plant growth medium	Drainage, air capacity, water capacity, bulk density, structure, penetration resistance	Playing fields, gardens, parks or sports fields
Prevention of erosion and runoff	Infiltration, drainage, structural stability	Foot paths, effluent disposal, parks or sports fields

Source: Adapted from Mullins, C.E., in *Soils in the Urban Environment*, Bullick, P. and Gregory, P.J., Eds., Blackwell Scientific Publications, Oxford, 1990, pp. 87–118.

Derelict sites pose some unique problems. In general, these sites have become by default disposal sites with unauthorized disposed goods and substances that include urban garden wastes and other kinds of household wastes. Perhaps the more dominant kinds of debris found in derelict sites are those items that are the result of demolished buildings or buildings in considerable distress. The debris generally found includes building materials such as pipes, pieces of foundation, tile, wood, plaster, rusting steel, and broken concrete slabs and structures.

4.3   Impact of Urbanization on WEHAB

4.3.1   Impact on Water

In urban and suburban areas, more and more land surfaces are covered by buildings, roads, and constructed parking areas using concrete, bituminous concrete, asphalt, or other impervious coverings. Because these covered surfaces prevent infiltration of rainwater and pond water into the subsurface, replenishment of the underlying groundwater is denied and groundwater levels may consequently be lowered. Another effect of such covered surfaces is to allow surface flow (i.e., streaming) of rainwater into collecting areas. It is not uncommon to find pollutants in the surface flow water or streaming water because of noxious substances deposited onto the covered surfaces. Since these waters will eventually find their way into the receiving waters, they can be considered to be a nonpoint source pollution of lakes and rivers.

An extensive underground system of parking areas, sewers, pipes, deep building foundations, and tunnels to depths of 100 m can also significantly affect the underground terrain. Water leakage from buried and degrading water supply pipes leads to excessive water accumulation. In some cities, due to a lack of maintenance, more than 40% of the water supply is lost because of leaking pipes.

In many cities, groundwater abstraction for consumption can be excessive, particularly in regions where available surface water is difficult to access and transport, in arid regions, and in regions where the quality of surface water supply is deemed unsafe for consumption. This will lead to severe problems of ground subsidence, soil erosion, aridification,

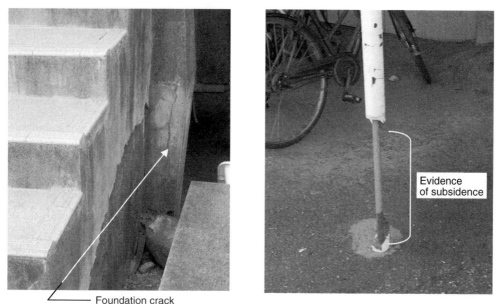

Evidence of subsidence

Foundation crack

FIGURE 4.2
Subsidence and foundation impairment resulting from excessive groundwater abstraction.

and salinization. In coastal regions in particular, high rates of groundwater abstraction can lead to land subsidence and flooding during high rainfall periods. Seawater intrusion into the aquifers can also occur as a result of groundwater abstraction from the aquifers, resulting in contamination of shallow aquifers. If vertical communication exists between shallow and deeper aquifers, contamination of the deeper aquifers will eventually occur (Yong et al., 1994, 1995). Land subsidence will also result in damage to structures with foundations affected by differential settlement (Figure 4.2). This problem has been discussed in detail in the previous chapter in respect to the impact of high rates of groundwater abstraction in regions where cities are founded on Quaternary sediments.

Urban discharges can substantially degrade groundwater quality. For example, in China, phenols, cyanides, mercury, chromium, arsenides, and fluorine have been found in more than 50 urban regions. Urban effluent infiltration can also increase levels of nitrates, sulfates, and chlorine up to 90 m in depth (Chamley, 2003). Groundwater pollution causes many health problems and restricts human use.

Salinity is also increasing in the urban environment from discharges of domestic wastewater, seawater intrusion into aquifers, and contamination from road salt. Detergents, soaps, and salts are added to domestic wastewater, which is subsequently sent to wastewater treatment plants. The salts, sodium, and chloride, however, are not removed in the treatment process, therefore resulting in increasing salinity of the water — as schematically illustrated in Figure 4.3. If the treated water is used for irrigation for agriculture, as for example in areas of the United States, Israel, and Jordan (Vengosh, 2005), groundwater and soil quality may be impaired. The addition of road salt for de-icing also contributes to the salinization of groundwater. Powdered road salt used for de-icing may be subject to wind dispersion, resulting in contamination of the surrounding area. Although calcium chloride can be used as a de-icing salt, it is more expensive and is not as effective as sodium chloride. The combination of sodium chloride with calcium chloride can reduce the sodium-to-chloride ratio. To reduce the stress on the land and water resources, processes such as reverse osmosis and nanofiltration will be required for desalination.

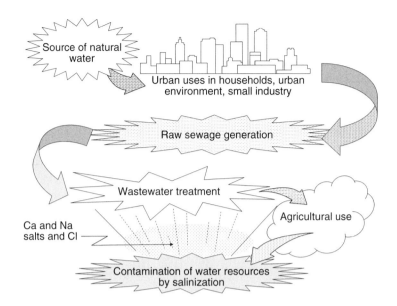

FIGURE 4.3
Cycle of salinization of water due to urban use. (Adapted from Vengosh, A., in *Environmental Geochemistry*, Vol. 9, Lolar, B.S., Ed., Elsevier, Amsterdam, 2005, pp. 333–366.)

4.3.2 Effect of Traffic and Energy Use

The impacts on the environment from traffic in urban areas are in the form of airborne, land, and water pollution. Understanding these problems and finding solutions for reduction of emissions are of paramount importance for the health and welfare of society and the geoenvironment. The occurrence of air pollutants from vehicular emissions at elevated levels can lead to serious health problems for the population living in high-traffic areas. Pedestrians or passengers in vehicles that do not stay in the area long will not be affected, but there could be significant impacts for those that live in the area. Carbon monoxide (CO) is a major pollutant from vehicles. In particular, intersections where vehicles are stationary for a period of time can substantially increase the level of carbon monoxide released into the air. A study by Croxford et al. (1995) shows that CO levels can reach 12 ppm. The most prevalent air pollutants from vehicular emissions are nitrogen oxides (NO_x), hydrocarbons, and CO.

In the past, lead in gasoline led to accumulation of lead on surface soils for many years. Page and Ganje (1970) showed that 15 to 36 $\mu g/g$ over a period of 40 years accumulated in high-traffic areas, compared with negligible amounts in low-traffic areas. Even though lead in gasoline has been eliminated in many countries, the low mobility of lead ensures that the lead remains in the ground as a contaminant for many years.

In terms of grams of emissions per mile, for the period from 1996 to 2003, emissions of hydrocarbons were 0.15 g/km; CO, 2.0 g/km; and nitrogen oxides, 0.2 g/km (U.S. EPA, 1998). Although emissions have decreased substantially over the years, the combination of poorly maintained cars with more cars and other types of heavy-duty vehicles that consume more gasoline and emit higher emissions (such as sports utility vehicles (SUVs), light trucks, vans, and pickup trucks) ensures that reduction in noxious emission will not be readily accomplished. Not only do vehicles emit various pollutants, but they also contribute more than 20% of the carbon dioxide emissions from gasoline. Other petroleum fuels add another 12%. Reduction in vehicle use or the use of alternate fuels or other means will be required to substantially reduce the emission of this greenhouse gas. Without entering into the lively debate (see Section 11.5 in Chapter 11) on global warming and climate change and (1) human factors causing such changes, (2) reliability of available data, (3) how far back in time one needs for data scrutiny, and (4) viability and applicability of climate change models, it is pertinent to note that the accumulation of greenhouse gases will contribute to other factors that are responsible for climate change.

Paved road surfaces share the same problems as covered impervious areas in cities, i.e., rainwater cannot infiltrate into the soil, but will accumulate on roads and will wash away the pollutants on the streets. This can include motor oil, grease, antifreeze, metals, phosphorus, and other pollutants. These are then washed into local waterways and rivers by streaming flows, resulting in oxygen depletion and pollution of the waters — killing fish, plants, and other aquatic life. Public health is threatened when the contaminated water and contaminated fish are ingested.

Soil quality is another issue of importance. Air pollutants are deposited on the ground from precipitation passing through the airborne pollutants. These will find their way onto surface soils and into the subsoil, and also into rivers and groundwater. Human exposure with the pollutants can be through inhalation, contact with the soil, and ingestion of water and crops grown in contaminated soil. Ingestion of vegetables grown in urban gardens and children playing in exposed sand boxes and bare soil landscapes are good examples of human exposure to deposited air pollutants. While awareness of the potential hazards of such a form of water and soil pollution exists, the same cannot be said for information and data on pollutant concentrations and distributions from such types of deposited airborne pollutants.

Soils and sediments contiguous to roads can exhibit high levels of pollution. In Germany, levels of lead, cadmium, chromium, nickel, vanadium, and zinc in soils near roads have been found to be up to 5 times higher than in soils located some distance from roads. In the case of polycyclic aromatic hydrocarbons (PAHs), these were found to be up to 100 times higher than in soils distant from the roads. The source for these pollutants can be traced to vehicular exhausts (Münch, 1992). Cuny et al. (2001) report that elimination of lead in gasoline has substantially reduced lead levels in parking and rest areas near the motorways in France.

Acid rain is a direct result of precipitation through sulfur and nitrogen oxides emitted from fossil fuel combustion in coal thermal power plants. Sulfuric and nitric acids are deposited on (1) the soil surface and into the underlying soil, and perhaps into the groundwater, (2) surface water courses, and (3) other surfaces. Northeastern U.S. and eastern Canada have experienced rain with pH values ranging from 4 to 5. Deforestation by acid rain has been significant in the nineteenth and early twentieth centuries in North America and Europe. Forests affected by acid rain, in particular, have less ability to retain water and protect against wind erosion. Heavy metals in contaminated soils may also be released by the acid rain. Fish and other aquatic organisms are susceptible to many metabolic disorders when the pH of lakes and other water bodies drops to 5 and below. Although not directly a geoenvironmental land problem, it is pertinent to note that respiratory problems in smog events are significant human health problems.

4.3.3 Implications on Health

The pathways of exposure of urban soils include inhalation and ingestion of soil and dusts through respiration, consumption of homegrown foods, and contact with the soil. Urban gardening is widely practiced. Thorton and Jones (1984) have reported on various tests conducted in the United Kingdom regarding radishes and lettuce grown in typical urban gardens with soils containing different concentrations of zinc, copper, and lead. Measurable values of lead in the radishes and lettuce were obtained, and it was concluded that both soil splash and foliar uptake contributed to the measured lead levels.

4.3.4   Impact of Land Use

Urban land can become degraded chemically and physically. As noted previously, contaminants degrade soil quality through release into the soil via spills, runoff, and other additives. Roads, sidewalk, parking lots, and other structures seal the land and reduce water infiltration into the groundwater. The installation of cables, sewers, foundations, and other underground structures disrupts the physical structure of the soil. Introduction of softer soils or wastes into the natural soil changes its characteristics.

Greenfields are lands that have not been disturbed. They have the capacity to maintain their biodiversity, ecological functions, and soil quality, and can renew their groundwater resources. Brownfields, on the other hand, have been degraded by contamination from various sources — primarily industrial and manufacturing facilities such as refineries, rail yards, gas stations, warehouses, dry cleaners, and other commercial enterprises using or storing hazardous chemicals. Table 4.2 provides a list of the various industries, activities, and contaminants that lead to land contamination, mainly in urban areas. Abandoned urban lands (brownfields) are clear indications of failure in complying with the principles of sustainable development as envisaged in the Brundtland Report (World Commission on Environment and Development, 1987), as well as that of the Club of Rome.

4.3.5   Impact of Urban Waste Disposal

Improper waste disposal has a major negative impact on the land. Wastes can be in solid forms as municipal solid waste, in liquid forms as sewage and wastewater, or in gaseous forms from vehicular emissions. Solid wastes can originate from (1) households as food and yard wastes, paper, chemicals, wood, and so on; (2) urban businesses that generate wastes similar to household wastes; (3) industrial wastes; and (4) construction wastes. Hospital wastes will be discussed in more detail in Chapter 7. As there are many factors that influence production of wastes, computer models have been developed to estimate waste production rates.

Various hydraulic computer models have been developed to calculate contaminant infiltration rates. Some of these models include (1) the hydraulic evaluation of landfill performance (HELP) model (Schroeder et al., 1994), (2) the water balance analysis program (MBALANCE) (Scharch, 1985), (3) the leaching estimation and chemistry model (Hutson and Wagenet, 1992), and (4) the unsaturated soil and heat flow model (UNSAT-H) (Fayer and Jones, 1990). HELP has been the most frequently used model for final cover and leachate collection system design and is particularly useful for humid and semihumid areas. UNSAT-H is more appropriate for arid and semiarid regions for landfill cover design.

Wastes can also be classified as demolition, nonhazardous (municipal), and hazardous wastes. Inert or nonleachable wastes include many types of construction wastes, such as soil, bricks, concrete, tile, and gypsum board. As long as they are not contaminated, these materials can be reused as backfill material, subgrade and road materials, and even as building materials. Organic and inorganic wastes have been discussed previously in Section 2.3.

Wastes can be classified according to the physical state, origin, degree of hazard, or ability to be recycled or transformed. Wastes are classified as *hazardous* if they have any of the following characteristics:

- Ignitability: Potential for fire hazard during storage, transport, or disposal under standard temperatures and pressures.

- Corrosivity: Potential for corrosion of materials in contact with candidate waste, resulting in environmental and health threats due to a pH less than 2.0 or greater than 12.5.

- Reactivity: Potential for adverse chemical reactions when in contact with water, air, or other wastes.

- Toxicity: As per the Toxicity Characteristics Leaching Procedure (TCLP). The leaching test system for application of the TCLP is shown in Figure 4.4, and the regulatory levels of the compounds in the leachate are given in Table 4.3.

TABLE 4.2

Urban Land Uses and Activities Leading to Contaminated Land

Industry	Activity Leading to Soil Contamination	Type of Contaminant
Airports	De-icing and fire control runoff, servicing, fueling	Acids/alkalis, asbestos, solvents, herbicides, PCBs, fuels, de-icing agents, fire-fighting chemicals
Animal slaughterhouses	Leaking tanks, pipework, spillages	Acids/alkalis, organic compounds, pathogens, metals, metalloids
Auto repair and refinishing	Leaking tanks, spills, sprays, solid wastes	Metals, dust, VOCs, solvent, paints and paint sludges, scrap metal, waste oils
Battery recycling and disposal	Discarded batteries	Pb, Cd, Ni, Cu, Zn, As, Cr
Incinerators	Solid wastes, gaseous emissions	Dioxin, ash, metals, wastes
Laundries and dry cleaning	Spillage of solvents and other contaminants	Organic compounds, including solvents (chloroform, TCE), PCE, PCBs, fuels, asbestos
Landfills (municipal and industrial)	Leachates and gaseous emissions	Metals, VOCs, PCBs, ammonia, methane, household products and cleaners, pesticides, wastes, hydrogen sulfide
Paper and printing works	Leakage from drums and other contaminants, may be buried on-site, spillages of solvents and other materials	Metals, inorganic compounds, acids, alkalis, solvents, inks, degreasing solvents, fuels, oils, PCBs, inorganic ions
Railway yards and tracks	Maintenance and repair of tracks, engines, coal storage	Fuel oils, lubricating oils, PCBs, PAHs, solvents, ethylene glycol, creosote, herbicides, metal fines, asbestos, ash, sulfate
Sewage treatment	Disposal of sludges, stones, and solid matter in landfills and other places	Metals, PCBs, PAHs, solvents, pathogens, acids/alkalis, inorganic compounds
Transport depots	Fueling, vehicle washing and maintenance, storage of tires and wastes, leaks from split drums	Metals such as Pb, Cr, Zn, Cu, and vanadium, acids/alkalis, solvents, PAHs, asbestos
Waste disposal	Landfill leachates, waste transfer station spills	Inorganic chemicals, oils, metals, PCBs, PAHs, solvents, acids/alkalis, inorganic compounds detergents, asbestos

Source: Adapted from Syms, P., *Previously Developed Land, Industrial Activities and Contamination*, 2nd ed., Blackwell Publishing, Oxford, 2004; U.S. EPA, *Road Map to Understanding Innovative Technology Options for Brownfields Investigation and Cleanup*, 2nd ed., EPA 542 B-99-009, U.S. Environmental Protection Agency, Office of Solid Waste and Emergency Response, Washington, DC, 1999b.

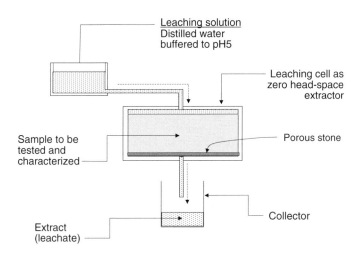

FIGURE 4.4

Typical leaching cell used as a zero-head extractor for application of TCLP. The results of chemical analysis of the extract (leachate) should be compared with the regulatory values shown in Table 4.3. (Adapted from Yong, R.N., *Geoenvironmental Engineering: Contaminated Soils, Pollutant Fate and Mitigation*, CRC Press, Boca Raton, FL, 2001.)

TABLE 4.3

TCLP Compounds and Regulatory Levels in Extract

Compound	Level (mg/l)	Compound	Level (mg/l)
Arsenic	5.0	Hexachloro-1,3-butadiene	0.5
Barium	100.0	Hexachloroethane	3.0
Benzene	0.5	Lead	5.0
Cadmium	1.0	Lindane	0.4
Carbon tetrachloride	0.5	Mercury	0.2
Chlordane	0.03	Methoxychlor	10.0
Chlorobenzene	100.0	Methyl ethyl ketone	200.0
Chloroform	6.0	Nitrobenzene	2.0
Chromium	5.0	Pentachlorophenol	100.0
o-Cresol	200.0	Pyridine	5.0
m-Cresol	200.0	Selenium	1.0
p-Cresol	200.0	Silver	5.0
1,4-Dichlorobenzene	7.5	Tetchloroethylene	0.7
1,2-Dichloroethane	0.5	Toxaphene	0.5
1,1-Dichloroethylene	0.7	Trichloroethylene	0.5
2,4-Dinitrotoluene	0.13	2,4,5-Trichlorophenol	400.0
Endrin	0.2	2,4,6-Trichlorophenol	2.0
Heptachlor	0.008	2,4,5-TP (Silvex)	1.0
Hexachlorobenzene	0.13	Vinyl chloride	0.2

In common with wastes generated from the various industries and manufacturing facilities, municipal solid wastes also will ultimately find their way into one or all of three disposal media: (1) receiving waters, (2) atmosphere, and (3) land. Land disposal of waste products and waste streams appears to be the most popular method for waste containment and management. The various impacts arising from this mode of disposal and containment include degradation of land surface environment and ground contamination by pollutants. Because of the inhomogeneous nature of wastes, stability problems on the surface of the landfill, slope stability, and settling of the landfill can occur. Chemical and biochemical

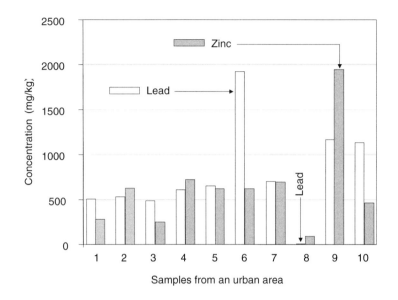

FIGURE 4.5
Lead and zinc concentrations of soil samples taken from various locations in the Montreal area (seven samples from communal garden and three samples from locations near it). (Data from Huang, Y.-T., Heavy Metals in Urban Soils, M.A.Sc thesis, Concordia University, Montreal, Canada, 2005.)

reactions within the landfill and water infiltration will inevitably change the mechanical properties of the landfill.

Both the wastes themselves and the products they produce (such as leachates and emissions) are health and geoenvironmental threats. Disposal of wastes in the ground, illicit dumping, leaking underground storage tanks, and others are all causes for concern. A recent sampling of backyards in Montreal, Canada, where wastes had been previously dumped, indicated elevated levels of the heavy metals lead and zinc (Figure 4.5).

Water entry into waste piles in landfills, together with dissolution processes, results in the generation of waste leachates. A liner and leachate collection system such as that shown in Figure 1.10 and Figure 3.10 in Chapters 1 and 3, respectively, is required to protect the groundwater from fugitive leachates. A cap system is also required to prevent rain from entering the waste pile and producing leachates. Typical leachates from landfills include organic chemicals (e.g., benzene, toluene, ethyl benzene, and xylene [BTEX], PAHs, phthalates, ketones, dioxins, phenols, pesticides, solvents, etc.) and inorganic components (e.g., mercury, cadmium, chromium, copper, zinc, lead, nickel, etc.). The composition of leachates varies significantly depending on the age and type of waste and landfill technology used to contain the waste. The groundwater level and the nature of the soil under the landfill are important factors in managing leachate pollution risks to the groundwater. Concentrations of both organic and inorganic components can be high, as shown in Figure 4.6.

In the past, quarries and other pits without proper barrier and liner systems were used for waste disposal, as for example the Gloucester landfill in Canada (Lesage et al., 1990). From 1969 to 1980, hazardous wastes, including laboratory organic chemicals, were disposed in this landfill, which was on glacial outwash deposits with a semiconfined aquifer. The chemicals reacted with the explosive charges that were also in the waste pile, and the leachate plume of more than 300 m in the aquifer in 1990 was found to contain the unaltered and transformed chemicals from the waste pile.

Gaseous emissions from biodegradation of organic materials in landfills also occur. These gases generally contain both carbon dioxide and methane. More than 350 m^3 of

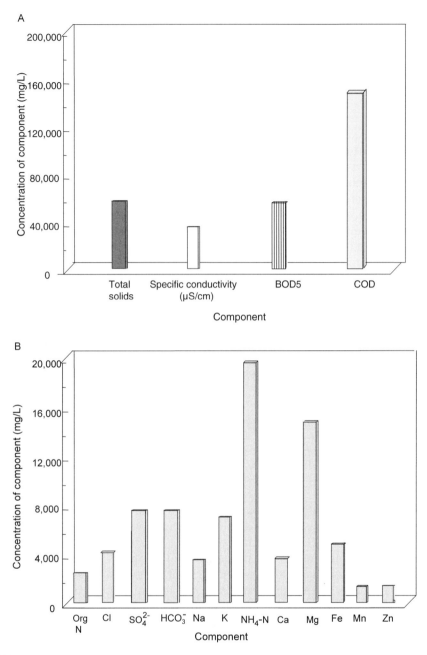

FIGURE 4.6
(A) Range of selected components found in landfill leachate. (Adapted from data reported by Kjeldsen, P. et al., *Crit. Rev. Environ. Sci. Technol.*, 32, 297–336, 2002.) (B) Range of selected components found in landfill leachate. (Adapted from data reported by Kjeldsen, P. et al., *Crit. Rev. Environ. Sci. Technol.*, 32, 297–336, 2002.)

biogas is produced by a tonne of municipal solid waste (MSW) (Genske, 2003). Volatile solvents, paints, and other chemicals, such as toluene, phenol, ethylbenzene, naphthalene, vinyl chloride, methylene chloride, xylene, and chloroform, will also evaporate. Dissolution processes involve chemical reactions between the various constituents in a waste pile, the end result of which will be transformed products and leachates. Hazardous waste dumping in the Love Canal in Niagara Falls, NY, in the 1940s and 1950s, and subsequent use of the site as a hazardous waste landfill, resulted in a situation (LaGrega et al., 1994) where the public was exposed to dioxin and other chemical fumes. Investigation of the many illnesses arising from the exposure showed that the chemicals causing the illnesses were neurotoxins, teratogens, fetotoxins, carcinogens, pulmonary toxins, and hepatoxins (Bridges, 1991).

4.3.6   Greenhouse Gases

Activities associated with urban living contribute significantly to the production of greenhouse gases. The combustion of fossil fuels is a major source of carbon dioxide production. Wood burning is another significant source. Methane released from anaerobic decomposition of organic waste in dumps and other sources contributes about 15% of the greenhouse effect. Chlorofluorocarbons (CFCs) used as refrigerants in air conditioners and as propellants were produced in the 1950s. Their use has decreased significantly in developed countries, with the exception of developing countries, after the introduction of measures for their reduction. Concentrations of ozone, produced in internal combustion engines, have also increased. Levels have tripled in Europe and North America in the troposphere. The evidence indicates that human activities during urbanization have clearly increased these levels (greenhouse gases and CFCs), and that it is very likely that these will affect the geoenvironment in the future. If global warming comes to pass, sea levels will rise as a result of ice cap melting and flooding of coastal regions will occur. This will disrupt coastal ecosystems and habitats. Permafrost melting will have an impact on arctic ecosystems and will also cause considerable distress to physical structures because of terrain instabilities.

4.3.7   Impact on Ecosystem Biodiversity

The multitude of microorganisms and their diversity in the soil ecosystem is essential in maintaining and developing a healthy soil. Organic matter and nutrient recycling, mineralization, and decomposition are essential processes that have not been studied extensively in urban soils, particularly when compared with agricultural soils (Harris, 1990). Soil organisms include microbiota (bacteria, fungi, algae, protozoa), mesobiota (arthropods, nematodes, springtails, etc.), and macrobiota (earthworms, mollusks, larger arthropods, enchytraeids, etc.).

   The impact of urbanization on soil properties and attributes includes great variability, compaction, bare top soil that is often water repellent, altered pH conditions and restricted aeration and drainage, altered nutrient cycling of the soil organisms, presence of other materials and contaminants in the soil, and altered temperature profiles (Craul, 1985). These changes all significantly have an impact on the soil organisms. For example, trampling by humans in urban forests can significantly reduce the numbers of earthworms. More information is needed, however, on species diversity and numbers in urban soils and the impact of recreation, disturbance, compaction, and contamination on these numbers. This microbial community can be used as an indicator of soil quality.

   Chemicals from wastes, leaks, and emissions enter the soil ecosystem. They can accumulate in organisms via bioaccumulation as higher animals on the food chain eat lower, contaminated organisms. As the contaminants increase in the species, the species may become compromised, causing an imbalance in the whole system. Anecdotal evidence indicates that fewer organisms, less biomass, and fewer species of organisms are found in urban soils. Assuming the evidence to be valid, this would be an indication of the stress on these organisms caused by the impact of human activities on urban soils. While direct measures of the level of degradation or impairment of the quality of urban soils are not available, an argument could be made in support of the use of sensitive soil fauna and microflora as indicators for urban soil quality.

4.4 Impact Avoidance and Risk Minimization

As cities grow, the problems and stresses created by human urban activities on the land environment will escalate. The requirements to satisfy the classical food, shelter, clothing, and recreation needs of the community impose several demands and stresses on the urban environment and its contiguous regions. To a very large extent, the main challenges for an urban society are linked to (1) sensible land management, (2) energy utilization, consumption, and management, (3) nonenergy resource (food, minerals, building materials, etc.) consumption and management, (4) waste management, (5) reduction of pollution to air and water, and (6) water resources management.

4.4.1 Waste Management

4.4.1.1 Pollution Management and Prevention

Waste generation (urban and industrial) is a large problem. New land must be acquired for classical waste disposal techniques involving landfills, since older landfills become filled and are decommissioned. Modern requirements for environmentally sensitive waste disposal and treatment require considerable resources. The essence of these requirements is embodied in the 4Rs: recycle, reuse, reduce, and recover. Reduction of waste entering landfills is the stated objective of almost all, if not all, communities and municipalities. Currently, Europeans on the average generate 400 kg of household waste every year, while the United States produces twice that amount (Genske, 2003). In Switzerland, a family of four generating 100 tonnes of household waste would require 200 m³ of landfill volume. Reduction in land-required waste landfill can be obtained through implementation of the other 3Rs (recycling, reuse, and recovery), together with incineration. Historically, landfills used to be located on the outskirts of cites. However, with the expansion of cities and suburbs in particular, these landfills are now part of the urban–suburban landscape. Although incineration is cleaner and energy efficient and can substantially reduce the volume of waste, incinerator facilities without the benefit of modern burning systems and efficient scrubber units can emit pollutants into the air and produce toxic materials and ash that require disposal. Hazardous wastes such as solvents, oils, medical wastes, and highly contaminated soils are often treated by incineration. With the present capabilities and efficiencies in incineration and scrubber systems that can clean properly captured pollutants before emission discharge, incinerators are being considered the disposal system of choice, since they can be located in or near cities. The system in the city of Vienna, Austria, is an example.

Pollution prevention is a method to reduce damage to the environment for future generations. Practices such as pesticide addition to lawns, city parks, etc., should be modified or eliminated. New methods to control pests should be introduced, such as biological control. Many cities have banned the use of pesticides for private use. The use of less harmful or hazardous products can reduce pollution through substitution. A key example where this should have been practiced is the case of chromated copper arsenate (CCA)-treated wood. While it has only recently been banned for use in many countries, wood previously treated still exists in service. CCA is a major wood preservative that was

TABLE 4.4

Results of Leaching Tests with Acetic Acid under Various Conditions

Conditions	Chromium (mg/l)	Copper (mg/l)	Arsenic (mg/l)
TCLP regulatory level	5.0	Not on the list	5.0
Experimental data at 35°C, pH 4			
5 days	0.4	1.3	2.5
10 days	0.7	1.2	3.0
15 days	0.8	1.4	3.2

Source: Adapted from Mulligan, C.N. and Moghaddam, A.H., *Int. J. Housing Sci. Appl.*, 28, 89–102, 2004.

used in North America for many years for lumber treatment against insects and microorganisms because (1) it was inexpensive, (2) it left a dry, paintable surface, and (3) it bonded well, and was thus relatively leach resistant. Primary utilization was for decks and playground equipment. Even today, it would not be unusual to find outdoor wooden facilities with CCA treatment. CCA associated with playground equipment is particularly problematic because children are directly exposed through physical contact and subsequent oral contact by ingesting food with hands that had previously been in contact with CCA-contaminated wood. The U.S. Consumer Product Safety Commission (CPSC) has stated that this might even cause cancer in children (*Green Building News*, February 2003 issue, oikos.com/news/2003/02.html). There is increasing concern about potential environmental contamination from leaching of Cu, Cr, and As from treated wood in service and from wood removed from service and placed in landfills. The life cycle of treated wood is estimated to be about 25 years, and the wood is then discarded as waste. By 1995, more than 90% of 67 million kg of utilized waterborne preservatives was CCA treated (Solo-Gabriele and Townsend, 1999). The quantity of removed treated wood from services is estimated, by the Forest Products Laboratory (FPL) in Madison, WI, to increase from 6 million m^3 to 16 million m^3 by the year 2020 (Cooper, 1993).

Mulligan and Moghaddam (2004) tested CCA-treated Gray Pine species wood for leaching using a modified TCLP method to determine the leaching of the three metals under various conditions. To obtain the results, samples of ground wood were soaked in acetic acid (0.1 N) at pH values of 3, 4, or 5; temperatures were 15, 25, or 35°C; and leaching time was 5, 10, and 15 days. The amounts of chromium as chromium oxide (CrO_3), copper as copper oxide (CuO), and arsenic as arsenic oxide (As_2O_3) leached from the wood were determined to be 49% CrO_3, 34% As_2O_3, and 17% CuO. The study also examined the effects of pH and temperature on the leaching of the three metals, from the wood, for a 5-day period, and found measurable amounts of chromium, copper, and arsenic in the leachates. Arsenic was found to be the least resistant metal to leaching when the temperature increased, and chromium was the most resistant. In addition, there was more leaching of all three elements as the pH decreased. The effect is shown for pH 4 in Table 4.4. The results of the study showed that there is the risk of soil, water, and environmental contamination by chromium, copper, and arsenic, wherever chromated copper arsenate-treated wood is used or disposed of in a landfill. Chromium leached the least despite being present in the greatest proportion. Disposal must be in a lined landfill to avoid contamination of the groundwater.

Since CCA-treated wood was exempt from the Toxicity Characteristics Leaching Procedure (TCLP) developed by the Environmental Protection Agency (EPA), few reports concerning tests on these substances are available. Wilson (1997) tested CCA-treated wood and found that it failed for arsenic and barely passed for chromium. Since the ash also failed the test, one would conclude that CCA-treated wood cannot be burnt in incinerators because the metals will remain in the ash. Arsenic can vaporize and be captured by air control equipment or escape into the atmosphere, but chromium and copper will stay in the ash. Mixing the CCA wood with mulch for landscape purposes is also problematic because of the potential for arsenic leaching. Wood-burning power plants cannot accept this type of wood if they use their ash for application onto agricultural fields. The ash becomes hazardous if 10.7% or more of the wood is CCA treated. While landfilling in municipal landfills is an option, it is not uncommon to find the waste (ash) sent to unlined *construction and demolition* (C&D) landfills — a practice that leads to contamination of groundwater.

Attempts in removal of the CCA treatment in the wood before reusing the wood have not been very successful. However, disposal of these products in the future is still uncertain. As landfilling in lined landfills is the only current option, the production of CCA-treated wood will be limited and was phased out from consumer application at the end of the year 2003 in the U.S. — as an agreement with manufacturers and the EPA. European countries had already banned this type of treated wood. While other alternatives, such as ammonium copper quatenary (ACQ) and copper boron azole (CBA) copper-based preservatives could be used, they have yet to be fully evaluated.

4.4.1.2 Waste Reduction

Reduction of the wastes or reduction of the source of waste is the key to reducing emissions from landfills and other waste management techniques. Life cycle analysis (LCA) has been identified as a tool to help achieve sustainable consumption, as it accounts for the emissions, and resource uses during production, distribution, use, and disposal (ISO, 1997). Three steps are involved: (1) the processes of the life cycle, (2) the environmental pressures of the processes (Figure 4.7), and (3) the environmental impact of the use, including the use of impact indicators. While ISO 14040 defines the inventory analysis and impact assessment steps, the other steps, involving definition of the process and the interpretation of the results, are not necessarily simple. Several databases such as EcoInvent (Frischknecht et al., 2005) and SimaPro are now available for some materials, but data are lacking. In

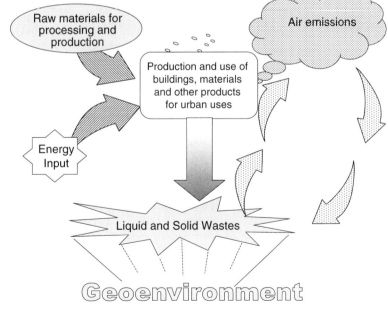

FIGURE 4.7
Elements of a life cycle assessment.

addition, life cycle impact assessment methods are being developed for minerals, land use, and toxic chemicals. A *Life Cycle Initiative* has been formed to promote the development of standard methods and exchange of information (UNEP, 2004). For example, waste management experts can determine which wastes are causing problems and inform the public on alternatives or methods for waste reduction; however, most of the current information for LCA methods has been in regards to energy analysis (Hertwich, 2005).

4.4.1.3   Recycling

Recycling, reuse of wastes, and incineration reduce the volumes of waste entering landfills. Application of waste limits could be implemented as a "polluter pays" mechanism. While recycling and reduction of the wastes, and thus the sources of pollution, have made inroads, particularly in advanced countries, a lot more progress is required. Although difficult to achieve, 100% recycling should be the goal for all materials, including those for construction (asphalt, concrete, wood, etc.). Collection of paper, plastics, glass, trees, and other materials for recycling is practiced to reduce energy, material requirements, and landfill spaces. Backyard composting can substantially reduce organic waste transport, collection, processing, and disposal, while providing a fertilizer for home gardens.

Sewage sludge, or biosolids, is produced by wastewater treatment plants. The amount of sludge depends on the amount of water treated by the city, town, or municipality. The sludge is typically treated to remove water, and stabilization is by thickening, digestion, conditioning, dewatering, drying, and incineration. After drying/incineration, the sludge can be reused as a powder for manufacturing bricks, or mixed with Portland cement and stabilized ash. Land application for agricultural uses or for land reclamation is also possible. Pathogen and heavy metal concentrations must be removed to protect groundwater quality and public safety.

Sewage sludge has been used as an amendment to urban and agricultural gardens for organic carbon and nutrients. The sludges generally contain about 45% organic matter, 2.0% nitrogen, 0.3% phosphorus, and 0.2% potassium (Bridges, 1991). These sludges have been known to contain up to 3000 ppm of zinc, 2000 ppm of chromium, 1400 ppm of copper, 385 ppm of nickel, 240 ppm of lead, 60 ppm of cadmium, and 60 ppm of arsenic. Repeated application of this sludge leads to an accumulation of heavy metals in the soil.

Glass is dense and takes up significant landfill space. Increasing recycle rates is required. However, separation of colored glasses is difficult and must not be contaminated with other materials, such as ceramics. Crushed glass can be processed to reach the characteristics of gravel or sand. It can thus replace aggregate in backfill, road construction, and retaining walls. The use of mixed glass as glassphalt is another possibility to replace aggregate in asphalt. More engineering tests (compaction, durability, skid resistance) will need to be performed on the properties of glassphalt and other materials. TCLP tests would need to be undertaken to determine the leachability of heavy metals from the glass. Lead has been shown to leach below the U.S. EPA levels (CWC, 1998).

Plastics are used in many everyday items such as containers, packaging, and trays. Polystyrene, polypropylene, and polyvinyl chloride (PVC) are some of the plastic materials. These can constitute around 10% of municipal solid waste (MSW). There are numerous problems with disposal, as it is not very biodegradable. It is a hazard in the marine ecosystem and occupies substantial volumes in landfills. Recycling can be achieved with generic plastic and mixed plastics to regenerate raw materials. New products can be formed from recycled polypropylene and polystyrene by injection moulding, blow moulding, and foam moulding. The construction industry uses PVC for piping and baseboard mouldings. Polyethylene can be used for trash containers, book binders, bags, and other household applications.

TABLE 4.5

Applications of Recycled Scrap Tires

Tire Form	Application
Tire chips	Embankments, aggregate materials retaining walls, bridge abutment backfills
	Reduction in frost penetration due to low thermal conductivity
	Adsorption of organic chemicals in leachate collection systems, for gas migration control trenches, gas collection and venting layer in caps, leachate recirculation trenches, and drainage layer in covers in landfills
Whole tires	Low-height retaining structures
	Highway applications for stabilization of road shoulders or noise barriers
Rubber	Rubberized asphalt concrete with long life and decreased thermal cracking
Crum rubber	Rubber is moulded to form panels for railway crossings that fit between tracks instead of timber crossings
Tire shreds	VOC movement reduction in groundwater by addition to bentonite slurry walls
Ground tires	Sorption of VOCs from wastewater in wastewater treatment plants

Source: Adapted from information reported in Sharma, H.D. and Reddy, K.R., *Geoenvironmental Engineering*, John Wiley & Sons, Hoboken, NJ, 2004.

Scrap tires are environmental threats (1) as mosquito hazards from accumulation of rainwater and (2) as fire hazards, since they are difficult to extinguish. Landfilling is difficult since landfills (1) are not compactable, (2) require hundreds of years to decompose, and (3) occupy substantial amounts of space. Contaminants may also leach into the soil. Heavy metals, PAHs, and total petroleum hydrocarbons (TPHs) have leached out under various conditions. The type of tire shredding can substantially affect the leaching results. Various applications for reused tires are shown in Table 4.5. Since there will be restrictions for tire-derived fuel in the future, the engineering applications for tires are the most promising.

Demolition debris and concrete are wastes from infrastructure renewal and building demolition. It has been estimated by the U.S. Environmental Protection Agency (2002) that 122 million tonnes of building-related waste were generated in 1996 in the United States. Approximately 43% of this was of residential origin. It was also estimated that 48% was generated by demolition, 44% by renovation, and 8% by new construction. Approximately up to 75% of the wood, concrete, masonry, metals, and drywall is potentially recyclable. According to the United Nations *Agenda 21* (1993), the promotion of environmentally sound waste disposal and treatment methods for construction debris is highly desirable and is one of the programs. Concrete can be recycled into aggregates of concrete, base foundation, new pavement, road shoulders, or backfill. Scrap wood can be used for landscaping (wood chips, mulch, ground cover), and wood-based geotextiles for landfills as fuel or in building products (fiberboard products, rigid boards, plastic lumber). Pentachlorophenol (PCP) leaching from treated wood may be a concern. Cardboard, dry wall, and rubble can be used as aggregate. The publication *WasteSpec* (Triangle J Council of Governments, 1995) provides specifications for waste reduction, material recovery, and reuse and recycling of construction waste. *WasteSpec* includes waste reduction techniques during construction, reusing waste material on-site, salvaging waste material at the site for reuse or resale, returning unused material for credit, and delivering waste material for recycling.

Composting is one of the simplest processes that can be used for treatment of organic wastes. It can also be very sophisticated. At the present time, it is used mainly for food wastes. However, composting materials can include garden and vegetable cuttings, paper and cardboard, garbage, and any decomposable organic matter. It is applicable for home owners, individual institutions, and even communities. It has increased in popularity significantly due to consumer awareness in decreasing the amount of wastes going to landfills. Moisture and oxygen levels, pile temperature, and odor must be monitored throughout the process. Carbon-to-nitrogen ratios are the other important factors that must be optimal to ensure the success of the process. Composted products can be used as soil conditioners. The use of in-vessel systems or biofilters with aerated piles will reduce odor levels considerably.

Anaerobic digestion is the microbial stabilization of organic materials without oxygen to produce methane, carbon dioxide, and other inorganic products. It is the most common biological treatment method used for high-strength organics in the world. At present, the technique is used for industrial, commercial, and municipal sludges (Mulligan, 2002). It will continue to be popular since it produces methane that can be recovered and used for energy. Solids concentrations up to 35% can be processed. In addition, less carbon dioxide is produced in comparison to aerobic processes. This could become an important consideration due to the reduction requirements of the Kyoto Protocol — an international treaty on global warming. This amendment to the United Nations Framework Convention on Climate Change (UNFCCC) reaffirms sections of the UNFCCC to reduce emissions of carbon dioxide and five other greenhouse gases from countries that ratify this protocol.

Solids from anaerobic digestion can be composted. Supernatants from the sludge often have high organic contents and need to be treated further. Anaerobic digestion of municipal solid waste (MSW) organics is performed at solids concentrations of 4 to 10%. Higher concentrations must be diluted. Gas production is 1.5 to 2.5 m^3/m^3 reactor or 0.25 to 0.45 m^3/kg of biodegradable volatile solids. Retention times are approximately 20 days. Most reactors operate under mesophilic conditions.

Aerobic digestion is a highly stable process that can be operated so that nitrification can also occur. Because of aeration requirements, it is more suitable for small and medium-size plants. Capital costs are less than costs for anaerobic digestion, and operation is safer since there are no explosive gases produced. Supernatants are also of higher quality. Thermophilic digestion is becoming increasingly popular due to pathogen destruction, and low space requirements and high sludge treatment rates. As an example, International Bio-Recovery Corp. (North Vancouver, BC) has developed a technology named autogenous thermophilic aerobic digestion (ATAD) that converts organic waste (food waste, sewage sludge, and animal manure) into single-celled protein organic fertilizer after 72 h.

4.4.2   Water Resources Management

As up to 90% of human sewage in developing countries is untreated; this is a growing threat to the health of surface water and groundwater. The capacity of rivers to absorb these pollutants without irreparable impairment of water quality is being exceeded. A combination of regulations and consumer incentives is needed to remedy the situation. Infrastructures must also be improved to reduce leakage from water supply systems. Since water quality monitoring is nonexistent in many countries, the impact of urbanization on water quality and water ecosystems is largely unknown — a situation that must be remedied.

The lack of clean water on one hand and flooding on the other hand are problems that beset many cities worldwide. Development of dams upstream instead of policies for water sustainability has led to water supply problems for downstream cities. We have previously discussed some of these problems in Chapter 3. Reuse of wastewater to reduce contaminant discharges into water systems may also be beneficial. Introduction of taxes or tax credits can also promote reduction in water use. Various other approaches have been proposed by König (1999), including the use of plant roofs to adsorb roof runoff with a combination of seepage or delayed drainage areas to allow rainwater to return to the groundwater. Roof runoff could also be collected for use in toilets, as shown in Figure 4.8.

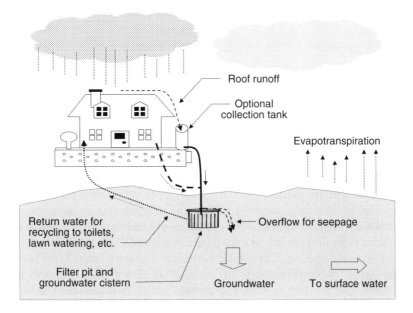

FIGURE 4.8
Recycling of roof and wastewater to replenish groundwater and provide nonpotable water for housing. (Adapted from information in König, K.W., in *Cities and the Environment: New Approaches for Eco-Societies*, Inoguchi, T. et al., Eds., United Nations University Press, Tokyo, 1999, pp. 203–215.)

4.4.3 Reduction in Energy Usage, Ozone Depletion, and Greenhouse Gases

As the burning of wood, coal, petroleum, and natural gas leads to substantial production of greenhouse gases and promotion of acid rain, it is obvious that reduction in the use of these energy sources is essential — as a step to fulfilling the requirements of the 1997 Kyoto Protocol. While many different other steps are needed to meet the requirements, and whereas this is an essential step in the right direction, it is recognized that combustion of fossil fuels is not a sustainable practice. Other sources of energy, such as solar, wind, geothermal, hydrothermal, and hydroelectricity, produce less greenhouse gases than the above-mentioned fuels. In the case of acid rain, the choice of fuels with lower sulfur and nitrogen oxide contents can also minimize acid fallout and damage to soil, water, fauna, and people. Nuclear energy is a clean energy. However, the problems of spent fuel rods and other radioactive waste materials are substantial and have yet to be fully resolved. A recounting of the repository disposal of high-level nuclear wastes and the technological tools for managing such wastes can be found in Pusch and Yong (2005). Biogas from waste dumps should be recovered and used for heating purposes. For example, the St. Michel Environmental Complex, in Montreal, Canada, recovers, at the landfill, approximately 20 MW of electricity, which is enough to power 12,000 homes.

CFCs can be replaced with other gases. Reforestation and increases in forest growth will increase carbon dioxide sink capacity. Sustainable energy generation and low-energy-consuming practices for transportation, heating, and so on, must be practiced, although they are not without political implications. Green buildings for houses and commercial use to reduce energy use are now being constructed. Eco-building codes and architecture will need to be developed and improved. Santa Monica, CA, for example, replaced city building electricity with geothermal electricity in 1999, and Austin, TX, offers a Green-Choice option for consumers that includes renewable electricity for wind, solar, or biogas (Portney, 2003).

Biodiesel is gaining popularity as a renewable fuel. It is made from the oil of various crops, including palm, mustard, rapeseed (canola), sunflower, and soybeans (Weeks, 2005). Rapeseed, which produces the highest yield, is the most common source in Europe, whereas soybean is the most common in the United States. A waste-frying oil can also be used. All emissions (with the exception of NO_x) are reduced when biodiesel is burnt instead of petroleum diesel. Spills are of less impact since biodiesel degrades faster in the environment. In addition, wastewater and solid waste generation rates are lower.

4.4.4   Minimizing Impact on Biodiversity

The encroachment of cities into natural landscape areas has a detrimental impact on biodiversity. To mitigate this impact, one needs to have access to a base inventory of the natural places and biodiversity existing within the target areas so as to develop a basis for determination of the nature and extent of potential impacts. This base inventory also forms the basis for planning for urban expansion that is sensitive to the need to (1) prevent introduction of invasive species, (2) protect endangered species, and (3) protect the natural habitats. Cities such as San Francisco are looking at ways to understand biodiversity, protect and restore the natural ecosystems, protect sensitive plants and animals through education, purchase green spaces, develop pest management programs, and enforce environmental regulations (Portney, 2003).

4.4.5   Altering Transportation

While some might argue otherwise, there is a consensus that automobiles have facilitated transport and contributed to economic progress. However, there are many environmental and sustainability issues related to transportation that need to be addressed. The method of delivery of urban transportation needs (buses, cars, trucks, etc.) requires modification if ozone production is to be reduced. Cities should be planned in an optimal manner to minimize transportation requirements. Bicycles and walking should be facilitated instead. Walking is a completely sustainable transportation mode. Copenhagen was one of the earliest cities to promote pedestrian streets (Vega, 1999). Numerous cities within Europe are promoting pedestrian areas, and others such as Amsterdam are promoting cycling. Public transportation will also need to be improved.

4.4.6   Brownfield Redevelopment

New policies are being developed in various countries to reduce greenfield consumption. In the United Kingdom, regulations require that 60% of all housing must be on brownfields by 2015. Germany is aiming to reduce greenfield consumption by 75% by 2020 (Genske, 2003). The U.S. EPA made special efforts since 1998 to encourage brownfield redevelopment — in line with the goals of sustainability (U.S. EPA, 1999a). Three cities, Baltimore, MD, Portland, OR, and Chattanooga, TN, were selected as EPA pilot cities. Aspects of these programs included (1) an inventory of the brownfield sites, (2) development of awareness programs, (3) cleanup and redevelopment criteria, and (4) finding financing mechanisms for cleanup. The aims of these programs were to (1) catalyze business, (2) create jobs, (3) clean up the environment, (4) rebuild the community, (5) promote land use, and (6) reduce urban sprawl as a means to reduce pollution.

An overview of the action plan devised by the U.S. EPA (1999b) regarding brownfield development is shown in Figure 4.9. The first part of the evaluation is the collection and assessment of the environmental information for the site. The data will determine what

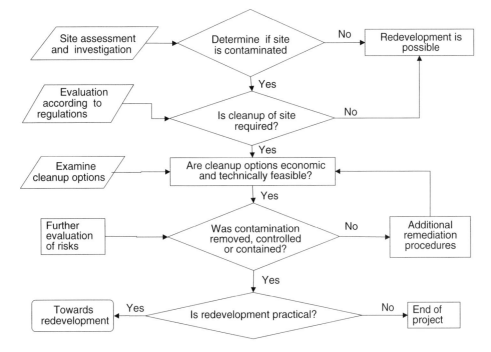

FIGURE 4.9
Evaluation of a brownfield project. (Adapted from U.S. EPA, *Road Map to Understanding Innovative Technology Options for Brownfields Investigation and Cleanup*, 2nd ed., EPA 542 B-99-009, U.S. Environmental Protection Agency, Office of Solid Waste and Emergency Response, Washington, DC, 1999b.)

cleanup may be required, depending on the intended use of the land. These data will need to be compared with regulatory levels to determine if the contaminant levels in the site are above or below regulatory limits. The ASTM (2003) has developed guidelines for all stakeholders for developing solutions to brownfield redevelopment. The source, nature, and extent of contamination must be determined in a site investigation. Exposure pathways for soil and dust include direct contact, ingestion, and inhalation. Water and air exposures are through ingestion and inhalation. Once the risks and contaminants are known, cleanup options will need to be evaluated. Depending on the type, location, and extent of contamination, various options are available, including bioremediation, chemical treatment, soil flushing, natural attenuation, phytoremediation, soil vapor extraction, air sparging, thermal desorption, permeable active barriers, and others. Many of these were briefly discussed in Chapter 3 and will be developed further in subsequent chapters as appropriate.

Once a cleanup option has been chosen, the technology will be implemented and monitored to determine if cleanup levels are reached. Additional contamination may be found that was not initially discovered. Long-term monitoring, particularly for natural attenuation, may be required. The site can only be developed when the regulatory levels have been achieved.

The U.S. EPA (1999a) organized various elements into a model framework — illustrated in Figure 4.10. Many of the principles of sustainability are incorporated into this process to enable the project to be sustainable for the community in the future. The framework is designed to assist municipalities, planners, and developers to undertake brown field projects. The technologies used should incorporate resource conservation, materials reuse, public safety and mobility, and information availability. Factors to measure sustainability should be incorporated in the future, and projects should be reevaluated every few years to monitor progress and failures.

FIGURE 4.10
Elements to be incorporated into a sustainable brownfield redevelopment. (Adapted from U.S. EPA, *A Sustainable Brownfields Model Framework*, EPA 500-R-99-001, U.S. Environmental Protection Agency, Office of Solid Waste and Emergency Response, Washington, DC, January 1999a.)

4.4.7 Sustainability Indicators for Urbanization

Indicators are useful for determining progress and comparisons with other cities or practices. There has been considerable effort in recent years to create indicators as a measure of the sustainability of cities. Most of the indicators presently available deal with easy-to-obtain data, such as recycling information. Cities such as Seattle (Portney, 2003) use a wide range of indicators, such as air quality, biodiversity, energy, climate change, ozone depletion, food and agriculture, hazardous materials, human health, parks and open space, economic development, environmental justice, education, etc. (Figure 4.11). Environmental indicators are generally associated with such elements or parameters as water, energy, solid wastes, air, and land use. For example, in regard to waste, one is interested in quantities (1) generated, (2) disposed, (3) recycled, and (4) composted. Other indicators regarding landscape and land use are (1) how much urban land is used for agriculture, (2) the quantity or number of contaminated sites, (3) the nature of contaminants in the contaminated sites, and (4) the number of brown field sites and their regeneration. Water quality indicators track (1) wastewater treatment facilities, (2) contamination of drinking water sources, such as lakes, rivers, and aquifers, and (3) quality of drinking water. For energy, the indicators are generally concerned with (1) use of energy and its form (less polluting energy sources are favored) and (2) amount of energy used. Air quality indicators seek to provide controls on (1) toxic releases, (2) emissions due to transportation, industry, and the energy sector, and (3) the acidity of precipitation.

4.5 Mitigation and Remediation of Impacts

4.5.1 Mitigation of Impact of Wastes

Proper daily cover of wastes deposited in active landfills can mitigate odor, dust, fires, and pests. A common technique is to use a granular-type soil fill material cover that is

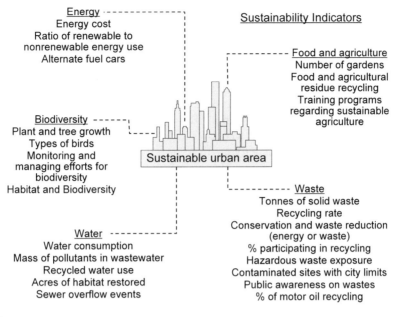

Energy
Energy cost
Ratio of renewable to
nonrenewable energy use
Alternate fuel cars

Sustainability Indicators

Food and agriculture
Number of gardens
Food and agricultural
residue recycling
Training programs
regarding sustainable
agriculture

Biodiversity
Plant and tree growth
Types of birds
Monitoring and
managing efforts for
biodiversity
Habitat and Biodiversity

Sustainable urban area

Waste
Tonnes of solid waste
Recycling rate
Conservation and waste reduction
(energy or waste)
% participating in recycling
Hazardous waste exposure
Contaminated sites with city limits
Public awareness on wastes
% of motor oil recycling

Water
Water consumption
Mass of pollutants in wastewater
Recycled water use
Acres of habitat restored
Sewer overflow events

FIGURE 4.11
Selected sustainability indicators for the city of San Francisco that affect the geoenvironment (www.ci.sf.ca.us/environment/sustain/Indicators.htm, August 2005).

compacted on the daily load of waste brought to the landfill. Techniques for placement of wastes in landfills, using compartments, for example, have been developed and are well illustrated in many handbooks dealing with disposal of wastes in landfills. The details of the interactions between wastes and the land environment are discussed in Chapter 9.

For landfill surface closure, a waterproof cover system, as shown previously in Figure 1.10 in Chapter 1, will prevent water infiltration and reduce the requirement for treatment of the leachates. This is the technique of landfill construction and closure that is called a dry garbage bag system. The intent of this system is to keep the material entombed in the land with liners surrounding the entire waste that are designed to be impermeable to water. Since water is the carrier for contaminants (i.e., contaminants cannot be transported into the surrounding ground without water), there is a school of thought that argues that denial of water will not only obviate dissolution processes, but also deny production of leachates. Along this line of reasoning, a dry garbage bag system will presumably keep the contained wastes in a dry state forever. However, reality forces one to accept the fact that engineered liners and barriers have a life span for secure containment that sometimes will have flaws, the result of which will admit water to the system and ultimately generate leachates that will find their way into the surrounding ground. In recognition of that fact, and in support of the thesis that if one could generate a bioreactor system with water entering the waste pile, leachate recycling into the waste pile would accelerate dissolution of the waste material in the landfill. The outcome of this kind of strategy is to obtain faster dissolution of the wastes in the landfill and a quicker return of the landfill to more fruitful and beneficial land use.

For closures that are designed to prevent moisture entry into the waste pile, a vegetative cover on top of the impermeable barrier can assist in reducing dust generation. The roots of the plants need to be short to avoid penetrating the waste that would inhibit their growth. Biogases at active landfills should be collected for energy recovery — as done, for example, in the landfill in the city of Montreal. Other gases generated in the waste pile will need to be collected to avoid problems — as seen in the Love Canal problem discussed previously.

Wastes are treated in many ways according to the regulations in the country where they are generated. In many ways, unsorted municipal solid wastes can be treated as a resource containing discarded bottles, paper, glass, and cans that can be recovered and recycled. Construction wastes contain many items that are recyclable, as for example broken concrete elements crushed and used as granular fill, and scrap lumber and wood chips that can be converted to particleboard and paper into cardboard. Organic wastes can be converted via composting or anaerobic digestion (Mulligan, 2002). One tonne of green waste produces 0.3 tonne of compost and 0.12 tonne of biogas.

Mixed wastes can be very difficult to treat and thus may require incineration. However, this reduces the weight of the waste to one third and the volume to one fifth of the original volume. In addition, 1 tonne of household waste can produce an equivalent energy value of 120 l of heating oil or 200 kg of coal (Genske, 2003). Despite these positive aspects, wastes generated from incinerators include gaseous emissions (74%), bottom (23%) and fly (3%) ashes, and process water. These ashes must be stored and properly disposed since exposure to acidic solutions enhances metal leaching that will find its way into aquifers or surface water. If these ashes are recycled, the heavy metals must be immobilized. Volatile substances such as dioxin are produced during incineration and can accumulate in animals, plants, and humans. Filtration and chemical methods are needed to reduce emissions of dioxins and other hazardous components, such as sulfur compounds that lead to acid rain and carbon and nitrogen oxides that can contribute to the greenhouse gas effect. The calorific value of the waste also must be sufficient to run the incinerator in a sustainable manner.

Treatment of liquid wastes is by physical and chemical means. Deep underground injection and storage is a method that has been used. However, in addition to possible compaction and ground subsidence, groundwater contaminant may also result. Modeling and experimental investigations to predict the consequences of long-term storage are required.

4.5.1.1 Fresh Kills Urban Dump, New York City

Fresh Kills Landfill is 7500 ha in area and was established in 1948 on Staten Island for the disposal of waste for the 20 million people living in New York (Chamley, 2003). Although more than 21,000 tonnes was dumped daily in the 1980s, this decreased to 13,000 tonnes per day by the 1990s, mainly due to sorting and recycling. The dump was closed in 2001, with the exception of the waste from the rubble from the September 2001 World Trade Center disaster. As the site was originally a wood, meadow, and saline–soil mixed coastal landscape, plans are underway to restore the site. The refuse is isolated by a layer of concrete and a network of leachate collection pipes. More than 3800 m^3 of liquid is treated each day. More than 280,000 m^3 of methane is produced from the waste degradation, which is used for heating buildings. Local vegetation has been planted on a new 100 m hill, with an ultimate objective of transformation of the site into a nature and leisure place.

4.5.1.2 Vertical Barriers and Containment

In Chapter 3, we have seen how vertical barriers can be used to isolate and contain the contaminants in groundwater. When a landfill is in contact with the groundwater or when a landfill leachate contaminates the groundwater, an available treatment option is to install vertical barriers to confine the contaminants. A low-permeability base is required to key in the cutoff walls. If this does not exist, an artificial base will have to be added. For older landfills where liners have degraded or do not exist, other means may be required to reduce the potential for contaminant leaching into the groundwater, as shown in Figure 4.12. Silicate gels, pozzolanics, or cements can be injected into the soil under the landfill

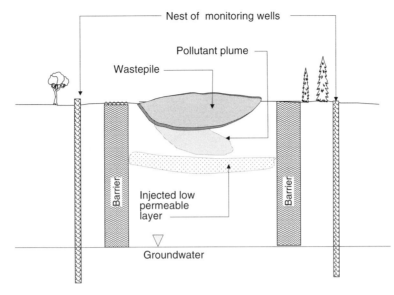

FIGURE 4.12
Vertical barriers and injections for an artificial low-permeable layer under a landfill. (Modified from Genske, D.D., *Urban Land, Degradation, Investigation and Remediation*, Springer, Berlin, 2003.)

to form a impermeable layer, thus immobilizing the plume. Groundwater levels may also need to be lowered by pumping to maintain their level below the landfill and the seal-off area. Pump-and-treat systems for treating the groundwater may also be necessary. Land settlement and decreased plant growth may result from lowering of the water table. Long-term monitoring is required to ensure that the system performs according to requirements specified by the indicators prescribed for safe performance.

4.5.1.3   Excavation

Excavation, partially or totally, to remove offending landfills can be a solution to decreasing contaminant generation. The waste can be recycled, incinerated, or dumped in a safe landfill. Handling and transport of these wastes is not without risk. Corroded drums are not easy to handle. The excavated landfill can be filled with clean or recycled soil and subsequently used for construction of buildings. Numerous measures have to be instituted to ensure worker safety and protection of the environment during the excavation work. Health and geoenvironmental threats arise from breaking bags, dust, emission of harmful gases or liquids, and unstable waste.

4.5.1.4   Landfill Bioreactor

Landfills traditionally have been operated without addition of liquid; i.e., they have been operated under the dry garbage bag concept. Increasing the moisture content of the waste, however, can increase waste degradation and methane production since most of the conversion processes are anaerobic. This concept is known as a landfill bioreactor and is becoming increasingly used worldwide. Experience with this type of landfill shows that gas recovery can be increased up to 90%, and that wastes can be stabilized within 10 years instead of 30 to 100 years (Block, 2000). Landfills can be designed to be aerobic, anaerobic, or anaerobic–aerobic bioreactors. Anaerobic bioreactors maintain moisture contents of 10 to 20% to optimize anaerobic degradation conditions. Aerobic reactors require injection of air or oxygen with vertical or horizontal injection wells. Although wastes can be stabilized

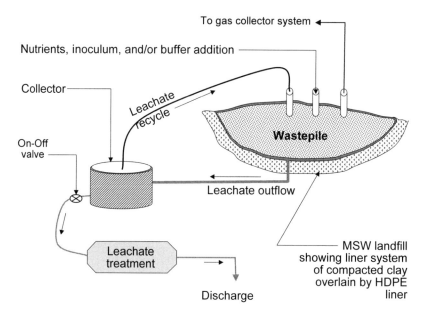

FIGURE 4.13
Schematic of a landfill bioreactor. Leachate can be directly fed back to the waste pile in the landfill without treatment or after treatment in the leachate treatment facility. The use of on–off valves will facilitate choice of treated or untreated leachate for recycling. (Adapted from Mulligan, C.N., *Environmental Biotreatment*, Government Institutes, Rockville, MD, 2002.)

in 2 years, by aerobic landfills, injection of the air can be costly and can lead to fires, and thus is not practiced often. The hybrid or anaerobic–aerobic bioreactor takes advantage of aerobic and anaerobic bacteria. The upper layer of waste is aerobically treated before burial and treatment with anaerobic bacteria. In all cases, moisture control is the most important parameter. Other factors include pH, waste pretreatment such as shredding, nutrient addition, settlement, cellulose content, leachate quantity and quality, and temperature control. Monitoring of these factors is essential. There is also the potential for mining the dredged waste for humic material and other recyclables. Figure 4.13 shows the major elements of a waste landfill bioreactor system. In addition to recycling of the collected leachate back into the waste pile, the option of addition of inoculum, nutrients, and other dissolution aids is provided. Gas generated from the waste pile is collected in a collection system for treatment.

In the United States, according to the EPA subtitle D rule, landfill bioreactors can only be used for landfills with composite liners such as 0.61 m of clay covered with a high-density polyethylene (HDPE) liner, as shown in Figure 4.13. Most installations increase the moisture content through recirculation of the leachate. Recirculation assists biodegradation by enhancing the transport of nutrients through the waste, redistribution of methane bacteria, buffering, dilution of inhibitory components, and retention of the constituents of the leachate in the landfill. It can be accomplished by spray application, infiltration ponds, horizontal or vertical injection wells, or trenches. Accumulated amounts of leachate will depend on (1) regulatory requirements, (2) how much is needed for each landfill compartment, and (3) minimization of quantity of leachate requiring disposal. Older areas in the landfill can serve as seeds for new areas. Ideally, other liquids will have to be added to increase the moisture content to 30 to 40%. However, a balance must be made such that organic acids or other components do not accumulate and inhibit methane production (Pohland and Kim, 2000). Leachate and gas monitoring are integral to the success of the process. In terms of heavy metal concentration in the leachate, pH is a major factor in the mobility of the metals. Mobility of the metals increases as the pH decreases. In addition,

TABLE 4.6

Conversion or Transport of Inorganic and Organic Components in a Landfill Bioreactor

Component	Transport or Conversion Mechanism
Heavy metals (Cd, Cu, Cr, Fe, Hg, Ni, Pb, and Zn)	Reduction of Fe, Cr, and Hg Complexation with organic or inorganic components and mobilization Precipitation as hydroxide (Cr) or sulfides (Cd, Cu, Fe, Hg, Ni, Pb, and Zn) after sulfate reduction Sorption and ion exchange with waste Precipitation under alkaline conditions
Halogenated aliphatics, including PCE, TCE, and dibromomethane	Volatilization and mobilization in leachate due to high vapor pressure and solubility
Chlorinated benzenes such as hexachlorobenzene, trichlorobenzene, and dichlorobenzene	Volatilization and sorption on waste due to low solubility and high k_{ow}
Phenols and nitroaromatics such as dichlorophenol, nitrophenol, and nitrobenzene	Low volatility, vapor pressure, and k_{ow}, with high solubility in leachate
PAHs and pesticides (lindane and dieldrin)	Low volatility and mobility due to low vapor pressure and high k_{ow}

Source: Adapted from Mulligan, C.N., *Environmental Biotreatment*, Government Institutes, Rockville, MD, 2002.

organic and inorganic agents can serve as ligands, promoting metal transport. However, the mechanisms of precipitation, encapsulation, and sorption ensure that the heavy metals are captured in the waste.

Solubility, volatility, hydrophobicity (k_{ow}), biodegradability, and toxicity influence the behavior of organic contaminants in the landfill. Compounds such as dibromomethane, TCE, 2-nitrophenol, nitrobenzene, pentachlorophenol (PCP), and dichlorophenol tend to be highly mobile, and thus are found in the leachate and gas phase (Pohland and Kim, 2000). They will also be biologically altered by reduction, complexation, and complete or partial degradation. Other compounds, such as hexachlorobenzene, dichlorobenzene, trichlorobenzene, lindane, and dieldrin, are more hydrophobic, and thus will remain in the waste and are available for biodegradation (Table 4.6).

In summary, the development of landfill bioreactors will continue due to its advantages over conventional landfills. Further efforts will be necessary to optimize leachate recirculation, gas generation, and removal of recalcitrant compounds. There are still many challenges, including regulator reluctance, ability to wet the waste uniformly, and availability of design criteria. Slope stability and settlement will differ from traditional landfills due to the increased moisture contents and degradation rates.

4.5.1.5 Natural Attenuation

Although we discussed natural attenuation and its application for groundwater in Chapter 3, and will be discussing this in greater detail in Chapter 10, we want to turn our attention here to the use of monitored natural attenuation (MNA) of soil to reduce the toxicity and concentration of contaminants in a soil–water system. This subject has been treated in detail in Yong and Mulligan (2004). This concept of passive remediation has gained acceptance by many jurisdictions and regulatory agencies. To ensure effectiveness, monitoring guidelines and criteria have been (or are being) established. The soil and the contaminants must be compatible to ensure the reactions of effective and optimum partitioning of the contaminants with the soil solids and the reactions and interactions to reduce the toxicity of the contaminants. Target concentrations must not be exceeded to protect the environment and biotic receptors.

To a certain extent this is true because the soil attenuation layer of an engineered barrier system is often composed of soil materials that are chosen for their attenuation capability

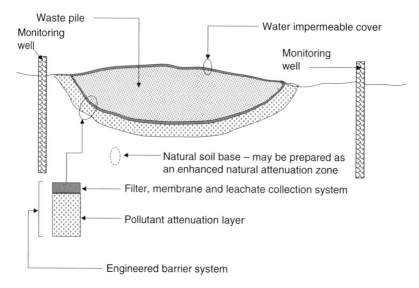

FIGURE 4.14

Pollutant attenuation layer constructed as part of an engineered barrier system. The dimensions of the attenuation layer and the specification of the various elements that constitute the filter, membrane, and leachate collection system are generally determined by regulations or performance criteria. (From Yong, R.N. and Mulligan, C.N., *Natural Attenuation of Contaminants in Soils*, Lewis Publishers, Boca Raton, FL, 2004.)

— a designed soil–water system that is generally called an *engineered clay barrier*. Figure 4.14 shows a general view of an engineered barrier system used for containment of a waste pile. The details of the filter, membrane, and leachate collection system and the nature and dimensions of the pollutant attenuation layer are specified by regulatory *command and control* requirements or by performance requirements. The nature of material comprising the engineered clay barrier that underlies the synthetic membrane is determined on the basis of a maximum permissible hydraulic conductivity performance expressed in terms of the Darcy permeability coefficient k. There is an implied understanding (not always well founded) that the minimum specified hydraulic conductivity is somehow related to the attenuation capability of the engineered soil.

The basic idea in the design details of the engineered barriers is that if leachates inadvertently leak through the high-density polyethylene (HDPE) membrane and are not captured by the leachate collection system, the pollutants in the leachate plumes will be attenuated by the engineered clay barrier. The engineered clay barrier serves as the second line of defense or containment. Even though the specifications refer only to a maximum permissible k value (generally in the order of 10^{-9} m/sec) for the engineered clay barrier material, it is prudent to conduct additional tests of the material. These tests, which determine the pollutant assimilation capability of the clay material, are part of the protocol in the *evidence of engineered natural attenuation (EngNA) capability* that assesses the capability of the engineered clay barrier material to attenuate the pollutants in the leachate plume. This will be discussed in the next section.

The EngNA as a direct link from NA is also used as the foundation base for the double-liner barrier system for landfill containment of hazardous waste. There are several options for the foundation base seen in Figure 4.14. Since a fully compacted foundation base is a standard requirement — to provide support for the material contained above — one has the option of working with the native material if it has the proper assimilative potential, or with imported fill material. Once again, the purpose of the foundation base is to provide attenuation of pollutants should leakage of pollutants through the double-liner system

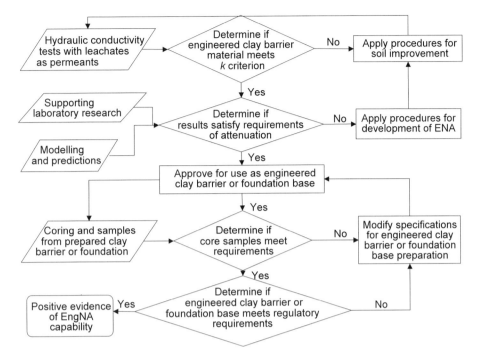

FIGURE 4.15
Protocol for determination of positive evidence of EngNA capability as required for use of engineered clay barriers or as foundation base material for double-liner systems. (From Yong, R.N. and Mulligan, C.N., *Natural Attenuation of Contaminants in Soils*, Lewis Publishers, Boca Raton, FL, 2004.)

occur. This in essence constitutes a third line of defense against pollutant transport into the subsurface soils.

One of the significant benefits in applying the protocols for evidence of EngNA capability is the determination of the required thickness of the pollutant attenuating layer and the engineered clay barrier shown in Figure 4.15. This specification can be obtained from the calculations from fate and transport models using the results of supporting laboratory research on partitioning and other attenuating phenomena. Since the dimensions for the engineered clay barriers specify "greater than or equal to" designations, application of the protocols for evidence of EngNA capability will likely provide dimensions that should satisfy regulatory requirements.

Water quality in the vicinity must remain acceptable in spite of contaminant dilution, degradation, and sorption processes. According to the action plan for Switzerland, *Sustainable Development* (FOEFL, 1997), these controlled leachate landfills fall within the concept of sustainable development because few resources are consumed. Contact with the waste is not necessary. As with all bioreactor-type landfills, a proper monitoring program is necessary — to ensure that fugitive leachates are captured and treated effectively with natural attenuation processes. To date, although little hard documentation exists regarding the use of monitored natural attenuation (MNA) for remediation of fugitive leachates from landfill sites, the limited options available favor its use. Examples of its use are shown in Table 4.7. Christensen et al. (2000) has suggested that there are five critical factors:

1. Local hydrogeological conditions
2. Size of the landfill and the variable nature of the leachate plume or plumes
3. Complexity of the leachate plumes

TABLE 4.7

Application of Natural Attenuation at Landfill Sites

Location	Geology	Chemicals	Electron Donors	Microbial Process	Studies
Farmington, NH, landfill (1995–present)	Silty sand (up to 20 m bgs) Bedrock	TCE, DCE, VC, trace ethane, DCM, TEX, ketones	TX, DCM, ketones	Acetogenesis, methanogenesis, cometabolic oxidation	NA investigation in groundwater, laboratory studies
Niagara Falls, NY (1994)	Overburden, fractured bedrock	TCE, DCE, VC, DCA, CA, CT, CF, DCM, CM, ethene ethane	Landfill leachate, other chemicals	Methanogenesis, sulfate reduction	NA investigation in groundwater
Cecil County, MD (1995–1996)	Sand and fill over fractured saprolitic bedrock	VC release	VC	Aerobic oxidation, anaerobic oxidation	NA investigation in groundwater

Source: Adapted from information reported in Sharma, H.D. and Reddy, K.R., *Geoenvironmental Engineering*, John Wiley & Sons, Hoboken, NJ, 2004.

4. Long timeframe for evaluation of the attenuation capacity of the soil
5. Demonstration of the effectiveness of natural attenuation based on a mass reduction basis

4.5.2 Remediation of Urban Sites

There are numerous benefits to restoring contaminated urban and brownfield sites. They included reducing sprawl, providing tax revenue, improving land and public health by improving air quality, removing threats to safety, and reducing greenhouse gas emissions (NRTEE, 1998). The United States and United Kingdom have national efforts in place for brownfield redevelopment. In addition, transportation costs can be reduced by up to $66,000/ha/year if brownfields are redeveloped, compared with greenfields, by reducing urban sprawl (NRTEE, 2004). Also, 4.5 ha of green land can be preserved for every hectare of brown field restored. Land use is more compact, thus increasing city competitiveness. The Revi-Sols program in Montreal and Québec has led to the cleanup and development of 153 projects for a total of 220 ha of land by the year 2004. The tax revenues in Montreal increased by $25.6 million, and 3400 new housing units were established.

One example of the restoration of a former contaminated site is the Angus Shops in Montreal (NRTEE, 2004). Between 1904 and 1992, the site was used as an area for railway and military maintenance and the construction of new equipment. Approximately one third of the sites (124 of 496 ha) were contaminated with heavy metals, petroleum hydrocarbons, and PAHs. All hazardous wastes were disposed of off-site. Recyclable materials, all debris, and contaminated soil were removed. Backfill consisted of clean on-site soil. The cleanup cost a total of $12 million. Five hundred houses and a supermarket and industrial mall have been built, with a biotechnology center under construction. Property taxes have increased to $2.2 million a year, and more than $391 million has been invested by private parties.

To restore urban contaminated sites, various parameters need to be considered. These include the characterization of the soil (mineral, texture, geochemical characteristics), the factors influencing fate and mobility of the contaminants, such as dilution, sorption/desorption, biodegradation, and transformation, and other factors, such as climate, hydrology, and microorganisms present. As a first step in the remediation of land for redevelopment, an investigation will be required to determine underground heterogeneity and hidden objects and obstacles. This can be accomplished by studying historical records and conducting geophysical surveys. Sewers, cables, underground tanks, pipelines, foundations, etc., must all be identified. They should be grouped according to their ease of extraction. For example, large foundations cannot be easily extracted and may require blasting, whereas tanks and other similar objects are not difficult to remove. Another factor is the type of material. Wood piles can be cut off, brick foundations may be extracted, but concrete foundations require special procedures. Various strategies could also be used to reduce redevelopment costs and promote sustainability. These are summarized in Figure 4.16.

As noted in Chapter 3, a variety of remediation techniques exist. They include excavation, contaminant fixing or isolation, incineration or vitrification, and biological treatment processes. *In situ* processes include (1) bioremediation, air- or steam-stripping, or thermal treatment for volatile compounds; (2) extraction methods for soluble components; (3) chemical treatments for oxidation or detoxification; and (4) stabilization/solidification with cements, limes, and resins for heavy metal contaminants. Phytoremediation is a developing technique (Mulligan, 2002). The most suitable types of plants must be selected based on pollutant type and recovery techniques for disposal of the contaminated plants.

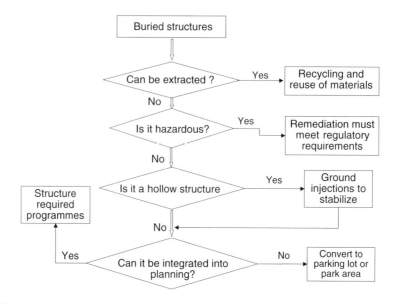

FIGURE 4.16
Schematic protocol for managing underground structures in a redevelopment plan. (From Genske, D.D., *Urban Land, Degradation, Investigation and Remediation*, Springer, Berlin, 2003.)

In Copenhagen, a soil treatment plant of 45,000 tonnes/year was recently established — mainly for oil-contaminated soils — with a requirement for reduction of levels of 700 to 2000 mg/kg to below 50 mg/kg (Cooper, 1999). The treatment procedures call for green waste to be added to the soil in a 50:50 ratio and placed in a 1.5-m-high windrow. Turning every week is required, to maintain the temperature above 15°C, even in the winter. Because of better air permeability characteristics, sandy soils are much easier to treat than clay soils. If the treatment is successful, the remediated soil can be used to grow vegetation in embankments or in construction projects. If the treatment fails or does not meet the required criteria, the soil must be sent to a special landfill.

Regardless of the origin of the contaminants and pollutants in the area, an evaluation of the threats to human health and the environment must be undertaken before the remediation process. Both the potential exposure time and level must be considered. Figure 4.17 gives a flowchart that illustrates a simple procedure for evaluation and treatment. Techniques such as selective sequential extraction are useful in determining the likelihood that the heavy metals are mobile. Selective sequential extraction studies were performed on nine soil samples (Huang, 2005). Figures 4.18A and 4.18B show the results for lead and zinc, respectively. It can be seen that both lead and zinc have different affinities toward different soil fractions. Both Pb and Zn have higher affinities toward the soil fractions of organic matter and oxides. Only a small fraction of both metals is associated with the exchangeable fraction. Metals bound to the exchangeable fraction of soil are mostly physically adsorbed (by electrostatic force) to the soil surfaces, and thus the bonding is weaker compared with other binding mechanisms. The moderate to high degree of leaching by rainfall and the competition from other cations present in the leachate solution possibly explains why only a limited amount of Pb and Zn was retained by this soil fraction. There is a high degree of association of Pb and Zn with soil oxides and organic matter. The metals associated with oxides are particularly susceptible to oxidation–reduction reactions and solubilization upon a decrease in pH by acid rain.

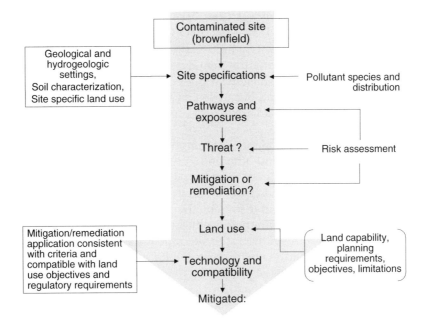

FIGURE 4.17
Simple protocol for rehabilitation of a contaminated site. (Adapted from Yong, R.N., *Geoenvironmental Engineering: Contaminated Soils, Pollutant Fate and Mitigation*, CRC Press, Boca Raton, FL, 2001.)

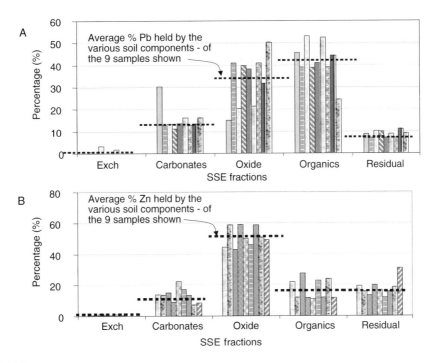

FIGURE 4.18
Selective sequential extraction characterization of nine different soil samples for (A) Pb and (B) Zn retention by the different soil components and mechanisms. (Data from Huang, Y.-T., Heavy Metals in Urban Soils, M.A.Sc thesis, Concordia University, Montreal, Canada, 2005.) Note that *Exch* represents Pb or Zn held by cation exchange mechanisms; *carbonates*, *oxides*, and *organics* are soil components responsible for retaining the metals; and *residuals*, what is remaining in the soil matrix.

4.5.2.1   Case Study of a Sustainable Urban Area

As part of the Olympic bids for Sydney, Australia (Sydney Olympics 2000 Bid Ltd., 1993), all tenderers for the Homebush Bay Olympic Site had to incorporate environmental concerns into their proposals, supply environmental documentation regarding performance, and conceive an environmental management plan.

As the land at Homebush Bay had been used as an industrial site for such facilities as (1) an abattoir, (2) a brickworks, (3) an armaments depot, and (4) an uncontrolled domestic and industrial waste disposal site, the area (soil, sediment, groundwater, and surface water) was contaminated with more than 70 compounds from food waste, heavy metals, oil products, asbestos, pesticides, and other industrial chemicals. Dioxin levels of approximately 2 ppb in soils, under 5 to 9 m of fill material, were of particular concern. Leachate collection systems were installed, and the leachate was sent to a waste treatment plant (CH2M Hill, 1997). It was decided to maintain all waste on the site, but to consolidate the wastes into four landfill areas and eliminate groundwater and water course contamination. The landfills were capped with soil from the site if possible. Native plants were planted to reduce water entering the landfill and to create habitats. Monitoring of the water, runoff, and leachates was included to observe progress of the remediation and recovery processes.

For water conservation, all fittings and devices were required to be energy efficient. Plants for landscaping were to be water efficient and native to Australia. Storm water and treated sewage were to be collected and reused for irrigation, flushing of toilets, and other uses. It was estimated that this would reduce the requirement for potable water by 50%. Energy use was minimized through design of buildings in terms of their orientation, insulation, and use of five-star energy-efficient appliances. Solar panels for electricity and water units helped to reduce energy requirements by 75%. The Olympic stadium used natural gas to produce electricity and hot water. This reduced greenhouse gases by 40%. The use of air conditioning and artificial lighting was also minimized.

Materials were chosen according to their environmental impact. Life cycle assessment and material eco-ratings were used as the basis for these evaluations. Renewable materials such as timber were chosen over nonrenewable ones, such as steel. Recycled materials such as concrete and brick rubble were incorporated into roads as a foundation base and other pavements.

Regarding biodiversity, it had been determined that the creeks and wetlands in the Homebush area had been contaminated with leachate from land disposal practices. The abundance and diversity of the fauna and flora were substantially reduced (Berents, 1996). To counter these problems, contaminated leachate inputs were removed. Barriers were also removed to allow free movement of the fish. Reeds were replanted to form a habitat for the birds. The creek was stabilized and isolated from the waste by a clay bund and impermeable bentonite liner. The wetlands were able to provide a habitat for flora and fauna and improved water quality by removing nutrients and organic material from storm water that flowed into the wetlands. Ongoing monitoring will determine changes and impacts on the flora and fauna in the area.

4.6   Concluding Remarks

Cities have a significant impact on the geoenvironment. Increasing urban population will increase pressures on the geoenvironment in the years to come. All aspects of the geoenvironment need to be protected. Optimal design of landfills is one of the methods to protect the urban environment. Natural attenuation of the landfill leachate plumes is a sustainable method of remediation. The operation of a landfill bioreactor has potential to increase methane production rates and waste degradation rates, which subsequently increases the life of the landfill. Although waste reduction and recycling programs exist in many cities, substantial improvements are needed to decrease wastes going into landfill and increase recycling rates. Life cycle analysis of products can be used to assist in determining which products have less impact on the environment. However, more data (particularly regarding emissions and leachates during production and disposal) are required to facilitate the comparison process.

Cities need to redevelop brownfields through remediation and other planning initiatives. This will substantially reduce the need for greenfields for future development projects. Few efforts have been made to protect the diversity of plants, animals, and microorganisms (i.e., biodiversity). Substantial efforts can be made to conserve resources such as energy. Electricity from geothermal and other renewable sources can be chosen. Buildings and their materials must also be energy efficient. Efforts must also be made regarding transportation to reduce its impact on the geoenvironment.

References

ASTM, (2003), *Standard Guide for Brownfields Redevelopment*, E1984-03, Vol. 11.04, ASTM International, West Conshohocken, PA.

Berents, P.B., (1996), *Wetlands and Benthos, Homebush Bay Ecological Studies 1993–95*, Vol. 1, Olympic Coordination Authority, CSIRO Publishing, Melbourne.

Block, D., (2000), Reducing greenhouse gases at landfills, *BioCycle*, 41:40–46.

Bridges, E.M., (1991), Waste materials in urban soils, in *Soil in the Urban Environment*, Bullock, P. and Gregory, P.J. (Eds.), Blackwell, Oxford, pp. 28–46.

Chamley, H., (2003), *Geosciences, Environment and Man*, Elsevier, Amsterdam, 450 pp.

CH2M Hill, (1997), *Olympic Site, Homebush Bay Dioxin Review Report*, report by CH2M Hill Australia Pty. Ltd., Olympic Coordination Authority, Sydney, 22 July.

Christensen, T.H., Bjerg, P.L., and Kjeldsen, P., (2000), Natural attenuation: a feasible approach to remediation of groundwater pollution at landfills?, *GWMR* 20:69–77.

Cooper, J., (1999), Solid waste management in Copenhagen, in *The Challenge of Environmental Management in Urban Areas*, Atkinson, A., Davila, J.D., Fernandes, E., and Mattingly, M. (Eds.), Ashgate, Aldershot, U.K., pp. 139–150.

Cooper, P.A., (1993), Disposal of treated wood removed from service: the issues, in *Proceedings, Environmental Considerations in Manufacture, Use and Disposal of Preservative Treated Wood*, Carolinas-Chesapeake Section, Forest Products Society, Madison, WI, pp. 85–90.

Craul, P.J., (1985), A description of urban soils and their desired characteristics, *J. Arboricult.*, 11:330–339.

Croxford, B., Penn, A., and Hillier, B., (1995), Spatial Distribution of Urban Pollution: Civilizing Urban Traffic, paper presented at the Fifth Symposium on Highway and Urban Pollution, May 22–24.

Cuny, D., Van Haluwyn, C., and Pesch, R., (2001), Biomonitoring of trace elements in air and soil compartments along the major motorway in France, *Water Air Soil Pollut.*, 125:273–289.

CWC, (1998), Crushed Glass Cullet Replacement of Sand in Topsoil Mixes, Final Report GL-97-10, CWC, Seattle, February.

Fayer, M.J. and Jones, T.L., (1990), *UNSAT-H Version 2.0, Unsaturated Soil Water and Heat Flow Model*, prepared for the U.S. Energy, Pacific Northwestern National Laboratory, Richland, WA.

FOEFL, (1997), *Sustainable Development: Action Plan for Switzerland*, report of the Federal Office of Environment, Forests and Landscapes, Bern.

Frischknecht, R., Jungbluth, N., Althaus, H.-J., Doka, G., Dones, R., Heck, T., Hellweg, S., Hischier, T., Nemecek, T., Rebitzer, G., and Spielmann, M., (2005), The ecoinvent database: overview and methological framework, *Int. J. Life Cycle Assess.*, 10:3–9.

Genske, D.D., (2003), *Urban Land, Degradation, Investigation and Remediation*, Springer, Berlin, 331 pp.

Harris, J.A., (1990), The biology of soils in urban areas, in *Soils in the Urban Environment*, Bullick, P. and Grogory, P.J. (Eds.), Blackwell Scientific Publications, Oxford, pp. 139–152.

Hertwich, E.G., (2005), Life cycle approaches to sustainable consumption: a critical review, *Environ. Sci. Technol.*, 39:4673–4684.

Huang, Y.-T., (2005), Heavy Metals in Urban Soils, M.A.Sc thesis, Concordia University, Montreal, Canada.

Hutson, J.L. and Wagenet, R.J., (1992), *Leaching Estimation and Chemistry Model, LEACHM*, New York State College of Agriculture and Life Sciences, Cornell University, Ithaca, NY.

ISO, (1997), *ISO 14040: Environmental Management — Life Cycle Assessment — Principles and Framework*, International Organization for Standardization, Geneva.

König, K.W., (1999), Rainwater in cities: a note on ecology and practice, in *Cities and the Environment: New Approaches for Eco-Societies*, Inoguchi, T., Newman, E., and Paoletto, G. (Eds.), United Nations University Press, Tokyo, pp. 203–215.

Kjeldsen, P., Barlaz, M.A., Rooker, A.P., Baun, A., Ledin, A., and Christensen, T.H., (2002), Present and long term composition of MSW landfill leachate: a review, *Crit. Rev. Environ. Sci. Technol.*, 32:297–336.

LaGrega, M.D., Buckingham, P.L., and Evans, J.C., (1994), *Hazardous Waste Management*, McGraw-Hill, New York, 1146 pp.

Lesage, S., Jackson, R.E., Priddle, M.W., and Riemann, P.G., (1990), Occurrence and fate of organic solvent residues in anoxic groundwater at the Gloucester landfill, *Can. Environ. Sci. Technol.*, 24:559–566.

Mulligan, C.N., (2002), *Environmental Biotreatment*, Government Institutes, Rockville, MD, 395 pp.

Mulligan, C.N. and Moghaddam, A.H., (2004), Sustainable building materials for urban development, *Int. J. Housing Sci. Appl.*, 28:89–102.

Mullins, C.E., (1990), Physical properties of soils in urban areas, in *Soils in the Urban Environment*, Bullick, P. and Gregory, P.J. (Eds.), Blackwell Scientific Publications, Oxford, pp. 87–118.

Münch, D., (1992), Soil contamination beneath asphalt roads by polynuclear aromatic hydrocarbons, zinc, lead and cadmium, *Sci. Total Environ.*, 126:49–60.

National Round Table for the Environment and Economy (NRTEE), (1998), *Greening Canada's Brownfield Sites*, NRTEE, Ottawa.

National Round Table for the Environment and Economy (NRTEE), (2004), *Cleaning up the Past, Building the Future: A National Brownfield Redevelopment Strategy for Canada*, NRTEE, Ottawa.

Page, A.L. and Ganje, T.J., (1970), Accumulation of lead in soils for regions of high and low motor vehicle traffic density, *Environ. Sci. Technol.*, 4:140–142.

Pohland, F.G. and Kim, J.C., (2000), Microbially mediated attenuation potential of landfill bioreactor systems, *Wat. Sci. Technol.*, 41:247–254.

Portney, K.E., (2003), *Taking Sustainable Cities Seriously*, MIT Press, Cambridge, MA.

Pusch, R. and Yong, R.N., (2005), *Microstructure of Smectite Clays and Engineering Performance*, Taylor & Francis, London, 335 pp.

RSU, (2000), Environmental report (in report). Report of the German Rat von SchVerständigen für Umweltfragen, Wiesbadeng, Germany.

Scharch, P.E., (1985), Water Balance Analysis Program for the IBM-PC Microcomputer, Department of Natural Resources, Madison, WI, May.

Schroeder, P.R, Lloyd, C.M., and Zappi, P.A., (1994), *The Hydraulic Evaluation of Landfill Performance (HELP) Model: User's Guide for Version 3*, EPA/600/R-94/168a, Office of Research and Development, U.S. EPA, Washington, DC, September.

Sharma, H.D. and Reddy, K.R., (2004), *Geoenvironmental Engineering*, John Wiley & Sons, Hoboken, NJ.

Solo-Gabriele, H. and Townsend, T., (1999), Disposal: end management of CCA-treated wood, in the *95th Annual Meeting of the American Wood Preservers' Association*, Ft. Lauderdale, FL, May 16–19, pp. 65–73.

Sydney Olympics 2000 Bid Ltd., (1993), *Environmental Guidelines for the Summer Olympic Games*, Sydney, September.

Syms, P., (2004), *Previously Developed Land, Industrial Activities and Contamination*, 2nd ed., Blackwell Publishing, Oxford.

Thorton, I. and Jones, T.H., (1984), Sources of lead and associated metals in vegetables grown in British urban soils: uptake from the soil versus air deposition, in *Substances in Environmental Health XVIII*, Hemphill, D.D. (Ed.), University of Missouri, Columbia, pp. 303–310.

Triangle J Council of Governments, (1995), *WasteSpec: Model Specifications for Construction Waste Reduction, Reuse and Recycling*, Triangle J Council of Governments, Research Triangle Park, NC, July.

UNEP, (2004), *UNEP/SETAC Life Cycle Initiative*, United Nations Environment Program, Paris.

United Nations Department of Public Information, (1993), *Agenda 21: Program of Action for Sustainable Development, Rio Declaration on Environment and Development*, United Nations, New York.

U.S. EPA, (1998), *Office of Mobile Sources*, U.S. Environmental Protection Agency, Washington, DC, www.epa/oms.

U.S. EPA, (1999a), *A Sustainable Brownfields Model Framework*, EPA 500-R-99-001, U.S. Environmental Protection Agency, Office of Solid Waste and Emergency Response, Washington, DC, January.

U.S. EPA, (1999b), *Road Map to Understanding Innovative Technology Options for Brownfields Investigation and Cleanup*, 2nd ed., EPA 542 B-99-009, U.S. Environmental Protection Agency, Office of Solid Waste and Emergency Response, Washington, DC.

U.S. EPA, (2002), *WasteWise Update Building for the Future*, EPA 530N-02-003, Office of Solid Waste and Emergency Response, Washington, DC, February, www.epa.gov/wastewise.

Vega, M., (1999), The concept and civilization of an eco-society: dilemmas, innovations and urban dramas, in *Cities and the Environment: New Approaches for Eco-Societies*, Inoguchi, T., Newman, E., and Paoletto, G. (Eds.), United Nations University Press, Tokyo, pp. 47–70.

Vengosh, A., (2005), Salinization and saline environments, in *Environmental Geochemistry*, Vol. 9, Lolar, B.S. (Ed.), Elsevier, Amsterdam, pp. 333–366.

Weeks, J., (2005), Building an energy economy on biodiesel, *Biocycle* 46:67–68.

Wilson, A., (1997), Disposal: The Achilles' Heel of CCA-Treated Wood, *Environmental Building News*, 6, March, http://www.buildinggreen.com/features/tw/treated_wood.html.

World Commission on Environment and Development, (1987), *Our Common Future*, Oxford University Press, Oxford, 400 pp.

Yong, R.N., (2001), *Geoenvironmental Engineering: Contaminated Soils, Pollutant Fate and Mitigation*, CRC Press, Boca Raton, FL, 307 pp.

Yong, R.N. and Mulligan, C.N., (2004), *Natural Attenuation of Contaminants in Soils*, Lewis Publishers, Boca Raton, FL, 319 pp.

Yong, R.N., Turcott, E., and Gu, D., (1994), *Artificial Recharge Subsidence Control, Bangkok, Thailand*, IDRC Canada Report File 90-1020-1.

Yong, R.N., Turcott, E., and Maathuis, H., (1995), Groundwater abstraction-induced land subsidence prediction: Bangkok and Jakarta case studies, in *Land Subsidence*, IAHS Publication 234, Barends F.B.J., Brouwet, F.J.J., and Shröder, F.H. (Eds.), IAHS, Oxfordshire, UK, pp. 89–97.

5

Natural Resources Extraction and Land Use Impacts

5.1 Introduction

Natural capital in the geoenvironment can be classified into two major categories: (1) renewable natural resources and (2) nonrenewable natural resources. Renewable natural resources consist of two subcategories: (1) living resources, such as forests and fishes and other aquatic species, and (2) nonliving resources, such as soil and water. Biodiversity, as a natural resource (see Section 1.2.2), is not within the purview of this book, and is therefore not included in this discussion of the natural capital in the geoenvironment. The key to classification as a living natural resource is the ability of the resource to regenerate or renew itself, and obviously, if the resource is not overharvested, it will be a sustainable resource. Farming and agricultural production are not included in the renewable resources group. The prominent nonrenewable natural resources include minerals and fossil fuels (coal, oil, and gas). The discussion in this chapter will be confined to industrial activities associated with the extraction of nonrenewable mineral, nonmineral, and energy mineral natural resources (uranium and tar sands). The materials that constitute these resources are extracted or harvested by primary industries devoted to such activities as mining, excavation, extraction, processing, drilling, and pumping. The outputs from these upstream industries are raw materials for their respective downstream industries (see Chapter 7).

Resource extraction and processing industries use the geoenvironment as a resource pool containing materials and substances that can be extracted and processed as value-added products. The common characteristic of the industries in this group is *processing of material extracted from the ground*. Included in this group are (1) the metal mining industries; (2) those industries involved in extraction and processing of other resources from the ground, such as nonmetallic minerals (potash, refractory and clay minerals, phosphates); (3) the industries devoted to extraction of aggregates, sand, and rock for production of building materials; and (4) the raw energy industries, such as those involved in the extraction of hydrocarbon-associated materials and other fuels (natural gas, oil, tar sands, and coal). Included in this list is the extraction and recovery of uranium for the nuclear power industry. Mineral resources are used in all fields of human activity. Worldwide consumption is in the order of a billion tonnes per year for iron ore and 2500 tonnes per year of gold. Metal extraction increased by 20% from 1980 to 2000 (Chamley, 2003). Some developed countries in Europe and Japan have completely depleted their underground resources.

Activities associated with the mining, extraction, and on-site processing of the extracted natural resource material (mineral and nonmineral) contribute significantly to the inventory of potential impacts to the terrestrial ecosystem. In the United Kingdom, for example, more than 200 years of the mining of coal and iron ore have left 500 million tonnes of

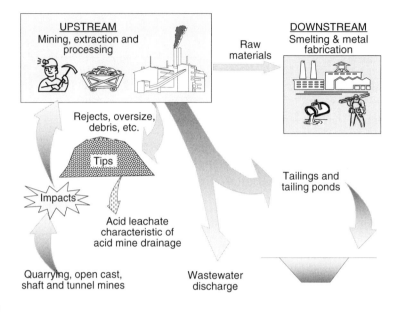

FIGURE 5.1
Illustration of some of the major features associated with mining and natural resource extraction. The example shown is typical of metalliferrous mining extraction operations.

residual waste from coal and iron and steel manufacturing as sterile waste tips (Barr, 1969). These tips are the result of removing more than 52 million tonnes of coal from the underground. Ground subsidence, water pollution, and toxic gas emissions are other impacts. Not only are these tips a blight on the landscape, but they are also a hazard when slide failures occur. There are historic reports of such failures that have resulted in human casualties. Figure 5.1 shows a simple generic illustration of operations associated with mining and extraction of metal mineral resources. Individual mineral extraction or beneficiation processes will differ between different types of minerals and their host ores. Specific examples will be discussed in a later section.

Activities mounted in conjunction with energy resources, such as the development of oil-producing wells, natural gas wells, recovery of bitumen from tar sands, and dams and hydroelectric facilities, also contribute their share to the potential impacts list. Not all the potential impacts are directly due to the discharge of wastes and pollutants in leachate streams or wastewaters. The discussion in this chapter will focus primarily on the mineral and nonmineral (including tar sands) mining and processing upstream industries — industries dealing with exploitation of nonrenewable natural resources.

5.1.1 Sustainability and Land Use

The nonrenewable natural resources discussed in this chapter are not going to satisfy sustainability from the viewpoint of renewal of supply of the natural resource material. Extraction of these resources from the ground will deplete them, and will therefore fail a key sustainability issue — replenishment or renewal of supply. The primary geoenvironmental concern in this respect is to achieve the slowest depletion rate possible. This requires initiatives from governments, industry, and the consumer. While a discussion of these initiatives is not within the purview of this book, it would useful to point out that these initiatives should be directed toward developing (1) incentives, legislation, and directives from governments, (2) support for basic research and new technology, (3) consumer awareness and citizenship, (4) alternative materials using renewable raw materials, (5) conservation of the extracted resource through minimal use and use of additives and supplements, (6) recovery, reuse, and recycle procedures as part of the conservation effort, and (7) replacements and alternatives.

The sustainability issues for nonrenewable resource extraction operations fall within the jurisdiction of land use — sustainable land use. At the present time, activities associated with extraction of the nonrenewable natural resources begin with operations designed to extract the host ore from the ground through open-pit or underground mining (tunnel and shaft mining). Beneficiation, ore dressing, and mineral extraction will generate waste materials in addition to processing wastes and discharges that find their way into the land environment. The nature of these and their impacts on the land environment, particularly with land use, constitutes the major concern in this chapter. The danger that one faces is the cascading or domino effect generated by these activities and discharges.

5.2   Sulfide Minerals and Acid Mine Drainage

Ores extracted from base metal mines include, in alphabetical order, copper, gold, iron, lead, molybdenum, platinum, silver, uranium, and zinc. The predominant base metal ores are those that contain copper, iron, lead, and zinc. These are primarily obtained from lode deposits using both open-pit and underground mining techniques. The sulfide minerals in the host rock, in the mines, and in the ores — such as iron sulfides, pyrites, arsenopyrites, chalcopyrites, pyrrhotites, sphalerites, and maracasites — present severe environmental problems when they are exposed to water and oxygen. Figure 5.2 gives an illustrative example of what happens when pyrites (FeS_2) are exposed to oxygen and a

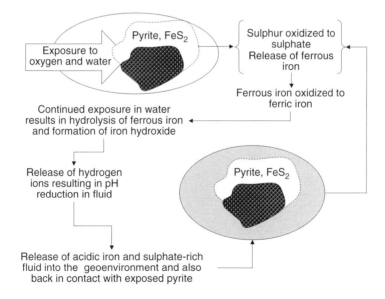

FIGURE 5.2

Effect of exposure of pyrite to oxygen and water. Note that continued exposure to water will result in the generation of iron hydroxide (yellowboy) and acidic solution that will be harmful to aquatic plants and animals, and will also release heavy metals previously held by the soil. Similar reactions shown in the diagram will also occur for sulfides of copper, lead, arsenic, cadmium, and zinc.

source of water. The example cycle of acid contact and oxidation of the pyrite shown in the diagram continues as long as oxidation processes can proceed. Release of the acid fluid into the land environment within the mining site will allow the fluid to come in further contact with other exposed pyrites. The generation of acidic leachate rich in iron and sulfate (known as yellowboy) is characteristic of the outcome of the various processes that accompany oxidation of the sulfur and iron in the pyrite. As long as there is a source of these in the host rock and ores, and as long as these continue to be exposed to water, generation of yellowboy will continue unabated. The presence of sulfate-reducing bacteria in soil and water will exacerbate the problem. These bacteria are anaerobes that use sulfate as electron acceptors.

While the generation of the acidic leachate, commonly known as acid mine drainage (AMD) or acid rock drainage (ARD), constitutes a major negative impact from mining operations, the cascading or domino effects that accrue from AMD can be severe. The domino effects arising from discharge of the leachate into the environment include (1) severe health threats to aquatic species, native habit, and plant life, (2) pollution of groundwater and drinking water, (3) deterioration of soil quality, and (4) release of trace metals and heavy metals previously retained by the soil solids in the ground. Information reported in the Interstate Mining Compact Commission study (IMCC, 1992) showed that for five western states in the United States, there were (in 1992) about 130,000 inactive and abandoned mine sites, and that in these same states, more than 2300 miles of streams and waterways were polluted with acid mine drainage.

The extent of acid generation at mine sites (underground mines, openings, leach ores, spent ores, etc.) is a function of several factors. These include: (1) type and concentration of sulfide minerals in the host ore and in the spent ores and leach piles, (2) the host rock, (3) availability of oxygen, (4) site hydrogeology, (5) pH of the water in the system, and (6) presence or absence of bacteria, e.g., *Thiobacillus ferrooxidans*.

Oxidation of the mineral sulfides may result in the solubilization of trace metals and heavy metals — effectively releasing them and allowing them to be mobile in the liquid phase. It is not uncommon to find evidence of arsenic, cadmium, cobalt, copper, lead, manganese, nickel, and zinc as released metals. Arsenic poisoning of groundwater and aquifers has been reported in many parts of the world. This problem has gained considerable publicity and has been reported as "the greatest mass poisoning of mankind" in relation to the poisoning of the tube wells in Bangladesh and West Bengal, as previously mentioned in Chapters 1 and 3. In this particular case, available information points to the presence of both arsenopyrites and arseniferrous iron oxyhydroxides in the substrate material as the immediate source materials for the arsenic. If oxygen is available in the groundwater, oxidation of the arsenopyrites will release the arsenic. Some reports have speculated on the use of tube wells as a means for introduction of oxygen into the subsoil strata. In the absence of oxygen, the processes associated with reductive dissolution of the arseniferrous iron oxyhydroxides will release arsenic while increasing the bicarbonate concentrations. This will result in arsenic pollution of the groundwater.

Mine tailings and effluents usually contain high concentrations of arsenic and are of concern as potential sources of environmental contamination. Arsenic occurs naturally in a wide range of minerals in soils in several forms of inorganic compounds. The most common arsenic-containing minerals are arsenopyrite or mispickel (FeAsS), realgar (AsS), and orpiment (As_2S_3). It was reported that the annual total fluvial input of arsenic to Moira Lake in Ontario, Canada, was approximately 3.5 tonnes due to local mining and mineral processing (Azcue and Nriagu, 1995).

In mine tailings, arsenic occurs in various forms, such as arsenopyrite (FeAsS), arsenian pyrite (As-rich FeS_2), arsenates, and association with iron oxyhydroxides. Wang and Mulligan (2004) measured the arsenic contents of six Canadian mine tailings. Inductively

TABLE 5.1

Arsenic Concentrations Measured in Canadian Mine Tailings

Mine	Location	Concentration (mg/kg)	Reference
Copper	Murdochville, QC	500	Wang and Mulligan, 2004
Gold	Musselwhite, ON	63	Wang and Mulligan, 2004
Copper–zinc	Val d'Or, QC	270	Wang and Mulligan, 2004
Iron	Mont-Wright, QC	<0.70	Wang and Mulligan, 2004
Lead–zinc	Bathurst, NB	2200	Wang and Mulligan, 2004
Gold	Marathon, ON	270	Wang and Mulligan, 2004
Con	Yellowknife	25,000	Ollson, 1999
Giant	Yellowknife	4800	Ollson, 1999
Negus	Yellowknife	12,500	Ollson, 1999
Rabbit Lake	Northern Saskatchewan	56 to 9871	Moldovan et al., 2003

coupled plasma mass spectrometry (ICP-MS) analyses indicated that the highest arsenic concentrations reached 2200 mg/kg in tailings from a lead–zinc mine at Bathurst, Australia. These and others are shown in Table 5.1. Many other countries, such as Thailand, Korea, Ghana, Greece, Australia, Poland, the United Kingdom (Kinniburg et al., 2003), and the United States, have also experienced significant arsenic contamination associated with mining activities. McCreadie et al. (2000) have found arsenic concentrations up to 100 mg/l in the pore water extracted from tailings in the province of Ontario. Donahue and Hendry (2003) showed that dissolved arsenic concentrations within the tailings could vary from 9.6 to 71 mg/l.

5.3 Resource Extraction and Land Use

There are at least three categories in this grouping of resource extraction and processing industry operations. These are operations mounted to process the ores or liquid/gaseous resources recovered from the various forms of mining and drilling activities. The products obtained from these industries serve as raw materials for their associated downstream industries. The associated downstream industries are discussed in detail in Chapter 7.

The three categories of resource extraction and processing include those dealing with (1) ores from metalliferrous-type mining, (2) ores from nonmetal resource mining and processing, and (3) hydrocarbon resources extracted from the ground, including tar sands and fossil fuel. The first two can be considered jointly as *mining-related industries*. For these (mining-related) industries, there are typically three types of interactive contact with the geoenvironment. The issues of prime importance in respect to the geoenvironment and sustainability relate to the various waste products issued from mining and resource extraction operation. Most mining wastes are associated with recovery processes of host rock material for production of natural resources such as metal and nonmetal ores and products. These include aluminum, iron, copper, gold, lead, molybdenum, silver, tungsten, uranium, zinc, coal, asbestos, gypsum, barite, syenite, potash mineral, salt mineral, bitumen (from tar sands or shales) quartz, lime, sand, and gravel and stone. In addition to solid wastes, liquid wastes in the form of tailings and other process liquid waste streams associated with mining and milling operations need to be considered. Specific and detailed examples of various mining activities and milling processes can be found in mining–milling textbooks dedicated to the study of these subjects.

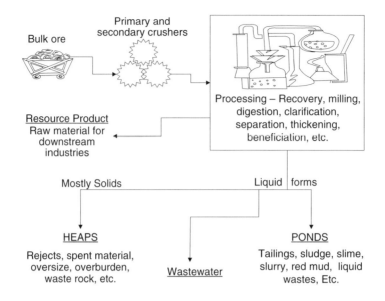

FIGURE 5.3
Generalized resource extraction recovery process. Specific details of additives, digestion and beneficiation, etc., will vary according to types of resources (minerals, nonminerals, hydrocarbon) being extracted. Discharges to the geoenvironment will take the general forms shown in the diagram.

5.3.1   Discharges from Beneficiation and Processing

Procedures and processes required to extract minerals from the ores obtained in mining operations vary according to the kinds of minerals being extracted. These procedures and processes fall under the categories of *beneficiation operations* and *mineral processing*. Operations included in beneficiation and mineral processing are (1) ore preparation — crushing, grinding, and washing; and (2) mineral recovery — including such processes as dissolution, filtration, calcining, roasting, leaching, concentration, separation, solvent extraction, electrowinning, and precipitation. The generalized procedure shown in Figure 5.3 applies to most of the processes used. Bulk ore transported to the processing and extraction plant needs to be crushed to smaller fractions where specific chemicals and additives are added. The resultant mixture is then subjected to the required mineral extractions processes, as required. The points of interest in regard to interactions of the mining–milling activities and industries with the geoenvironment shown in Figure 5.3 include debris from mining and liquid and solid waste materials from beneficiation and processing.

Both operations and the site itself can cause impact to the environment. Collapse and subsidence are major problems. The heap and tailings basins are also subjected to movement and have exhibited substantial damage in towns where the tailings have liquefied and flowed into towns, burying houses and people. Noxious and toxic releases from dissolution, bacterial activity, and subsequent runoff also cause damage. The acid water is carried into surface and groundwater, thereby threatening humans, animals, and the local flora. Dust inhalation may also lead to cancer and other illnesses, such as asbestosis and silicosis. Bioaccumulation of mercury, lead, and other heavy metals in the food chain may also be significant. Kobayashi and Hagino (1965) reported that the runoff of cadmium from a zinc mining waste in the 1950s led to accumulation in the Jintsu River that was used for drinking water and irrigation of rice fields, and that itai-itai disease afflicted the population living in the area. Secure disposal, cleanup, and management of drainage eliminated the disease.

5.3.1.1 Solid Waste Materials

The principal sources of solid waste materials issuing from extraction of the various minerals are waste rocks, ore spoils, and overburden. The nature of the host ores and the leaching solutions used to extract the minerals from the ores are perhaps the more significant sources of problems ascribed to solid wastes found on-site. Sulfide minerals are found in most of the host ores. Commonly used ores for extraction of lead and zinc ores are galena (PbS) and sphalerite (ZnS). In the case of iron, the ores are in oxide and sulfide forms. These include magnetite (Fe_3O_4), hematite (Fe_2O_3), goethite ($Fe_2O_3H_2O$), siderite ($FeCO_3$), and pyrite (FeS_2). Although copper sulfide minerals are found in such minerals as chalcopyrite ($CuFeS_2$), covellite (CuS), chalcocite (Cu_2S), and bornite (Cu_5FeS_4), the most common source of copper ore exploited in present mining operations is chalcopyrite (Simons and Prinz, 1973). Oxide minerals include chrysocolla ($CuSiO_3$), malachite (Cu_2CO_3), azurite ($2CuCO_3Cu(OH)_2$), tenorite (CuO), and cuprite (Cu_2O).

Using extraction of copper as an example, solid waste material discharges include the overburden material such as soil, debris and unconsolidated material, oversize material and waste rock (rejects), spoil heaps, leach ore obtained from leaching of copper oxide mineral ores, and other solid by-products and wastes issuing from the recovery process. It is estimated that on the average, about 100 tonnes of waste material are generated in the extraction of 1 tonne of copper. Using the 100:1 ratio of waste material for each tonne of recovered copper as an example, we can surmise that for the various mineral extraction industries in the world, this would mean that hundreds of millions of tonnes of waste materials are generated in a year. All of these materials will need proper management and disposal. This is particularly significant because the rejects and other ores in the piles contain trace amounts of sulfide minerals. These will contribute to the generation of acid mine drainage problems. Not all the solid heaps are waste heaps. In operations where sulfur recovery is obtained, open-air storage on-site is not an uncommon procedure. Figure 5.4 shows some of the more significant impacts to the geoenvironment and also impacts on land use. The significant impacts from the presence of solids as rejects, oversize, etc.,

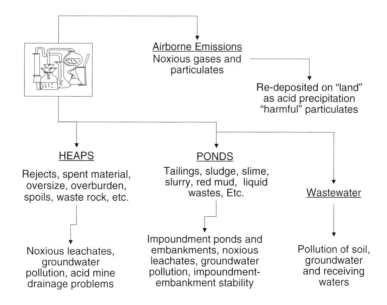

FIGURE 5.4
Discharges from resource recovery operations and some of their more significant impacts on land use and the geoenvironment. The ponds containing the slurry tailings could also be dammed-up valleys or even abandoned mine pits.

on land use are storage, noxious leachates, acid generation (AMD), and runoffs, leading to acid pollution of watercourses, groundwater, and land surface.

5.3.1.2   Liquid Waste Streams and Discharge

There are essentially two or three liquid waste streams associated with mining–mineral extraction industries. These include:

- Mine water and waste streams generated in a mining operation due to hydrologic drainage from the mining site, and percolation from waste rock and mill tailings piles, and surface runoffs. While containment of mine water and generated waste streams in abandoned mines has been previously practiced, present worries of acid mine generation have led to prevention of excess water entry into abandoned mine sites. Extraction pumping of mine water serves to remove a potential source for generation of acid in the mine. However, without benefit of capping and sealing of mine shafts and openings, it is inevitable that some water will be introduced into the mines, resulting ultimately in the problem identified as acid mine drainage.

- Liquid waste streams from the processing plants, such as solutions or liquids containing solvent extraction sludge, spent electrolytes, spent leaching solutions, spent solvents, and used oils. In general, the liquid wastes are contained in ponds and are often referred to as solution ponds, pregnant ponds, holding ponds, etc. Pregnant (solution) ponds are those that contain some of the minerals, and solution ponds are those that are presumably devoid of these minerals.

- Slurry tailings discharge consisting of water and a wide range of inorganic and organic dissolved constituents obtained from the remnants of reagents used in the recovery processes and fine fractions of the host ores. Impacts on land use consist of large open containment facilities such as dammed-up valleys and ponds containing sludges and slimes that pose health and safety threats to animals and humans, runoffs from the containment facilities, embankments for the impoundments used to contain the sludges and slimes, and leachates polluting ground, groundwater, and receiving waters. Most of the slurry tailings' containment facilities contain slurries and slimes that will not exhibit liquid–solids separation; i.e., the solids in these facilities will not sediment or settle to the bottom of the containment structure. These pose potential safety and health threats to the human population within the immediate area. The strategies for maintaining and operating containment facilities are discussed in the next subsection.

- Embankment stability can be a problem if the containment embankments become too high and are subject to drawdown pressures when large fluctuations occur over a very short period in the height of the pond. This has happened in instances when containment ponds are emptied quickly. This particular issue is a geotechnical problem that can be corrected with proper design and management techniques.

- Smokestack emissions of fugitive dust from the crushers and noxious airborne gases and particulates from the processing extraction plant. Treatment of smokestack emissions will be covered in Chapter 7.

5.3.2   Containment of Tailings Discharges

Central to the extraction of the resource material contained in the ores obtained in mining is the fine grinding of the ores — as the preliminary stage of the beneficiation process.

The common elements of the various techniques used in the beneficiation processes include both fine grinding and washing flotation and, finally, at some stage in the beneficiation process, separation of the resources from the finely ground material. Flotation and other methods of resource separation (e.g., magnetic) are required. The end result of all of these processes is the generation of liquid wastes containing a suspension of the finely ground material. This liquid waste is generally defined as *tailings waste slurry* or *slurry tailings*. It can be well appreciated that the generous quantities of tailings are reflective of the significant amounts of water needed in the processing of the finely ground ore material for beneficiation.

5.3.2.1   Containment of Tailings and Land Use

Slurries, slimes, sludges, and red mud ponds are all names that are given to the general class of slurry tailings discharge from beneficiation processes. These slurry tailings cannot be directly discharged into the land environment, not only because they contain suspended fines, but also because of their chemistry and potential detrimental impact on the land ecosystem. Some of the main reasons for containment of these slurry tailings are:

- Avoid pollution of land surface environment. This is one of the principal reasons for containment, and is generally coupled with other specific disposal containment strategies.
- Permanent containment of the tailings. This strategy generally includes several kinds of scenarios, ranging from *permanent ponds* to totally reclaimed solid land surfaces.
- Recovery of water for reuse in the beneficiation processes as process recycle water or other mine site requirements. This requires implementation of treatment of the supernatant — assuming that liquid–solids separation is effective in producing sedimentation of the suspended fines in the slurry tailings.
- Secondary recovery pond storage. This strategy presumes that some residual resource is contained in the slurry tailings waste, and that secondary recovery of this resource can be obtained when appropriate technology becomes available, and when the economic climate is favorable.

There are at least three basic types of slurry tailings impoundment facilities: (1) abandoned, used-up, fully exploited mine pits; (2) dammed-up valleys, as illustrated in Figure 5.5; and (c) constructed ponds with containment embankments (Figure 5.5). The choice of containment facility depends on many factors, not the least of which is site conditions and company mining strategy. These are not within the scope of this book and will not be addressed. The engineering preparation and construction of these facilities can be demanding. Some of these aspects will be discussed in a later section dealing with sustainable land use.

There are two distinct categories of slurry tailings contained in the containment facilities or structures:

- Containment structures that contain sedimented solids and particulates discharged from mineral extraction processes. These structures will show a solids sedimented layer overlain by water, as illustrated by the containment pond shown in the top diagram in Figure 5.6. For such kinds of sediment slurries, treatment of the supernatant (liquid) may or may not be necessary, depending on the chemistry or toxicity of the supernatant. In some mining processes, such as

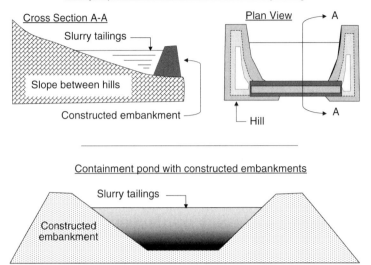

FIGURE 5.5
Illustration of valley impoundment for containment of slurry tailings (top diagram) and constructed pond (bottom diagram).

aggregate harvesting from transported surface soils, since water is the only agent used in the beneficiation process, the supernatant obtained is considered to be nontoxic. As an example, tin mining of placer deposits using the gravel pump and dredging method will leave behind slime ponds with well-developed sediments and clear supernatants.

- Containment structures that contain solids suspensions. The solids in these suspensions may or may not finally sediment. Using containment ponds as an example, the bottom diagrams in Figure 5.6 show the characteristics of these kinds of ponds. The dispersion stability of these types of slurry tailings will be discussed in the next section.

5.3.2.2   Dispersion Stability of Slurry Tailings

The common factors in all of the materials contained in the various kinds of containment facilities are (1) fine particles that may or may not exhibit tendency to remain suspended in the aqueous phase and (2) an aqueous phase chemistry that reflects the types of reagents used in the beneficiation process. Since the fine particles are obtained from the preparatory grinding–crushing processes required before application of techniques for removal of the mineral, they will have the same composition of the host ore. In some cases, the material may undergo transformation because of the chemical treatments used in the beneficiation process. These are the materials (fines and aqueous phase) that constitute the tailings that are discharged in the beneficiation process — generally in the flotation phase of beneficiation.

For suspended fines in slurry tailings that do not exhibit liquid–solids separation behavior, some common features can be identified. If one determines the solids concentration with depth of the slurry tailings, one will obtain at least four distinct zones: (1) clear supernatant liquid, (2) a transition zone where the solids concentration begins to register some small value that increases as one progresses in depth, (3) a stagnant zone where the solids concentration remains relatively constant or increases imperceptibly with depth,

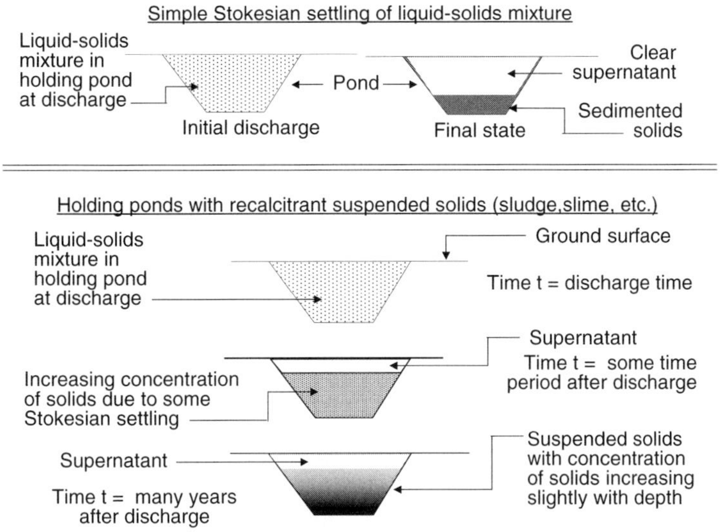

Simple Stokesian settling of liquid-solids mixture

Liquid-solids
mixture in
holding pond
at discharge

← Pond →

Clear
supernatant

Sedimented
solids

Initial discharge   Final state

Holding ponds with recalcitrant suspended solids (sludge,slime, etc.)

Liquid-solids
mixture in
holding pond
at discharge

Ground surface

Time t = discharge time

Increasing concentration
of solids due to some
Stokesian settling

Supernatant
Time t = some time
period after discharge

Supernatant

Time t = many years
after discharge

Suspended solids
with concentration
of solids increasing
slightly with depth

FIGURE 5.6

Two types of behavior of liquid–solids discharge in slurry tailings containment structures. The examples shown are containment ponds. The top diagram shows simple Stokesian settling of the solids in the liquid–solids mixture. The bottom diagram illustrates solids behavior over some time period, ending up with recalcitrant performance of suspended solids.

and (4) a sediment zone that contains the solids that have finally settled to the bottom of the containment structure. Figure 5.7 shows a typical solids concentration (sc) profile in the four zones that are typical of various kinds of sludges, slimes, slurries, etc. The sc values refer to the weight ratio of the suspended solids to aqueous phase that constitutes the suspension fluid. Figure 5.8 shows the solids concentrations in the stagnant zones for various types of slurry tailings. Leaving the suspended fines in the containment structures is not an acceptable land use option. Strategies have been developed to render the material

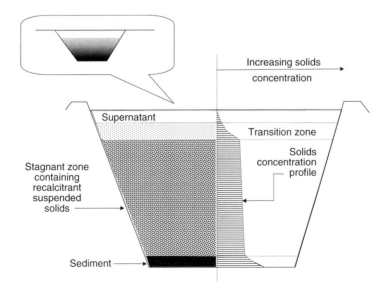

Increasing solids
concentration

Supernatant

Transition zone

Solids
concentration
profile

Stagnant zone
containing
recalcitrant
suspended
solids

Sediment

FIGURE 5.7

Illustration showing solids concentration profile for the recalcitrant suspended solids behavior pattern — typical of such sludges and slimes as red muds, tar sands sludges, and phosphatic and other clay slimes.

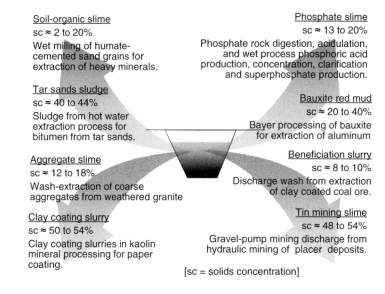

FIGURE 5.8
Some examples of slurries, slimes, and sludge found in holding ponds. The solids concentrations (sc) are obtained from results reported by Yong and his co-workers (see, for example, Yong, 1984). Details of the compositional features of the slurries, slimes, and sludge are given in Table 4.2.

in the containment structures to a state that would not pose a threat to the immediate environment and biota (Yong, 1983a, 1983b).

Studies on the nature of the solids suspensions constituting those slurry tailings facilities (ponds, etc.) where the solids remain in suspension for some considerable length of time show that the principal factors responsible for the dispersion stability of the suspended solids include (1) colloidal nature of the solid fines, (2) reactive surfaces on the fines, and (3) chemistry of the suspending fluid. The theoretical basis for the dispersion stability of the colloidal-type solid fines has been well developed and reported (Kruyt, 1952). The diffuse double-layer (DDL) model is a good fit with the types of suspended fines (e.g., montmorillonite, kaolinite, amorphous materials) found in many slurry tailings. It essentially provides one with a basis for determining (theoretically) the maximum volume of water or fluid in the diffuse ion layer that surrounds individual reactive suspended particles. The intensity of the interaction forces between the two particles resulting from the interpenetration (or overlapping) of the contiguous diffuse ion layers is a function of (1) the extent of the overlapping or interpenetration of the adjoining diffuse ion layers, (2) the nature of the reactive surfaces of the particles, and (3) the chemical composition of the suspending fluid. The electrostatic interactions of the ions in the diffuse ion layer and their relation to the surfaces of the reactive suspended particles are expressed as an electric potential ψ that decreases in an exponential manner as one departs farther from a particle surface. The DDL model provides one with the basis for computing the average electric potential ψ as a function of distance from the surface of the particle as follows (Yong and Warkentin, 1975):

$$\psi = -\frac{2\kappa T}{e} \ln \coth\left(\frac{x}{2}\sqrt{\frac{8\pi z_i^2 e^2 n_i}{\varepsilon \kappa T}}\right) \qquad (5.1)$$

where the negative sign on the right-hand side indicates that the potential ψ decreases as one departs farther away from each particle surface, and where κ = Boltzmann constant,

TABLE 5.2

Equilibrium Suspension Volumes Obtained from Soil Suspension Tests

Suspended Solids	Equilibrium Volume, cc/g	Void Ratio
Kaolinite	1.3	3.4
Illite	3.1	8.2
Montmorillonite	21.5	57
Amorphous Fe_2O_3	20.5	82
Gibbsite	1.0	2.6
Mica	3.0	7.9
Quartz	0.14	1.12

Source: Adapted from Yong, R.N., in *Sedimentation Consolidation Models: Predictions and Validations*, Yong, R.N. and Townsend, F.C., Eds., American Society of Civil Engineers, New York, 1984, pp. 30–59.

T = temperature, e = electronic charge, n_i and z_i = concentration and valence of the ith species of ions in the bulk solution, and ε = dielectric constant. A detailed treatment of the DDL theory and models can be found in Kruyt (1952). The development and application of the DDL models to soil mineral particles such as those found in slurry tailings can be found in Yong and Warkentin (1975) and Yong (2001a).

Calculations of the volume of water associated with a gram of soil particle in equilibrium in an aqueous phase, based on type of soil fraction and diffuse double-layer interactions, can be made using the DDL models. These can be compared with measurements of equilibrium solids concentrations obtained in soil suspension experiments. The results of solids suspension tests reported by Yong (1984) are shown in Table 5.1 for some typical soil solids found in slurry tailings. These results are expressed in terms of the equilibrium volume of water per unit weight of suspended solids, and the units are given in terms of cc/g of soil. The void ratios shown in the third column of Table 5.2 have been calculated from the measured equilibrium volumes.

Yong (1984) has shown good correlation between calculated and measured equilibrium solids concentration for the stagnant region of a slime pond using the equilibrium volumes shown in Table 5.2. In the actual cases examined, predicted solids concentrations were compared with actual solids concentrations obtained from samples in the stagnant zone (Figure 5.7 and Figure 5.8) for phosphatic slimes, aggregate slimes, tin mining slimes, beneficiation slurry (slimes), tar sand sludges, etc. Table 5.3 shows that except for the aggregate slimes obtained from aggregate recovery of aggregate loams, a comparison of the predicted solids concentration with actual measured values showed good accord. The ratio of predicted to measured solids concentration (predicted:measured) varied from 0.96 to 1.05. What this tells us is that colloidal dispersion of the suspended solids is responsible for the dispersion stability of the slurry tailings.

5.4 Sustainable Land Use

Since depletion of nonrenewable natural resources will not qualify under any category of sustainability, the discussion in this section will deal with some of the more noteworthy impacts on sustainable land use from mining operations and natural resources extraction.

TABLE 5.3

Composition of Slurry Tailings Solids and Suspension Fluid in Stagnant Zone of Slurry Tailings' Ponds

Type of Slurry Tailings (Location)	Suspended Solids Composition	Suspension Fluid, Dominant Ions	Measured sc, %	Computed sc, %
Phosphate slime (Florida)	Carbonate–fluorapatite, quartz, montmorillonite, attapulgite, wavellite, feldspar, dolomite, kaolinite, illite, crandallite, heavy minerals	Ca^{2+}, Mg^{2+}, Na^+, K^+, SO_4^{2-}, HCO_3^-	14	13.4
Aggregate slime (SE Asia)	Kaolinite, montmorillonite, illite	Ca^{2+}, Mg^{2+}, Na^+, K^+, SO_4^{2-}	14.3	10.5
Clay coating (SE U.S.)	Illite, montmorillonite, mixed-layer minerals, chlorite, quartz	CO_3^{2-}, HCO_3^-, Cl^-, SO_4^{2-}, Na^+, Mg^{2+}, Ca^{2+}	52.7	51.8
Tin mining slime (Malaysia)	Kaolinite, gibbsite, mica, quartz, other	Na^+, Ca^{2+}, K^+, Mg^{2+}	52	50.8
Beneficiation slurry (Western Canada)	Montmorillonite, illite, feldspar, kaolinite, chlorite	CO_3^{2-}, HCO_3^-, Cl^-, SO_4^{2-}, Na^+, Ca^{2+}, K^+, Mg^{2+}	8.7	9.1
Tar sands sludge (Western Canada)	Kaolinite, illite, chlorite, montmorillonite, mixed-layer minerals, feldspar, quartz, siderite, ankerite, pyrite, Fe_2O_3	Not available	41.9	42.2

Source: Adapted from Yong, R.N., in *Sedimentation Consolidation Models: Predictions and Validations*, Yong, R.N. and Townsend, F.C., Eds., American Society of Civil Engineers, New York, 1984, pp. 30–59.

There are at least three significant kinds of land use impacts: (1) mining excavations, pits, underground caverns, debris piles, waste rock, rejects, overburden, etc.; (2) acid generation or acid mine drainage from exposure of debris piles, exposed mine cavities, etc.; and (3) slurry tailings containment facilities. The last two concerns can be seen in Figure 5.4. Mining excavations and underground mining create situations where the excavated (empty) volumes present challenges that are beyond the scope of the material discussed in this book. The impact from debris discharge and heaping into tips has been briefly mentioned at the beginning of this chapter. In addition to the sterilization of the immediate landscape surrounding the tips as a result of the leachates emanating from the tips, possible instability of the tips is a question and problem that needs attention. Geotechnical engineers have, because of a history of instability problems experienced by some problem tips and heaps, developed guidelines and procedures that are designed to render spoil–debris heaps and tips stable.

5.4.1   Geoenvironmental Inventory and Land Use

By and large, a major proportion of mining and on-site resource extraction operations are initially in regions situated some distance from urban centers. Original land use, prior to the time of mining exploration in such regions, would be characterized by the local physiographic features, such as those discussed in Section 1.2.1. An environmental inventory, and more specifically a geoenvironmental inventory, prior to mining operations is needed to establish a base upon which decisions regarding impacts on land use, sustainability indicators, and restorative requirements can be sensibly made. The principal features of the geoenvironmental inventory, which is a baseline descriptor of the state of the various constituents of the local geoenvironment (*ab initio* condition), include:

FIGURE 3.2
Iron runoff from a coal mine into canal area.

Tin mining ponds

Rejects, sand and
alluvial debris

FIGURE 5.10
Tin mining slurry ponds obtained as a result of hydraulic dredging and pumping. Reclamation of pond is necessary to accommodate expansion of the housing units seen at the left of the picture (and left of the canal).

(a) Non-filtered   (b) Filtered

FIGURE 8.9
Lucite dishes showing (a) nonfiltered and (b) filtered seawater obtained with purification units in the purification vessel.

2005.2.21

2005.2.21

2005.3.16

2005.9.30

FIGURE 8.14
Countermeasure for coastal erosion by dumping and the action of erosion.

1. Regional controls such as climate and meteorological factors
2. Local terrain features, including linear features, physical attributes, topography, watershed, local hydrology, surface layer quality, vegetative cover, receiving waters, and water quality
3. Subsurface features such as geological and hydrogeological settings, soil subsurface system, and groundwater–aquifer regimes

Decisions on sustainability of potential land uses cannot be made without the inventory and without determination of the qualities necessary for various land uses. These are highly dependent on whether (1) the mining and resource extraction operations remain isolated from urban communities, or (2) small urban communities are located contiguous to the mining site. In the first instance, i.e., mining operations remain isolated from habitable communities, return of the exploited land to original natural conditions is an obvious response to sustainable land use. Specification of *ab initio* land environment sustainability indicators will be guided by the geoenvironmental inventory established before mining operations. Historically, geoenvironmental inventories have not been made prior to or even during mining operations. Nevertheless, a study of the natural geoenvironment system contiguous to the mining operations will serve to provide the basis for establishment of *ab initio* sustainability indicators. While return of the land to its pristine original function, as a sustainability objective, may not be absolutely possible, one can return the land to a natural state. For example, one could seal shafts and openings, relevel and contour the landscape to conform to local topography, add organic fertilizers, and reseed with tolerant plant species (Davies, 1999). One recognizes that whereas one no longer has the natural resource that has been harvested, this mine closure procedure allows for recovery of the functionality of the affected ecosystem.

It is not uncommon for small urban communities to be developed within regions close to mining and resource extraction operations to service the mining operations. Land use and land restoration will depend on the requirements of the urban community and will undoubtedly be markedly different from *ab initio* land use conditions. The general procedure is to perform an assessment of impacts on land use from the mining extraction operations, and to determine what reclamation and land restoration requirements are needed to meet the needs of the community. The basic steps shown in Figure 5.9 include:

1. Determination of impacts to geoenvironment and to land use. This requires information from *ab initio* geoenvironmental inventory.
2. Compilation of community or regulatory land use schemes or plans, e.g., return to *ab initio* conditions, housing estates, parklands, schools, recreation facilities, natural landscape, waste dump site, industrial estate, agriculture, etc.
3. Determination of land quality requirements for land use schemes and plans.
4. Assessment of the required 4R technologies — remediation, reclamation, restoration, and rehabilitation requirements to meet the required land use schemes or plans.
5. Assessment of quality of rehabilitated lands.
6. Matching land quality and land suitability with community or regulatory land use plans and requirements.

In the final analysis, land suitability is dependent on the quality of the rehabilitated land. Rehabilitated land use will be dependent not only on the rehabilitated land suitability, but also on regulatory or community land use plans. For rehabilitated land use that does not return the land to initial conditions, as for example reclamation of land for

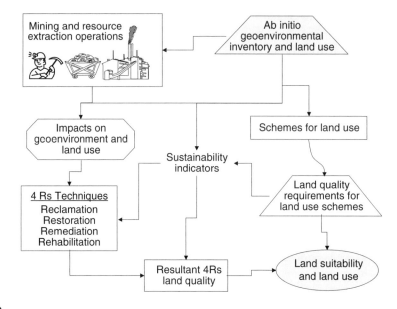

FIGURE 5.9
Basic steps of rehabilitation of lands affected by mining extraction operations. Implementation of schemes for land use are dependent on the restored land quality.

housing developments, it is important to prescribe sustainability indicators and requirements according to the new land use role.

5.4.2 Acid Mine Drainage Problem

The phenomenon of acid generation from heaps with trace amounts of sulfide minerals and from exposed sulfide minerals in mined-out caverns and pits, etc., has been discussed in Section 5.2. The magnitude of the sets of cascading problems attributed to the actions of acid leachate cannot be overstated. Cleanup of all the affected areas, sites, waterways, etc., is estimated in the billions of dollars in North America. The threat expressed is not only in terms of acid leachates finding their way onto the land environment and into receiving waters, but also in terms of release of trace metals into the geoenvironment. The leachates and released metals will negatively affect the functionality of the various ecosystems that comprise the geoenvironment — causing them to be eventually unsustainable. In terms of sustainable land use, the obvious protection against such agents of unsustainability is to remove conditions and circumstances favorable for generation of acid leachates — on the assumption that acid generation has not occurred. This means denying access to oxygen and water. The principle of *keep it dry* is good to practice at mine sites. This is an easy statement to make, but in reality is a very difficult and an almost impossible principle to adhere to. Since water is an essential element in mining extraction processes, keeping it dry requires one to provide protective covers, isolation barriers, and pump discharge operations. Implementation of these techniques is dependent on site-specific and operation-specific conditions. Adherence to the "keep it dry" principle is one of the basic requirements for minimization of impacts to land use, and to the 4Rs technique (Figure 5.9).

For conditions where acid generation has occurred and acid leachate has found its way to the land environment and its receiving waters, the two courses of required action are (1) protection of the affected land receptors and water bodies from accepting further acid leachates and (2) treatment of the affected land and water bodies. It has been suggested that once pollution of the receiving waters such as those described in Section 5.2 has occurred, destruction of aquatic habitat will render these waters to be bereft of aquatic life for a very long time. Nevertheless, treatment of these waters is necessary. Methods for treating polluted water have been discussed in Chapter 3.

Prevention of further reception of acid leachate requires mitigation strategies to be implemented at the mine site. This also applies to historic abandoned mine sites. The essential element here is to capture the leachate for treatment —an expense that will be ongoing for an interminable period since the source of acid generation will most likely be almost inexhaustible. Considerable work is presently being undertaken by the mining industries to apply mitigation strategies. These include (1) diversion of waters around the mine sites as part of the "keep it dry" (keep it as dry as possible?) strategy, (2) channeling the generated leachates through constructed aerobic and anaerobic wetlands as a neutralization procedure, (3) complete inundation of mined-out sites — to deny access to oxygen, and (4) capturing and channeling the generated leachates for active treatment before discharge.

5.4.3 Slurry Tailings Ponds

Slurry tailings ponds are by far the major type of containment facilities for slurry tailings. Their use has been discussed in the previous section. Their presence in the landscape degrades land quality and considerably reduces land use capabilities. Such kinds of ponds are shown in Figure 5.10 — for tin mining slurry ponds in Southeast Asia obtained as a result of alluvial tin mining hydraulic operations using the wash separation technique. Sand and debris are collected at the end of the sluice box, with the ponds serving as sedimentation facilities (Chow, 1998). It is not unusual for these ponds to develop a crust overlying a slime layer with solids concentration ranging from 50 to 60%. Encroachment of housing estates onto such kinds of ponds and debris, as seen, for example, in Figure 5.10, will pose limits on housing and introduce safety hazards. Reclamation of the ponds is necessary to allow for utilization of the reclaimed land for further urban development and other land uses consistent with an urban ecosystem.

Sustainable land use, in the context of slurry tailings ponds and their like, is not different in principle from the acid mine problem or mined-out caverns and pits. We will consider sustainability of land use in the physical landscape sense, i.e., in respect to the physical features and properties of the land and its utility. Figure 5.11 gives a summary view of the various options available for management of slurry tailings ponds. The first option is to leave the ponds in their disposal state. This is not a generally acceptable option, but nevertheless must be considered, in the event of negligent abandonment of such ponds. Health and safety concerns require implementation of (1) monitoring procedures for these ponds to be implemented and (2) health and safety measures to protect human and wildlife populations. Figure 5.11 also shows that three distinct categories of sustainable land use options for the slurry tailings ponds are available: (1) return of land to *ab initio* physical condition, (2) reclamation of the ponds in the *keep it wet* state, and (3) reclamation of the ponds in the *take it to dry* state. Returning the land to its *ab initio* landscape condition fulfills the landscape portion of sustainability requirements.

Two basic options are available for the "keep it wet" state: (1) freshwater pond and (2) wetlands. In the "take it to dry" category, the options available depend to a large extent on (1) quality of dry land obtained from the reclamation process and (2) regulatory and community requirements. The basic element of all the schemes for reclamation of slurry tailings ponds and other types of containment facilities must deal with the question of

Tin mining ponds ——————   Rejects, sand and ——————
                                            alluvial debris

FIGURE 5.10
(See color insert following page 142.) Tin mining slurry ponds obtained as a result of hydraulic dredging and pumping. Reclamation of pond is necessary to accommodate expansion of the housing units seen at the left of the picture (and left of the canal).

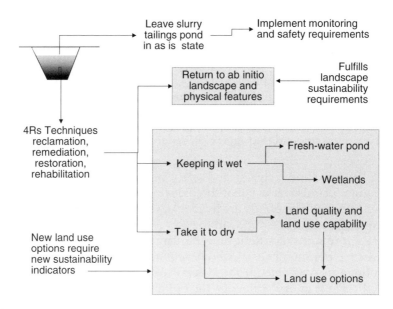

FIGURE 5.11
Sustainable land use in the context of slurry tailings pond. Options range from nonsustainable to enhanced land use and new sustainability scenarios.

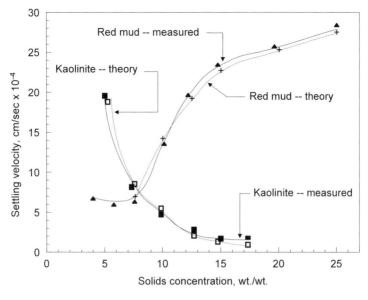

FIGURE 5.12
Calculated and measured settling velocities for a kaolinite soil suspension and bauxite red mud.

what to do with the stagnant layer. Liquid–solids separation and treatment of the released water are basic requirements for any of the pond reclamation options. Physical methods for liquid–solids separation include surcharging the top of the stagnant layer to achieve consolidation — a geotechnical process that provides compression of the solids skeletal matrix through the applied surcharge and drainage of the water in the skeletal structure. Other physical methods include removal of the stagnant layer for treatment and filling of the emptied pond with new fill material.

Chemical and physicochemical methods for increasing the sedimentation rate of the suspended solids in the stagnant layer include the use of polyacrylamides and polyelectrolytes. The basic intent of these kinds of flocculants is to overcome the domination of interparticle forces typical of colloidal interaction. Various kinds of flocculants and flocculating agents have been developed for such types of slimes and sludges (Yong and Sethi, 1982, 1983, 1989). They all have the aim of promoting aggregation of the particles into flocs, thus increasing the mass of individual groups of particles, and hence allowing for gravitational forces to dominate and sedimentation to occur. Calculations performed by Yong and Wagh (1985) using the two-particle collision theory to study the stability of the suspended solids in the stagnant zone have shown the effect of aggregation on the settling velocities of the suspended solids. Confirmation of their calculations has been obtained from experiments on a pure clay mineral suspension (kaolinite) and the red mud discharge from bauxite processing (Figure 5.12). Aggregation of the red mud particles with increasing solids concentration caused the increase in settling velocity. For the kaolinite soil suspension, increasing solids concentration in the soil suspension served to decrease the settling velocity — probably due to the hindrance effect posed by the proximal particles.

Another method for increasing the settling velocity of the suspended particles is to increase the zeta potential, ξ, of the particles. This is the potential that represents the charge at the shear layer between the suspended particle surface and the suspending fluid. We recall from Equation 5.1 that the potential ψ provides us with a means for determining the electric charge distribution as a function from the particle surface. This potential has two basic components: (1) the potential at the surface of the particle, represented by ψ_o, and (2) the potential ψ_s at the Stern layer boundary (double-layer boundary), where the shear action

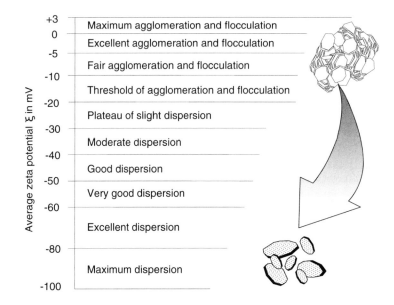

FIGURE 5.13
Relationship between zeta potential, ξ, soil microstructure (flocs and aggregations of particles), and dispersion stability for clay soils.

with the liquid medium occurs. This potential, which is commonly identified as the zeta potential, ξ, is a function of the nature of the surface charge possessed by the suspended particle, the ions in the suspending fluid, and the ions in the double layer. By changing the zeta potential, aggregation of the suspended particles in the stagnant zone can be obtained — with resultant increased settling velocities. Figure 5.13 shows the relationship between zeta potential, ξ, and the dispersion stability for clay soils reported by Yong (2001b).

5.5 Remediation and Treatment of Mining Sites and Wastes

Mining sites are characterized by piles of waste rock, huge pits, and acid mine drainage. Decommissioning procedures often must include remediation strategies — a requirement that had not been imposed in the past. Remediation through decontamination, isolation, or immobilization may be required for contaminated areas. For open pits, either the pit can be refilled for new land use options, or it can be harmonized with the surroundings. Before open pit filling, consultation is needed to ensure that the geological heritage is not destroyed. Maintaining formations is essential for research and teaching purposes or education of the public. Waste storage in these pits used to be a common practice. However, because the base of the quarries is often relatively permeable, groundwater contamination from waste leachates can result.

   If the pit is left for recreation or as a monument, slope stability must be ensured. Flooding of the pit can be quite beneficial. Waring and Taylor (1999) have proposed this as a method that is low cost and passive for preventing acid mine drainage by eliminating exposure of the sulfide minerals to atmosphere, and hence denying oxidation of these minerals. The reservoirs can be used for fishing or wildlife or recreation or restoring biodiversity. Vegetation may also be introduced. Mines could also be sealed against seepage of rainwater

TABLE 5.4

Reuses of Mining Wastes

Type of Residues	Application
Coal	Hard rocks for road and rail ballast
	Clay stone and mark for brick components and cement
	Lignite for soil conditioners, humus substitutes
	Clay stone for aggregates
Slates	For filling cavities at quarries
Phosphates	Additives for light aggregates, pipes, bricks
Kaolin	Filling up cavities, brick, tile, concrete, or artificial pozzolana making
Iron ores	Aggregates, building or roofing material
Copper ores	Road ballast, bitumen containers
Gold and uranium ores	Road building materials, additives to concrete or brick components
Aluminum	Thermal insulators, porosity testers, concrete additives, acids, soil conditions, pigments

Source: Adapted from Chamley, H., *Geosciences, Environment and Man*, Elsevier, Amsterdam, 2003.

and snow. Drainage systems divert the water away from the tailings, and thus reduce acid mine drainage problems.

5.5.1 Spoil or Tailing Heaps

Remediation of spoil heaps is also possible in different ways. The heaps can be potentially used for recreation, ecological purposes, or flora and fauna. Revegetation enables reintegration of the heap with the surrounding environment. A thorough site investigation will be required to optimize this process. According to Genske (2003), revegetation reduces erosion and dust and leachate production, while biodiversity is increased. Acidity, heavy metal content, and lack of nutrients and organic matter can decrease the ability to revegetate the heap. The permeability of the heap may be too permeable or compact. Treatments can include lime addition for pH adjustment, fertilizer addition to increase nutrients, or top soil addition for organic matter. Combustion may also be a hazard for coal heaps. The vegetation process may start slowly, but once it is initiated, it can develop quickly. Various applications are being developed for various mining wastes. Some of these are listed in Table 5.4.

5.5.1.1 Biohydrometallurgical Processes

Bacterial leaching of metals from mining ores, also bioleaching, is a full-scale process that can be performed by slurry reactors or heap leaching (Figure 5.14). Mining wastes include low-grade ores, mine tailings, and sediments from lagoons or abandoned sites. Low pH values lead to solubilization of the metals in the mining ores. Elemental sulfur or ferrous iron may be added as bacterial substrates. Reactors such as Pachuca tanks, rollings reactors, or propeller vessels have been used (Tyagi et al., 1991). Heap leaching is more common because it allows the large volume of wastes to be treated in place (Boon, 2000). To enhance this process, aeration can be forced through the pile. Alternatively, hydrophilic sulfur compounds can be added (Tichý, 2000). *Thiobacilli* bacteria are responsible for the oxidation of inorganic sulfur compounds. Applications include metal dissolution of low-grade sulfide ores, generation of acidic ferric sulfate leachates for hydrometallurgical purposes, and removal of gold by oxidation of pyrite by bacterial sulfide production. The extraction of metals from low-grade metals ores and refractory gold ores is a multibillion-dollar business worldwide (Rawlings, 1997). Bacterial solubilization by oxidation of the sulfide min-

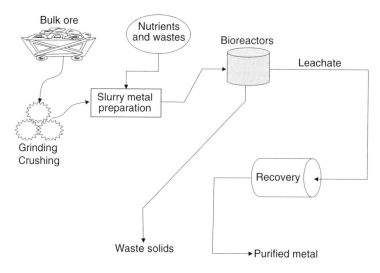

FIGURE 5.14
Flow sheet of a biometallurgical process.

erals, pyrite, and arsenopyrite enhances gold extraction by the traditional method of cyanidation. The solubilization mechanisms have been debated extensively, however.

Biohydrometallurgical processes are highly efficient and cause less environmental problems than chemical methods (Torma and Bosecker, 1982). For slurry processes, oxidation rate per reactor volume, pH, temperature, particle size, bacterial strain, slurry density, and ferric and ferrous iron concentrations need to be optimized. Bioleaching is very effective for recovery of gold from refractory gold pyrite and copper from chalcopyrite. The feasibility of the recovery of metals from a mining residue was shown by *Aspergillus niger*, which exhibited good potential in generating a variety of organic acids effective for metal solubilization (Mulligan and Kamali, 2003). Organic acid extraction effectiveness was enhanced when sulfuric acid was added to the medium. Different wastes such as potato peels, leaves, corn husks, and wood shavings were tested. In addition, different auxiliary processes were evaluated in order to either elevate the efficiency or reduce costs. Finally, maximum solubilizations of 68, 46, and 34% were achieved for copper, zinc, and nickel, respectively (Figure 5.15). Also, iron co-dissolution was minimized with only 7% removal.

Rhamnolipid biosurfactants (biodegradable and of low toxicity) have also been added to mining oxide ores, to enhance metal extraction (Dahr Azma and Mulligan, 2004). Batch tests were performed at room temperature. Using a 2% rhamnolipid concentration, 28% of the copper was extracted. The addition of 1% NaOH with the rhamnolipid enhanced the removal up to 42% at a concentration of 2% rhamnolipid, but decreased at higher surfactant concentrations. Sequential extraction studies were also performed to characterize the mining ore and to determine the types of metals being extracted by the biosurfactants. Approximately 70% of the copper was associated with the oxide fraction, 10% with the carbonate, 5% with the organic matter, and 10% with the residual fraction. After washing with 2% biosurfactant (pH 6) over a period of 6 days, it was determined that 50% of the carbonate fraction and 40% of the oxide fraction were removed by the biosurfactant.

Pintail Systems, Inc. (Aurora, CO) has developed a biological process (U.S. EPA, 1999) for the removal of cyanide in a heap leach process. Native microorganisms were extracted from the ore and tested for cyanide detoxification potential. Those identified were kept for bioaugmentation purposes. Pilot tests were performed to simulate pile conditions and determine detoxification rates, process parameters, and effluent characteristics. Once the test was complete, sufficient quantities of bacteria were applied to the heap. Metals were

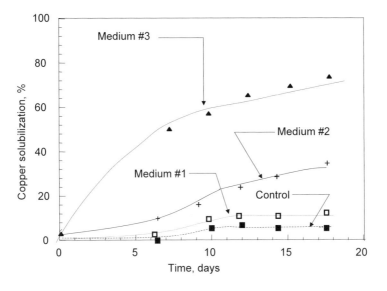

FIGURE 5.15
Effect of dilute acid pretreatment on leaves and sawdust on copper removal by *A. niger*. The control is dilute acid, with medium 1 (40 g/l potato peels and dilute acid), medium 2 (40 g/l sawdust and dilute acid), and medium 3 (40 g/l leaves and dilute acid). (Adapted from Mulligan, C.N. and Kamali, M., *J. Chem. Technol. Biotechnol.*, 78, 497–503, 2003.)

biomineralized during the leaching process. This process can be used for spent ore heaps, waste rock dumps, mine tailings, and process water from silver and gold mining operations. The technology has been evaluated in the SITE Demonstration Program, and two full-scale cyanide detoxification projects were completed.

An anaerobic treatment process was developed by Geo-Microbial Technologies (Ochelata, OK). The technology is called anaerobic metals release (AMR). Instead of using aerobic acidophilic bacteria that form acids and solubilize metal sulfides that can contaminate streams and lakes, the AMR technology uses *Thiobacillus* with a denitrifying culture at neutral pH. The anaerobic conditions were controlled by the nitrate levels in the leaching solutions, which allow solubilization of the metals. All nitrates were consumed in the process. The metals are removed from the leachate by standard methods and the effluent is recycled. Levels of sulfide-reducing bacteria and sulfides are kept to a minimum. The technology was demonstrated in a 1994 SITE Emerging Technology Program.

5.5.2 Acid Mine Drainage

Acid mine drainage (AMD) can also be managed. Active treatments that use wastewater treatment techniques are possible. They are expensive and must be maintained regularly, such as chemical or electrolytic treatments. Prevention of AMD is possible by addition of limestone or by sealing the area with fly ash grouts. For example, Bulusu et al. (2005) used a grout of coal combustion by-product to reduce AMD. The grout was durable and did not exhibit signs of weathering. Alkaline agents (calcium oxide, calcium hydroxide) and wastes (cement kiln dust, acetylene gas sludge) can be added to pools of AMD for neutralization. Alkaline trenches with limestone or soda ash are used to neutralize runoff. Bacteriocides have to be used to inhibit bacterial growth and hence pyrite oxidation. Natural treatment through carbonate formations may also be possible, as in the case of Cretaceous chalk underlying the coalfield slag heaps in the north of France (Chamley, 2003). Permeable reactive barriers may also be a solution. Constructed wetlands are a low-cost alternative for passive treatment, as described in the next section. Anoxic limestone drains are being evaluated for use because of their ease in maintenance and operation. However, they require large areas for effective application.

There is also a significant capacity for natural attenuation at mining sites. A particular example of this was the Falun Copper Mine in Sweden. During its operation of more than 450 years, it was estimated that a 0.5 to 1 megatonne of copper, lead, zinc, and cadmium was emitted into forest soils and streams in the area (Lindeström, 2003). Concentrations of 70 and 2000 µg/l were found in streams in the city of Falun after treatment of the mine water was initiated, but decreased thereafter. Soils in the area, however, were able to recover substantially faster. Most of the metals concentrated in the sediments in two lakes. Copper was 120 to 130 times normal background levels, while lead and zinc levels were 30 to 40 times. Cadmium accumulation was much less. The aquatic ecosystem was able to return due to the low bioavailability of the metals and possible interactions between the metals.

5.5.2.1   *Wetlands*

Natural wetlands as discussed in Chapter 3 are areas of land with the water level close to the land surface, thus maintaining saturated soil conditions and vegetation that includes plants, peat, wildlife, microbial cultures, cattails (*Typha* spp.), reeds (*Phragmites* spp.), sedges (*Carex* spp.), bulrushes (*Scirpus* spp.), rushes (*Juncus* spp.), water hyacinthe (*Eichhornia crassipes*), duckweeds (*Lemna* spp.), grasses, and others (Mulligan, 2002). Algae and mosses, together with the wet areas, can trap the heavy metals. Constructed wetlands have been specifically designed to include these species for the removal of biological oxyten demand (BOD), suspended solids, nutrients, and heavy metals for optimal performance. It has been reported by Reed et al. (1995) that 1000 managed wetlands are in operation throughout the world. In 1988, 142 North American wetland systems were used for acid mine drainage (Wieder, 1989).

Iron and manganese removal are often the key objectives in treatment of mine drainage. Because of possible clogging of subsurface systems due to precipitation of iron and manganese in subsurface systems, preference is usually given to surface systems, since they can be aerated more efficiently. Since the pH typically decreases from 6 to 3 in acid mine drainage phenomena, the Tennessee Valley Authority (TVA) has developed an anoxic limestone drain (ALD) for use as a treatment tool. This consists of a high-calcium limestone aggregate (20 to 40 mm) placed in a trench of 3 to 5 m wide and 0.6 m to 1.5 m depth (Brodie et al., 1993). The anoxic conditions in the trench are ensured by backfilling with clay. A plastic geotextile is placed between the clay and limestone. The inlet of the trench is placed at the source of the acid mine drainage. However, when the oxygen in the drainage is greater than 2 mg/l or the pH is greater than 6 and the redox potential is greater than 100 mV, use of the ALD is detrimental due to the formation of oxide coatings. Installation of a sedimentation pond before the wetland treatment with or without ALD is preferred since it is easier to remove iron precipitation from the pond than the wetlands. Sanders et al. (1999) have indicated that wetlands systems were used to remove Zn, Fe, Cu, Pb, and some other heavy metals from a moderate to severe acidic drainage from a mining complex in Montana. This treatment will be required for decades. Long-term monitoring will be needed, as shown in Table 5.5.

5.5.2.2   *Biosorption*

Biosorption is potentially an attractive technology for treatment of water containing dilute concentrations of heavy metals. Activated carbon is the currently recognized adsorbent for removal of heavy metals from wastewater. However, the high cost of activated carbon

TABLE 5.5

Monitoring Requirements for an AMD Remediation Process

Category	Parameter
Physicochemical	pH
	Redox potential
	Total dissolved solids
	Specific conductance
	Dissolved oxygen
Cationic and anionic species	Fe, Cu, Pb, Zn, Cd, Hg, As, SO_4^{2-}
Gases	O_2, CO_2, SO_2, H_2S
Flow	AMD flow rate, hydrostatic pressure
Meteorological conditions	Precipitation, temperature, sunlight, wind speed

Source: Adapted from Fytas, K. and Hadjigeorgiou, J., *Environ. Geol.*, 25, 36–42, 1995.

limits its use in adsorption. A search for a low-cost and easily available and renewable adsorbent has led to the investigation of wastes of agricultural and biological origin as potential metal sorbents (Hammaini et al., 1999). Biosorption is the ability of certain types of microbial biomass to accumulate heavy metals from aqueous solutions by mainly ion exchange mechanisms. A large number of microorganisms belonging to various groups, such as bacteria, fungi, yeasts, and algae, have been reported to bind a variety of heavy metals to different extents (Volesky and Holan, 1995).

The main requirement of an industrial sorption system is that the sorbent can be utilized as a fixed or expanded bed for use in a continuous process. Immobilization techniques have been developed, but the employment of immobilization procedures is expensive and complex (Liu et al., 2003). Two attempts to market two different types of immobilized microbial biomass, one by BV SORBEX and the other by the U.S. Bureau of Mines, were not commercially successful applications (Tsezos, 2001). The feasibility of anaerobic granules for industrial wastewater reactors was investigated as a novel type of biosorbent for the removal of cadmium, copper, nickel, and lead from aqueous solution by Al Hawari and Mulligan (2006). Results showed that a living biomass has a higher sorption capacity than a dried biomass, but due to the difficulties in maintaining the biomass, the dried biomass would be more suitable for industrial applications (Figure 5.16). Unlike most forms of biomass, immobilization or stiffening was not necessary prior to using the biomaterial. Anaerobic granules possess compact porous structures, excellent settling ability, and high mechanical strength. Even under aggressive chemical environments (acidic or basic conditions), the biomass demonstrated good stability with no visible structural damage — making this biomass more advantageous over other biosorbents.

5.6 Concluding Remarks

Harvesting of nonrenewable natural resources brings with it at least three major areas of geoenvironmental impacts: (1) depletion of the nonrenewable resource, (2) mining excavations on the surface and underground, and (3) discharges from processes associated with extraction of the natural resource. The biological components of the terrestrial ecosystem are not covered in the discussions in this book. We have focused specifically on the land use aspect of the geoenvironment in mining operations as an upstream activity. The use of the raw materials obtained from such mining operations by downstream

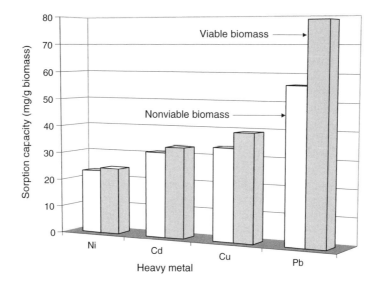

FIGURE 5.16
Comparison of a viable and a nonviable anaerobic granulated biomass for the biosorption of heavy metals.
(Adapted from Al Hawari, A. and Mulligan, C.N., *Bioresource Technol.*, 2006.)

industries is covered in Chapter 7. From the viewpoint of sustainability, we accept that depletion of the nonrenewable resource will not qualify for inclusion in any category of sustainable resources. This leaves us with the mining excavations and the discharges from extraction processes.

Underground mining excavations pose problems in two categories: (1) the physical aspect of excavations as empty chambers, tunnels, etc., together with the heaping of debris and other spoils as heaps and tips, and (2) the chemical problem represented by acid generation and the resultant acid mine drainage problem, because of the interaction of the exposed sulfide minerals in the mined-out areas to oxygen and water. If sustainable land use requires one to return the land to its *ab initio* condition, this would mean filling the empty mined-out chambers and excavations. This would, or should, avoid or minimize the problem of acid generation. Since practical and economic considerations have so far militated against this course of action, attention has been directed toward mitigating acid mine and acid rock drainage problems.

Measures undertaken by industry to better protect against acid generation in present mining and extraction operations are continually being improved. However, for historic and abandoned mine sites, the problem of acid mine drainage remains. The direct impact on land use from acid generation problems is not only in the immediate sense of acid interactions with the land environment, but also in the ripple or cascading effect. Pollution of groundwater and receiving waters has severe consequences to the native habitat and plant life and other receptors.

In regard to liquid discharges and containment facilities, sustainable land use is not different in principle from the acid mine problem or mined-out caverns and pits. Options for reclamation of the slurry tailings ponds and containment facilities have been summarized in Figure 5.11. Much depends on both regulatory and industry requirements.

Closure of mining sites requires restoration so that they may be used for other purposes and to prevent risk to the environment and humans. The reuse of mining residues, stabilization of mine areas, and neutralization of pollutants are some of the challenges. Numerous techniques are being investigated, such as replanting, wetland treatment, and biological treatments. As a more sustainable approach, waste products must be recycled as much as possible and should also be integrated into the treatment processes. An integrated approach for land, solids, and leachate management is highly desirable.

References

Al Hawari, A. and Mulligan, C.N., (2006), Biosorption of cadmium, copper, lead, and nickel by anaerobic granular biomass, *Bioresource Technol.*, 97(4):692–700.

Azcue, J.M. and Nriagu, J.O., (1995), Impact of abandoned mine tailings on the arsenic concentrations in Moira Lake, Ontario, *J. Geochem. Explor.*, 52:81–89.

Barr, J., (1969), *Derelict Britain*, Penguin, London, 240 pp.

Boon, M., (2000), Bioleaching of sulfide minerals, in *Environmental Technologies to Treat Sulfur Pollution: Principles and Engineering*, Lens, P. and Pol, L.H. (Eds.), IWA Publishing, London, pp. 105–130.

Brodie, G.A., Britt, C.R., Toamazewski, T.M., and Taylor, H.N., (1993), Anoxic drains to enhance performance of aerobic acid drainage treatment wetlands: experiences of the Tennessee Valley Authority, in *Constructed Wetlands for Water Quality Improvement*, Lewis Publishers, Chelsea, MI, pp. 129–138.

Bulusu, S., Aydilek, A.H., Petzrick, P., and Guynn, R., (2005), Remediation of abandoned mines using coal combustion by-products, *J. Geotechnol. Geoenviron. Eng.*, 131:958–969.

Chamley, H., (2003), *Geosciences, Environment and Man*, Elsevier, Amsterdam, 450 pp.

Chow, W.S., (1998), Studies of Slurry Slime in Mined-Out Ponds, Kinta Valley, Peninsular Malaysia, for Purposes of Reclamation, Ph.D. thesis, Department of Geology, University of Malaysia, Malaysia.

Dahr Azma, B. and Mulligan, C.N., (2004), Extraction of copper from mining residues by rhamnolipids, *Pract. Periodical Hazardous Toxic Radioactive Waste Manage.*, 8:166–172.

Davies, C.S., (1999), Derelict land, in *Encyclopedia of Environmental Science*, Alexander, D.E. and Fairbridge, R.W. (Eds.), Kluwer Academic Publishers, Dordrecht, pp. 120–123.

Donahue, R. and Hendry, M.J., (2003), Geochemistry of arsenic in uranium mine mill tailings, Saskatchewan, Canada, *Appl. Geochem.*, 18:1733–1750.

Fytas, K. and Hadjigeorgiou, J., (1995), An assessment of acid rock drainage continuous monitoring technology, *Environ. Geol.*, 25:36–42.

Genske, D.D., (2003), *Urban Land Degradation, Investigation and Remediation*, Springer, Berlin, 331 pp.

Hammaini, A., Ballester, A., Gonzalez, F., Blazquez, M.L., and Munoz, J.A., (1999), Activated sludge as biosorbent of heavy metals, Biohydrometallurgy and the environment toward the mining of the 21st century, in *International Biohydrometallurgy Symposium IBS'99*, Madrid, Spain, pp. 185–192.

IMCC, (1992), *Inactive and Abandoned Non-Coal Mines: A Scoping Study*, Cooperative Agreement X-817900-01-0, prepared for IMCC of Herndon, VA, by Resource Management Associates, Clancy, MT, July.

Kinniburg, D.G., Smedley, P.L., Davies, J., Milne, C., Gaus, I., Trafford, J.M., Burden, S., Huq, S., Ahmad, N., and Ahmed, K.M., (2003), The scale and causes of the groundwater arsenic problem in Bangladesh, in *Arsenic in Groundwater: Occurrence and Geochemistry*, Welch, A.H. and Stollenwerk, K.G. (Eds.), Kluwer, Boston, pp. 211–257.

Kobayashi, J. and Hagino, N., (1965), *Strange Osteomalacia by Pollution from Cadmium Mining*, Progress Report WP 00359, Okayama University, Japan, pp. 10–24.

Kruyt, H.R. (Ed.), (1952), *Colloid Science: Irreversible Systems*, Elsevier Publishing Co., Amsterdam, 389 pp.

Lindeström, L., (2003), *The Environmental History of the Falun Mine*, Stiftelsen Stora Kopparberget & ÅF-Moljöforskargruppen AB, 110 pp.

Liu, Y., Yang, S., Xu, H., Woon, K., Lin, Y., and Tay, J., (2003), Biosorption kinetics of cadmium (II) in aerobic granular sludge, *Process Biochem.*, 38:997–1002.

McCreadie, H., Blowes, D.W., Ptacek, C.J., and Jambor, J.L., (2000), Influence of reduction reactions and solid-phase composition on porewater concentrations of arsenic, *Environ. Sci. Technol.*, 34:3159–3166.

Moldovan, B., Hendry, M.J., and Jiang, D.T., (2003), Geochemical and mineralogical controls on arsenic release from uranium mine tailings, *Geophys. Res. Abs.*, 5:12538.

Mulligan, C.N., (2002), *Environmental Biotreatment*, Government Institutes, Rockville, MD, 395 pp.

Mulligan, C.N. and Kamali, M., (2003), Bioleaching of copper and other metals from low-grade oxidized mining ores by *A. niger*, *J. Chem. Technol. Biotechnol.*, 78:497–503.

Ollson, C.C., (1999), Arsenic Contamination of the Terrestrial and Freshwater Environment Impacted by Gold Mining Operations, Yellowknife, NWT, M.Eng. thesis, Royal Military College of Canada, Kingston.

Rawlings, E. (Ed.), (1997), *Biomining: Theory, Microbes and Industrial Processes*, R.G. Landes and Springer-Verlag, Heidelberg, 302 pp.

Reed, S.C., Crites, R.C., and Middlebrooks, E.J., (1995), *Natural Systems for Waste Management and Treatment*, 2nd ed., McGraw-Hill, New York, 448 pp.

Sanders, F., Rahe, J., Pastor, D., and Anderson, R., (1999), Wetlands treat mine runoff, *Civil Eng.*, 69:53–55.

Simons, F.S. and Prinz, W.C., (1973), Copper, in *United States Mineral Resources*, Geological Survey Professional Paper 820, U.S. Department of the Interior, Geological Survey.

Tichý, R., (2000), Treatment of solid materials containing inorganic sulfur compounds, in *Environmental Technologies to Treat Sulfur Pollution: Principles and Engineering*, Lens, P. and Pol, L.H. (Eds.), IWA Publishing, London, pp. 329–354.

Torma, A.E. and Bosecker, K., (1982), Bacterial leaching, *Prog. Ind. Microbiol.*, 16:77–118.

Tsezos, M., (2001), Biosorption of metals. The experience accumulated and the outlook for technology development, *Hydrometallurgy*, 59:241–243.

Tyagi, R., Couillard, D., and Tran, F.T., (1991), Comparative study of bacterial leaching of metal from sewage sludge in continuous stirred tank and air-lift reactors, *Process Biochem.*, 26:47–54.

U.S. EPA, (1999), *Superfund Technology Evaluation Program, Technology Profiles*, 10th ed., Vol. 1, *Demonstration Program*, EPA/540/R-99/500a, U.S. EPA, February.

Volesky, B. and Holan, Z.R., (1995), Biosorption of heavy metals, *Biotechnol. Prog.*, 11:235–250.

Wang, S. and Mulligan, C.N., (2004), Arsenic in Canada, in *Proceedings of the 57th Canadian Geotechnical Conference and 5th Joint IAH-CNS/CGS Conference*, Session 1D, Environmental Geotechnology I, Quebec City, Canada, October, pp. 1–8.

Waring, C. and Taylor, J., (1999), A new passive technique proposed for the prevention of acid drainage: GaRDS, in *Mining in the Next Century: Environmental Opportunities and Challenges*, Minerals Council of Australia, Townsville, pp. 527–530.

Wieder, R.K., (1989), A survey of constructed wetlands for acid coal mine drainage treatment in the eastern United States, *Wetlands*, 9:299–315.

Yong, R.N., (1983a), Method for Dewatering the Sludge Layer of an Industrial Process Tailings Pond, U.S. Patent 4,399,038.

Yong, R.N., (1983b), Treatment of Tailings Pond Sludge, U.S. Patent 4,399,039.

Yong, R.N., (1984), Particle interaction and stability of suspended solids, in *Sedimentation Consolidation Models: Predictions and Validations*, Yong, R.N. and Townsend, F.C. (Eds.), American Society of Civil Engineers, New York, pp. 30–59.

Yong, R.N., (2001a), *Geoenvironmental Engineering: Contaminated Soils, Pollutant Fate, and Mitigation*, CRC Press, Boca Raton, FL, 307 pp.

Yong, R.N., (2001b), Interaction in clays in relation to geoenvironmental engineering, in *Clay Science for Engineering*, Adachi, K. and Fukue, M. (Eds.), Balkema, Rotterdam, pp. 13–28.

Yong, R.N. and Sethi, A.J., (1982), Destabilization of Sludge with Hydrolyzed Starch Flocculants, U.S. Patent 4,330,409.

Yong, R.N. and Sethi, A.J., (1983), Decarbonation of Tailings Sludge to Improve Settling, U.S. Patent 4,414,117.

Yong, R.N. and Sethi, A.J., (1989), Methylated Starch Compositions and Their Use as Flocculating Agents for Mineral Wastes, Such as Bauxite Residues, U.S. Patent 4,839,060.

Yong, R.N. and Wagh, A.S., (1985), Dispersion stability of suspended solids in an aqueous medium, in *Flocculation, Sedimentation and Consolidation*, Moudgil, B.M. and Somasundaran, P. (Eds.), National Science Foundation, New York, pp. 307–326.

Yong, R.N. and Warkentin, B.P., (1975), *Soil Properties and Behaviour*, Elsevier Scientific Publishing Co., Amsterdam, 449 pp.

6

Food Production and the Geoenvironment

6.1 Introduction

We consider the primary sets of activities in food production to be agricultural based, and the industry identified with these sets of activities is the *agro-industry*. Agriculture (food production) is a basic activity for human beings. It is an essential component in the life support system of the human population. The importance of agriculture was highlighted by the Johannesburg World Summit on Sustainable Development (WSSD) in 2002 as one of the four components within sustenance and development, i.e., industrialization, urbanization, resource exploitation, and agriculture. To obtain a sustainable society and yet afford development that is compatible with the goals of sustainability, the activities and patterns associated with food production must be managed properly.

Agro-ecosystems are natural ecosystems that are manipulated by humans to produce food, fibers, and other products. The land is cleared, crops are planted, the soil is tilled, water is added (by irrigation, if necessary), nutrients are supplied, and weed and pests are controlled by various mechanical and chemical aids. Soil is a natural resource material and is considered a geoenvironment natural capital. Agricultural soil has a balance of inorganic and organic components. Both climate and the environment control the value of the soil. Factors such as water content, soil type (composition), soil thickness, salt content, and other physical, chemical, and biological properties are important determinants of soil quality. In regions where winters and cold temperatures are factors, the presence or absence of permafrost will also contribute to soil value. The combination of physical, chemical, and biological agro-aids, climate, soil management, and technology has rendered the soils in Europe and North and Central America favorable for agriculture. In contrast, the lack of many of the agro-aids, inadequate technology, poor soil management, and unfavorable climate have combined to produce nutrient-depleted soils in regions of Africa and Asia.

Hunger and nutrient deficiencies are experienced daily by more than one billion people. Nutrient deficiency is defined as insufficient levels of food proteins and caloric energy. As can be seen in the famines in many parts of the world, nutrient deficiency is a major problem. The United Nations found that from 1994 to 1996, 19% of the population in developing countries, or 828 million people, were undernourished (FAO, 1998). The Rome Declaration on World Food Security called for a decrease of undernourished people from the present amount to 400 million by the year 2015. If one factor is the increase in population up until that time, then this would mean a rate reduction of undernourished people of more than 50% (WFS, 1996).

The demands for food are immense, and as we have discussed previously in Chapter 1, according to the Malthusian model, food availability is linked with population growth (Malthus, 1798). It has been estimated (Daily et al., 1998) that by 2050, food demand (1) could double in response to population growth, (2) would increase in relation to per capita income, and (3) would increase in response to measures to reverse the undernutrition of the poor. Increased urbanization has decreased available agricultural land. To increase yields, the use of fertilizer and pest control chemicals has increased. Irrigated areas have expanded, and high-yield crops have been developed. Some of these agricultural activities designed to enhance crop yields have led to decreases in soil fertility and increases in soil and water contamination. Subsequent depletion of the soil resource and the presence of these contaminants could lead to a decline in per capita food production according to Meadows et al. (1992), which could be a threat to human survival (Commoner, 1971). More than 75% of the arable land in North and Latin America, 25% in Europe, and 16% in Oceania can be considered damaged. This threatens future food supplies (Fischer Taschenbuch, 1996). North America's breadbasket (the Midwest of the United States to the Great Plains in Canada) is particularly at risk due to wind and soil erosion and fertility loss. Loss of native habitat in Canada due to farming has been significant. The Canadian Biodiversity Information Network (CBIN) reports that more than 85% of short-grass prairie, 80% of mixed-grass prairie, 85% of aspen parkland, and almost all the native tall-grass prairie have been lost (CBIN, 1998).

Food production must be increased without increasing the impact on the geoenvironment. Improper irrigation can lead to waterlogging and soil salinity problems. Use of pest control chemicals has increased pest resistance and destroyed natural species (NRC, 1991). The many aspects of agricultural engineering and soil management practices are subjects that are well studied in soil science. Their attention to efficient food and crop production, together with research into the various issues of soil management and soil quality, has the aim of providing agricultural productivity without compromising the objectives of sustainable agriculture.

The focus of this chapter is from a geoenvironment perspective. Its concern is in respect to the impacts on the geoenvironment resulting from agricultural activities required for the agricultural production of the food supply necessary for human sustenance. It is not a soil science focus on agricultural production of food. Instead, it will examine the likely geoenvironmental impacts due to food production activities such as pesticide use, nutrient addition, waste management, and irrigation.

6.2 Land Use for Food Production

The requirement for sufficient food to sustain life has led humans to require substantial use of grasslands, forests, and freshwater. Factors such as excess or lack of water, drainage, soil quality and thickness, salinity, annual mean temperatures, and, in cold regions, the freezing index and the presence or lack of a permafrost area determine the appropriateness of the soil for agriculture. Richards (1990) reports that there has been a fivefold increase in agricultural lands in the past 300 years. This needs to be balanced with the large quantities of land lost from production due to erosion, salinization from irrigation, desertification, and conversion to roads and urban uses. The capillary rise of salts (chlorides, sulfates, and carbonates) due to groundwater extraction and irrigation is threatening the U.S. High Plains, Canadian prairies, and Australian soils. This rate of loss has increased in the twentieth century to 70,000 km^2/year (Rosanov et al., 1990). With the exception of a few high-value crops, the market value of the land for nonagricultural purposes is much higher — a factor that appears to drive conversion of these lands to urban uses.

The other consideration that is significant is the increasing productivity of agriculture in the past century — a factor that has decreased famine rates significantly (Pinstrup-Anderson et al., 1997). Innovations including high-yield crop varieties, application of fertilizers and pest control, and utilization of mechanized equipment in both developed and developing countries have substantially increased agricultural productivity.

More than 50% of the land's surface is involved in one way or another with forestry, agriculture, or animal husbandry. Pastures alone make up 6 to 8% of the land. This does not include land for grazing. Agriculture, in combination with urban areas, occupies up to 10 to 15% of the land (Vitousek and Mooney, 1997). More than 13 million ha of forests were eliminated for agriculture and wood harvesting between 1980 and 1995 (FAO, 1997). This has substantial implications in global warming since forest soils and removed trees account for recycling of much larger amounts of carbon dioxide from the atmosphere than agricultural lands. Deforestation also leads to increased risk of soil and wind erosion because of exposure to the elements. Although some forests have been allowed to regrow upon depletion of agricultural lands, reforestation rates are much smaller than deforestation rates.

Ploughing disrupts (1) soil horizons, (2) chemical weathering, and (3) soil horizon interchanges. Ploughing compacts soil and decreases soil permeability. This inhibits evaporation processes and plant germination. Cattle and sheep herds also compact the soil. In arid climates, the soil is particularly vulnerable to erosion due to mechanization of ploughing. Overgrazing and slash-and-burn cultivation also reduce soil cohesion, leaving many areas affected by soil degradation and loss of vegetation. More than 35% of all degraded land is due to overgrazing (Quendler and Reichert, 2002). Arid and semiarid lands are particularly susceptible in Oceania and Africa.

From the geoenvironmental perspective, one could ask, How sustainable is agriculture? Although most agree that agriculture is not sustainable under present practices, there is considerable debate concerning the means for measuring sustainability (FAO, 1995, 1996). While full consensus is not available, it is agreed that the factors that need consideration are the use of genetically identical plants, irrigation water, fertilizers, and pesticides, and the assortment of wastes produced. Intensive agricultural practices utilizing high levels of technology and mechanization are generally not kind to the soil environment. A simple summary of the major inputs and outputs of agricultural activities is depicted in Figure 6.1. The impact of agricultural activities on the land environment will be discussed in more detail in the next section.

6.3 Impact of Food Production on WEHAB

6.3.1 Water Quantity and Quality

6.3.1.1 Water Use

The availability and quality of water are two important water factors that impact significantly on the sustainability of a growing population, particularly in developing countries. Freshwater is a precious resource because of the limited amounts directly available for use. More than half of runoff water is used by humans, and 70% of this is for agriculture (Postel et al., 1996). Rivers are diverted to serve the needs of humans. Only 2% of the rivers in the United States have not been manipulated. As an example of the impact of excessive use of water for agricultural purposes, it is reported that the water levels of the Aral Sea, located in Central Asia in the lowlands of Turan, have been substantially reduced

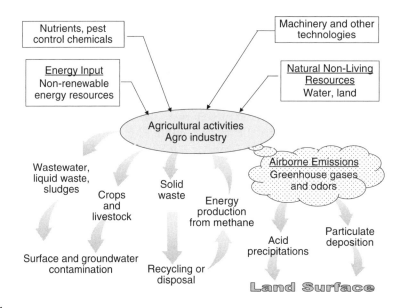

FIGURE 6.1
A summary of some of the major inputs and outputs related to agricultural activities.

by agricultural practices. The result of this has been (1) loss of native fish and biota, (2) creation of a source of windblown dust from the exposed salty sea bottom, (3) increase in the frequency of human diseases, and (4) creation of a drier local climate (Micklin, 1988).

The cost of energy influences one's ability to extract, pump, and irrigate abstracted groundwater. Increasing agricultural yields coupled with increasing land for agriculture result in corresponding increases in water demand. Currently, about 33% of the world's food is produced by irrigation (Postel, 1992), and 16% of all land is irrigated (WRI, 1992). Management for salinity and drainage is required to avoid decreases in agricultural yields. Falling water tables increase the costs of abstraction of groundwater — a factor that needs to be incorporated into management of irrigation.

Irrigation involves exploitation of rivers, aquifers, or other freshwater sources, causing a disruption of the natural hydrological cycle. If irrigation is poorly controlled, desiccation can occur between watering periods. This can lead to increasing rates of aridification and wind erosion. Overirrigation can deplete freshwater sources, and the soil can become salinated due to increased evaporation at the surface horizons. Substantial amounts of water are required for production of various crops, as shown in Figure 6.2. Corn, rice, and soybeans require substantially more water than wheat. Livestock requires 100 times more water than 1 kg of vegetable protein (Pimentel and Pimentel, 1996). Overall, about 1.3% of the total amount of water is used to raise livestock (Gleick, 1993). However, this does not include the 100 kg of hay and the 4 kg of grain required for production of 1 kg of beef. It has thus been estimated that up to 200,000 l of water is required to produce 1 kg of beef on a range (Thomas, 1987). Increase in livestock and crop production required to satisfy the needs of a growing population will continue to stress water resources (Giampi-etro and Pimentel, 1995).

Mining of groundwater is necessary because of demand for freshwater by the increasing population and the subsequent use of irrigated agricultural practices (Falkenmark, 1989). Up to 68% of the groundwater withdrawn in the United States is for agriculture (Solley et al., 1993). Water tables have fallen 3 to 120 cm/year in the United States in some areas (Sloggett and Dickason, 1986), and in China, at rates of 1 to 2 m/year (Postel, 1992). Some groundwaters cannot reach the ocean in the dry season.

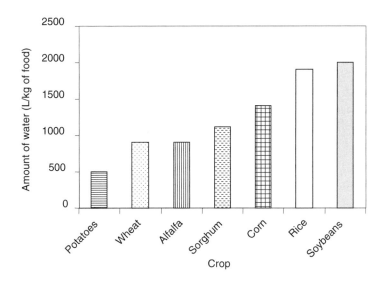

FIGURE 6.2
Amount of water required to produce one kilogram of food. (Data from Pimentel, D. et al., in *Ecological Sustainability and Integrity: Concepts and Approaches*, Kluwer Academic Publishers, Dordrecht, the Netherlands, 1998, pp. 104–134.)

Conflicts over water use due to irrigation have occurred in various parts of the world. The Egyptians have used the Nile for more than 5000 years, while the other nations in the upper drainage basin have not (McCaffrey, 1993). Recently, however, lack of water and increasing populations have made the other nations more dependent on the Nile. Construction of a dam on the Ganges River by India has led to riots and protests in Bangladesh, as the water needed for irrigation is now diverted (Kattelmann, 1990).

6.3.1.2   Water Quality

6.3.1.2.1   Nutrients

Agricultural practices can have a significant impact on water, as summarized in Figure 6.3. The plant nutrients, nitrogen and phosphorus, have significant impact on water quality. This can affect animal and human health. Sources of nutrients include inorganic fertilizers, animal manure, biosolids, septic tanks, and municipal sewages. In the province of Québec, Canada, in 1990 to 1991, an average of 190,000 tonnes of nitrogen and 120,000 tonnes of phosphorus were applied to agricultural lands in the form of fertilizers, or was present as livestock wastes. In 1990, almost 470,000 tonnes of commercial fertilizer was applied to crops in the province (0.47 tonnes/ha of farmland). In that same year, more than $43 million were spent on pesticides for crops (Statistics Canada, 1994). The U.S. Environmental Protection Agency (EPA) has established 10 mg/l of nitrate-N as the maximum contaminant level in groundwater, and a goal of 0.05 mg/l. It has also set a maximum of 0.1 mg/l for phosphate effluents that will enter a lake or reservoirs (U.S. EPA, 1987). Nash (1993) determined that 25% of the U.S. drinking wells had above 3 ppm levels of nitrate, and that other wells have nitrate levels that have reached as high as 100 ppm.

Soybeans, alfalfa, and other legumes fix more than 40 Tg/year of nitrogen fertilizer (Galloway et al., 1994). Depletion in nitrogen levels can haven an impact on soil fertility. Nitrate runoff from agricultural fields can decrease water quality. The nitrogen cycle can subsequently be altered through production of nitrogen oxides, which can affect human and ecosystem health (Figure 6.4). The nitrogen can travel from agricultural fields, to the rivers, streams, groundwater, and finally to the oceans. The diffusion processes for fertil-

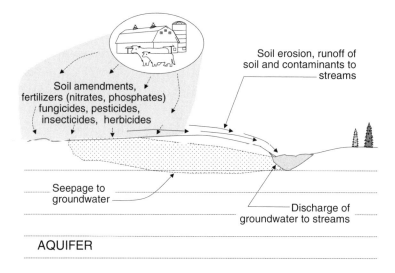

FIGURE 6.3
Schematic of pollutant transfer from agricultural activities to surface and groundwater — from a geoenvironmental perspective.

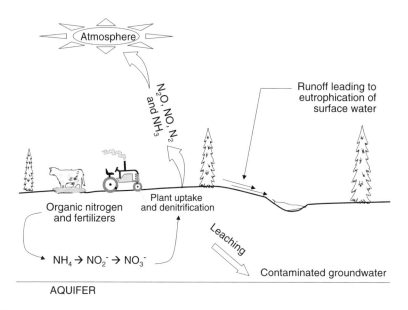

FIGURE 6.4
Illustration of the nitrogen cycle as it relates to the geoenvironment and agricultural practices.

izers (1 m/year) are slow. Consequently, groundwater contamination may not be seen
until a decade after a large amount of fertilizer is spread (Chamley, 2003).

The nitrogen from fertilizers can be converted to nitrate before uptake by plants, causing
pollution of rivers, lakes, estuarine, deltaic, or coastal waters. Toxic algal blooms (by the
dinoflagellate *Pfisteria*) can result from the eutrophication of estuaries and kill fish via
their toxins. Eutrophication processes also deplete the oxygen in the water, destroying
aquatic ecosystems. These phenomena affect the sustainability of fisheries. In the United
Kingdom, eutrophication decreased tourism and the values of waterfront properties by
$105 to 160 million per year in the 1990s, and has led to expenditures of $77 million per
year to fix the damage (Millennium Ecosystem Assessment, 2005). Humans can also be

affected by shellfish poisoning. High nitrate levels (above 10 ppm) can also lead to meth-emoglobinenemia (also known as blue-baby syndrome), abortions, and increased rates of non-Hodgkin's lymphoma. Levels above 5 ppm can affect young animals (Pimentel, 1989). Algal blooms initiated by the elevated nutrient levels lead to (1) disruption of the ecology through oxygen consumption, (2) accumulation of organic content, (3) reduction of water and sediments, (4) asphyxia, and (5) biota mortality, such as plankton and benthos (Chamley, 2003). Inorganic nitrogen levels in coastal areas have increased by a factor of 2.5, and phosphorus levels by 2, particularly in Western Europe. Dating of ^{210}Pb in Tahiti has allowed the determination of the accumulation of phosphorus in the sediments. The steps included exchangeable, iron-sorbed, carbonate, marine organic, and terrigenous phosphorus forms. In the 1950s, terrestrial phosphorus increased significantly due to both soil erosion and waste discharges.

The conversion of ammonia to nitrate lowers soil pH, particularly in northeastern Canada and Scandinavia (Chamley, 2003), where the bedrock formations are siliceous. The acidity increases the mobility of the toxic components, aluminum, and manganese. This can diminish vegetation and plant growth and increase soil erosion.

Due to its charge, phosphorus binds to soils, but it will leach from sandy soils with low levels of clay, oxides, and organic matter. Surface runoff will most likely cause contamination of streams and lakes, whereas groundwater contamination by nitrates is more likely, since these anions are held more weakly by negatively charged clayey soils than phosphorus. Nitrate diffusion in the groundwater is very slow, in the order of 1 m/year. Thus, the impact from excessive fertilization of the soil may only be seen in the groundwater one decade later.

Inorganic fertilizers and other soil amendments, such as animal manure and biosolids, contribute to elevated levels of N and P in the environment. Inorganic fertilizer use increased by 20- and 4-fold respectively from 1945 to 1980. Their use has since leveled off. Animal manure contributes 6.3 million tonnes of N and 1.8 million tonnes of P, compared with 10.8 million tonnes of N and 1.8 million tonnes of P from inorganic fertilizers (U.S. Geological Survey Circular 1225, 1999). About 15% of shallow groundwaters sampled beneath agricultural areas were above the acceptable levels for nitrates.

The government of Australia funded a project to study the movement of phosphorus in soils supplemented with piggery effluents (Redding, 2005). Soil samples were taken down to 5 cm in areas with and without the effluent amendment. Leaching was not significant if correct management procedures were used, but it did occur if application rates were excessive. Significant adsorption of the phosphorus occurred in the top 5 cm, and phosphorus runoff readily occurred from surface soil.

6.3.1.2.2 Pesticides

Pesticides have been used since the nineteenth century in the form of lead, arsenic, copper, zinc salts, and nicotine for insect and disease control. Since the 1930s and 1940s, with the introduction of 2,4-D and DDT, agricultural use has increased substantially. Worldwide use of pesticides is approximately 2.6 billion kg (Donaldson et al., 2002). The Database on Pesticide Consumption, which is maintained by the United Nations Food and Agriculture Organization (2003), provides information on the use of specific pesticides within each country.

Although pesticides have enhanced crop yields, concerns are increasing regarding their effects on the health of humans and animals and their transport in the environment. Levels of contamination of surface streams and groundwater increase with increased nutrient and pesticide use (USGS, 1999). Agricultural streams show the highest concentrations of pesticides. Herbicides are the most frequent pesticides found in agricultural streams and groundwater. Atrazine, deethylatrazine, metalochor, cyanazine, alachlor, and EPTC are

TABLE 6.1

Levels of Contaminants in Agricultural Areas

Contaminant	Streams	Shallow Groundwater
Nitrogen	Medium to high	High
Phosphorus	Medium to high	Low
Herbicides	Low to high	Medium to high
Currently used insecticides	Low to medium	Low to medium
Insecticides used in the past	Low to high	Low to high

Source: Adapted from U.S. Geological Survey, *The Quality of Our Nation's Waters: Nutrients and Pesticide,* U.S. Geological Circular 1225, 1999, water.usgs.gov.s

the most commonly detected chemicals — correlating well with their usage (USGS, 1999). Frequent pesticide and fertilizer use can alter the natural resistance of the plants and may also increase their resistance to parasites. This will make some soils unusable for agricultural purposes. Levels of contaminants typically found in agricultural areas are shown in Table 6.1 and Table 6.2. Overall, more than 2.3 million tonnes of pesticides are applied (Pimentel et al., 1992), resulting in the pollution of more than 10% of the rivers and 5% of the lakes in the United States Groundwater quality is also affected by pesticide use.

TABLE 6.2

Frequency of Detection of Herbicides and Insecticides in Streams and Groundwater in Agricultural Lands

Pesticide	Streams	Groundwater
Agricultural		
Atrazine	65	30
Deethylatrazine	35	29
Metalochlor	53	8
Cyanazine	25	2
Alachlor	28	2
EPTC	12	1
Urban Herbicides		
Simazine	44	12
Prometon	26	10
2,4-D	11	1
Diruron	7	2
Tebuthiuron	7	1
Insecticides		
Diazinon	10	1
Carbaryl	6	<1
Melathion	4	<1
Chlorophirofos	10	<1

Source: Adapted from U.S. Geological Survey, *The Quality of Our Nation's Waters: Nutrients and Pesticide,* U.S. Geological Circular 1225, 1999, water.usgs.gov.s

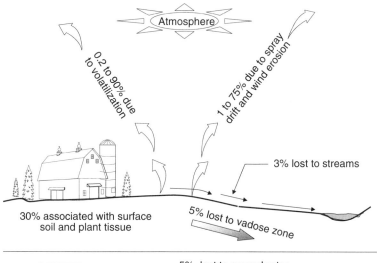

FIGURE 6.5
Estimates of the fate of pesticides after application to the soil for agricultural use. (Data from Barbash, J.E., in *Environmental Geochemistry*, Sherwood Lollar, B., Ed., Elsevier, Amsterdam, 2005, pp. 541–612.) Percentages of pesticide associated with plants, etc., and lost to atmosphere, vadose zone, and streams are dependent on highly climatic factors, soil management practice, type of pesticides used, and manner of application.

Estimated mass balances of pesticides have been used to evaluate the fate and transport of the pesticides in the environment, as illustrated in Figure 6.5. These estimates exhibit substantial variation in some cases. For example, atmospheric drift can vary due to weather conditions, the application method, and the properties of the pesticide. Unaccounted for pesticides can be due to the formation of covalent bonds with plant material or organic matter of soils (Xu et al., 2003) or biodegradation. The interactions of organic chemicals with soil organic matter has been briefly discussed in Section 2.5 in Chapter 2, and is further discussed in detail in Section 9.5.2 in Chapter 9. These discussions show that the association of pesticides with the organic matter of soil can be described by the relationship $C_{oc} = k_{oc} C_{aq}$, where C_{oc} is the concentration of solute contaminant sorbed onto the soil organic carbon, k_{oc} is the organic carbon–water partition coefficient, and C_{aq} is the dissolved concentration of the contaminant. The Freundlich adsorption isotherm for many pesticides can be described by the following relationship: $C_{oc} = k_f C_{aq}^{1/n}$, where k_f and $1/n$ are Freundlich parameters. These parameters are known for more than 60 pesticides (Barbash, 2005).

The sorption of an herbicide, triasulfuron, in Molinaccio clay loam, with and without humic and fulvic acids as organic materials, determined by Said-Pullicino et al. (2004), is shown in Figure 6.6. The Freundlich isotherm values, k_f ($\mu g^{(1-1/n)} ml^{1/n} g^{-1}$) were 0.14 for the soil, 0.18 with the compost, 0.20 with the humic acid, and 0.15 with hydrophobic dissolved organic matter. The $1/n$ values were all less than 1. The high concentration of compost from municipal waste led to an increase in sorption. However, in all cases, saturation was not reached and the herbicide preferred to remain in solution, and thus was highly likely to leach. The humic acids and the hydrophobic dissolved organic matter in the compost were responsible for the sorption characteristics of the compost.

Trace elements in the form of trace metals, heavy metals, metalloids, micronutrients, and trace inorganics are also reaching surface and groundwater from application as commercial fertilizers, liming materials, irrigation waters, and biosolids. Fertilizers often contain arsenic, cadmium, and lead. These can increase soil toxicity.

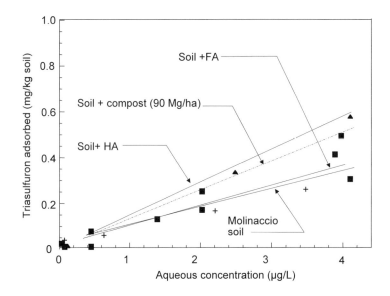

FIGURE 6.6
Adsorption of triasulfuron on a Molinaccio soil. HA, humic acid; FA, fulvic acid. (Adapted from information reported by Said-Pullicino, D. et al., *J. Environ. Qual.*, 33, 1743–1751, 2004.)

6.3.2   Energy Use

Agriculture, in general, uses nonrenewable energy sources. In the United Kingdom, 19.4% of the energy to produce a loaf of bread is from production (growing) of the wheat, and the remainder is due to production of the bread and its distribution. Farm-based production in the United Kingdom requires 2% of the primary energy from oil, gas, and electricity. If one were to assume that all countries use the same rates, this would amount to a total of 40% of global consumption (Ogaji, 2005) — a sum that would not be a sustainable use of energy. One could argue, however, that agriculture is also involved in energy production, and that this should offset some of the energy expenditure in calculations of the energy budget for farm-based production. For example, corn and wheat are used as substrates for the production of ethanol that can be mixed with gasoline. The trend toward increased biofuel production can lead to more conversion of land for production of energy-producing types of plants instead of food purposes.

6.3.2.1   Atmospheric Changes

At least three separate agricultural practices have some direct and indirect impact on atmospheric conditions. These are:

- Burning of farm materials associated with agricultural practices: This can have serious impact on human health and ecosystems through emissions of carbon, nitrogen, and sulfur gases. Subsequent chemical reactions can lead to the formation of ozone and acid rain, which can negatively impact agricultural yields.
- Methane production: Methane is a greenhouse gas that is produced during the degradation of stubble, such as from rice paddies, as it is ploughed into the soil (Yagi and Minami, 1990). The alternative, burning of the stubble, also increases air pollution. Thus, the risks of each alternative must be evaluated carefully.

- Removal of forests for agricultural purposes: The reduction in forest growth affects the carbon dioxide balance. Trees, shrubs, and other plants within forests are much better carbon sinks than agricultural plants. This subsequently increases the carbon dioxide emissions into the atmosphere, enhancing the greenhouse gas effect.

6.3.3 Impact on Health

Thirty new known diseases have emerged in the 1980s and 1990s (WHO, 1996). Although agricultural yields have increased, irrigation, land conversion, and disturbances of the ecosystems have increased the rates of even older diseases, such as malaria. Irrigation has increased the number of diseases, up to 30 diseases, including mosquito-borne diseases in Central and South America (WHO, 1996) and malaria and Japanese encephalitis due to rice paddy irrigation. Irrigation systems in hot climates are directly linked to schistosomiasis incidences.

Deforestation and increasing cultivation decrease the soil's ability to retain contaminants and nutrients. Chemicals such as mercury are normally stabilized by iron oxyhydroxide adsorption, and they can also bioaccumulate in fish. Erosion destabilizes mercury, thus increasing the release of mercury into the water supply. Freshwater fish can contain an average of 48 µg/l of mercury, a potential health hazard (Richard et al., 2000).

The decrease in water quality due to pollution from a variety of urban and industrial sources also can contaminate crops and lead to the poor health of farm workers and consumers of the contaminated crops. Agriculture cannot be viewed in isolation to the other sectors. Agriculture has been called both a cause and a victim of water pollution (Ogaji, 2005).

6.3.4 Impact on Biodiversity

One of the major impacts of land transformation for food production and other uses has been the extensive loss of biodiversity. The agricultural sector has led to the extinction of more species than any other sector. The rates of extinction have increased to 100 to 1000 times the level prior to the industrial revolution (PCAST, 1998). Decreases in the amounts of pollinating insects have negatively affected yields of particular crops (Nabhan and Buchmann, 1997). The diversity of soil organisms is also decreased due to the decreased opportunity for organic decomposition and increased nutrient content. Specialized predators and weeds will also develop due to the low diversity of species in agricultural lands. The change in land use has also affected the various cycles within the global system, including the carbon, water, biogeochemical, and biotic, to name a few. For example, lack of soil organic matter and soil organisms will affect the carbon cycle.

Biological diversity is the highest in forested areas. For example, intertropical forests make up 6% of the surface area, but half of all plant and animal species. Thus, conversion of these forested areas to agriculture can be highly detrimental to biodiversity. Modifying rivers and lakes for irrigation purposes can lead to extinction of the fauna, flora, and terrestrial organisms due to variations in water composition, temperature, and flow. Water pollution from runoff of pesticides, fertilizers, and salinization can influence biodiversity. Salinity levels in the Aral Sea have tripled due to cotton irrigation (WRI, 1992). This has led to the extinction of 24 species of fish (Postel, 1992). Also, erosion can be detrimental to fish and other aquatic organisms due to siltation of breeding grounds. Mulholland and Lenat (1992) estimated that there has been a decrease of up to 50% of the species in streams affected by agriculture. Increasing the use of surface water for agriculture can also damage

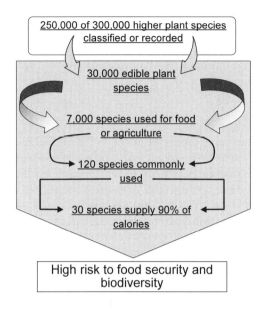

FIGURE 6.7
Risk to food security due to loss of agricultural biodiversity. (Data from El Bassam, N., in *Proceedings of the International Conference on Sustainable Agriculture for Food, Energy and Industry* (Joint Conference of the Food and Agriculture Organization (FAO), Society of Sustainable Agriculture and Resource Management (SSARM), and Federal Agriculture Research Center (FAC)), Braunschweig, Germany, June 1998.)

ecosystems in lakes and watersheds. The biodiversity losses or changes may not be evident initially, but in the long term can have significant impacts on the sustainability of agricultural practices.

New breeding methods have caused a significant loss of agricultural biodiversity. Some of these impacts are shown in Figure 6.7 which shows that although there is a large number of species available, only a small portion is utilized for food production. Uniformity and standardization of farming practices has led to this restricted biodiversity — a clear demonstration of the effect of human activities on biological diversity. The genetic base of crops needs to be widened to avoid dependence on a restricted genetic base that renders the world's food supply at risk to diseases, pests, and other dangers. New crops will also ease the demand for food in areas where it is scarce, and new management techniques will be required to incorporate new or underdeveloped crops to broaden the genetic base. To strive toward agricultural sustainability, minimization of the negative impacts on natural biodiversity is required — together with conservation of the available genetic resources. It is useful to bear in mind that agricultural fields are not isolated from the surrounding natural environment. Current agricultural policies concentrate on product yields, demographic changes, and land ownership (Fischer Taschenbuch, 1999). Genetic erosion will lead to increased risks to food security, as plants will be less able to adapt to changes in the environment. Agro-biodiversity is a vital factor in agricultural management practices.

Other impacts on biodiversity have been studied in the United Kingdom. Hedgerows were used extensively in the past as field boundary markers (Nature Conservancy Council, 1986). However, more than 16,000 km of hedgerows have been removed annually in the 1960s, and it has been estimated that the 20 species of mammals, 37 species of birds, and 17 species of butterflies that live in the boundaries could be threatened by the loss of the hedgerows. Biodiversity has been severely affected due to removal of the hedgerows and other intensified agricultural production processes. Other species, such as brown hares, arthropods, insects, bees, flowers, and bats, are also threatened.

6.4 Tools for Evaluating Sustainability in Agriculture

6.4.1 Agricultural Sustainability

To avoid or mitigate risks and protect soil quality, we need to develop a methodology for evaluation of the existing and potential risks of agricultural activities on agricultural and geoenvironmental sustainability. It has been estimated that 28% of soil degradation is due to faulty agricultural practices (Quendler and Reichert, 2002). There appears to be some significant room for improvement, as has been discussed previously. One needs to (1) examine the impact of the activity on water, energy, health, agriculture, and biodiversity (WEHAB), (2) seek methods to quantify the emissions, and (3) determine the fate and transport pathways of the harmful emissions. By including air, water, soil, and vegetation as part of the agro-ecosystem, one could seek measures in an integrated approach to (1) limit the harmful discharges and (2) curtail inefficient practices or emissions, as the first major step toward generation of sustainability. Some of these measures include (1) ecological footprint, (2) sustainable processing index (SPI), and (3) material intensity per service unit (MIPS) or land intensity per service unit (LIPS) (Quendler et al., 2002). Models and sustainability indicators are also tools that can be used to evaluate sustainability. The analysis of sustainability thus will rely on sustainability indicators, reference values, and an established evaluation method. A recounting of some indicator systems has been reported by the Organization for Economic Cooperation and Development (OECD) (1996).

Attempts have been made to define the components of agricultural sustainability. One of the most inclusive was formulated by Christen (1996), as depicted schematically in Figure 6.8. Ethical considerations include fairness between generations. Resources and

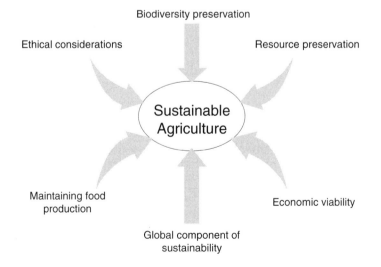

FIGURE 6.8
Components of sustainability for agriculture. (Information from Christen, P., *Berichte Landwirtschaft*, 74, 66–86, 1996.)

biodiversity need to be preserved without reducing production, but by minimizing environmental impact. Economic viability must also be ensured for both small and large enterprise farming units. Since soil will always be used for agricultural purposes, the impact of pollution and erosion must be determined through research and field studies.

Some authors have argued that the definitions of organic farming (Quendler and Schuh, 2002) follow the same principles as those used to establish sustainable agriculture. These principles include:

- Working with a closed system to draw on local resources
- Maintaining the fertility of the soil over the long term
- Avoiding all forms of pollution
- Producing high-quality nutritional food in sufficient quantities
- Reducing the use of fossil fuels to a minimum
- Giving livestock humane conditions
- Making it possible to allow agricultural producers to earn a proper living
- Using appropriate technologies for biological systems and decentralized systems for product processing, distribution, and marketing
- Creating aesthetically pleasing systems to all
- Preserving and maintaining wildlife and their habitats

The Food, Agriculture, Conservation and U.S. Trade Act of 1990 established that practice of sustainable agriculture is to

> satisfy human food and fibre needs; enhance environmental quality and the natural resource based upon which the agricultural economy depends; make the most efficient use of non-renewable resources and on-farm resources and integrate where appropriate natural biological cycles and controls; sustain the economic viability of farm operations and the quality of life for farmers and society as a whole.

Canada's International Institute for Sustainable Development agrees with this, but says that organic farming is not the only way to achieve sustainable agriculture. Technology needed to achieve sustainability is minimized in organic farming.

6.4.2 Model Development

As we have seen in the previous section, various agricultural practices can have a significant impact on WEHAB — highlighting the problems and difficulties in achieving the goals of agricultural sustainability. A pertinent question that can be posed is, What methods can we use to determine the impact of the practices, and how can we measure and optimize the improvements of these practices? The development of predictive models and sustainability indicators to measure sustainability progress is an ongoing process to accomplish the task needed to answer the question posed.

We have seen in Section 2.5 in Chapter 2 that soil can provide means to retain pollutants by natural attenuation of both organic and inorganic contaminants. This subject is discussed in greater detail in Chapter 10. For example, in Germany, levels of nitrate as high as 250 mg/l, aluminum as high as 0.64 mg/l, and potassium up to 60 mg/l have been found in the groundwater in agricultural areas (Houben, 2002). Acid rain has decreased the pH of the soil to 2.75, and that of the groundwater to 3.4. The soil buffering capacity has also been diminished. Cation exchange, autotrophic denitrification (reaction of nitrate with FeS_2), and other natural attenuation mechanisms have restricted the movement of the pollutants. Modeling, together with determination of the age of the groundwater, mass balances, and reactive transport, was undertaken using hydrochemical and geochemical data. The PHREEQC-2 geochemical model was used for the hydrochemical equilibrium modeling. Sorption and desorption column experiments with undisturbed samples of sandy sediments for magnesium, sodium, potassium, and aluminum ions were performed. Modeling was accurate for most ions with the exception of potassium. Competition cannot be accounted for in the mass balance approach. Due to the high velocity of fluid flow in the columns, there was insufficient time for the nitrate to react with the pyrite. The models indicated that the contaminants move a few centimeters per year.

A number of models have been developed to integrate scientific information to enable policy development and future management practices. Integrated assessment modeling in particular has been applied to determine the impacts and to predict climate change. Relevant models for agriculture with some of their features are summarized in Table 6.3. The term *forcings* used in the second column of the table refers to climate forcings. These are perturbations imposed on the energy balance of the earth — either naturally or through man-made events. Because energy flows in and out of the sun are in respect to visible wavelengths and long-wave infrared radiation, respectively, disruptions of these through, for example, airborne fines naturally emitted from volcanic eruptions or from smokestack emissions are considered forcing agents. A measure of the forcing energy perturbation is given as watts per square meter (W/m^2).

Other models focus on predicting pollutant transport and leaching in the soil. PSALM is the *phosphate solute and leaching model* developed at Rothamsted Research by Addiscott et al. (2002). It is used to simulate losses of phosphorus at plot and field scales from agricultural soils. The criteria for leaching from agricultural lands were based on soil extractions to determine 0.01 M $CaCl_2$-extractable P and 0.5 M $NaHCO_3$-extractable P (Olsen P) measurements on soil (Hesketh and Brookes, 2000). This extraction is able to predict the concentrations of Olsen P, above which further concentrations would be lost to drainage water. However, there were wide variations in results between soil types. To assess the impact of different management regimes and the relative importance of the various loss mechanisms of P in soils, a dynamic modeling approach is used for PSALM.

According to the U.S. Department of Agriculture, Agricultural Research Service (ARS), Great Plains Systems Research (Fort Collins, CO), the Nitrate Leaching and Economic Analysis Package (NLEAP) is a field-scale computer model developed to provide a rapid and efficient method of determining potential nitrate leaching associated with agricultural practices. It combines basic information concerning on-farm management practices, soils, and climate, and then translates the results into projected N budgets and nitrate leaching below the root zone and to groundwater supplies, and estimates the potential off-site effects of leaching. The NLEAP model was designed to predict leaching of nitrate. The processes modeled include movement of water and nitrate, crop uptake, denitrification, ammonia volatilization, mineralization of soil organic matter, nitrification, and mineralization (immobilization associated with crop residue, manure, and other organic wastes). It can be used with various geographic information systems (GIS).

CREAMS (Knisel, 1980) is a field-scale model for prediction of runoff, erosion, and chemical transport from agricultural management systems. It is suitable for application to individual storms or long-term averages over a period of 2 to 50 years. It is able to estimate runoff, percolation, erosion, and adsorption of plant nutrients and pesticides.

The Soil and Water Assessment Tool (SWAT) has been developed to model changing land use patterns and practices on nitrogen and phosphorus movement to surface and groundwater. Sediment transport, crop growth, and nutrient cycling are simulated. It has

TABLE 6.3
Summary of Agriculture Integrated Assessment Models

Model	Forcings[a]	Geographic Specificity	Socioeconomic Dynamics	Geophysical Simulation	Treatment Uncertainty	Treatment of Decision Making
AIM	Land use, CO_2, other GHG, aerosols	Countries, grids/basins	Economics, technology choice, land use, demographic	1- and 2-dimensional temperature and pressure change	None	Simulation
FUND	CO_2 and other GHG	Global	Economics	Global temperature change	None or uncertainty	Optimization
IIASA	CO_2	Global	Economics	1-dimensional temperature and pressure change	None	Optimization
IMAGE 2.0	Land use, CO_2, other GHG, aerosols	Grids/basins	Exogenous, technology choice, land use	2-dimensional temperature and pressure change	Uncertainty	Simulation
MIT	Land use, CO_2, other GHG, aerosols	Countries, grids/basins	Economics	2-dimensional temperature and pressure change and climate	Uncertainty	Optimization, simulation
PAGE	CO_2, other GHG	Continental, countries	Economics	Global temperature change	Variability	Simulation
ProCAM	Land use, CO_2, other GHG, aerosols	Countries, grids/basins	Economics, technology choice, land use, demographic	2-dimensional temperature and pressure change	Uncertainty	Simulation
TARGETS	Land use, CO_2, other GHG, aerosols, other	Global	Economics, technology choice, land use, demographic	2-dimensional temperature and pressure change	Cultural perspectives	Optimization, simulation

[a] Forcings refer to *climate forcing*, which is defined as an imposed perturbation of earth's energy balance. The units are given as watts per square meter (W/m^2).

Source: Adapted from Weyant et al., 1996.

also been used for pesticide transport (Rekolainen et al., 2000). It is integrated into ArcView geographic information systems (GIS) software and was developed by the USDA Agriculture Research Services. It is a combination of EPIC and Groundwater Loading Effects of Agricultural Management Systems (GLEAMS). GLEAMS was developed to determine the effect of various management practices on pesticide and nutrient leaching at, through, and below the root zone. CREAMS (Knisel and Davis, 1999) is a forerunner of SWAT and was also the basis for GLEAMS. The model has shown promise for the development of best management practices in reducing nonpoint source pollution in watersheds (U.S. EPA, 1997). SWAT has been used by decision makers and could be used to protect water quality, if it is adequately tested.

Other countries, such as the United Kingdom, have also developed nitrogen leaching models. MAGPIE (Modeling Agricultural Pollution and Interactions with the Environment) was developed by the Environmental Modeling and GIS Group at ADAS Wolverhampton. Simulations of nitrate leaching have been performed at catchment and field scales. Climate, land use, soil survey data, and crop and livestock information are included.

Other models have focused on pesticides. The European registration process uses the Pesticide Leaching Model (PELMO). PELMO estimates the leaching potential of pesticides through distinct soil horizons using an extended cascade model. Processes include estimation of soil temperatures, pesticide degradation, sorption, and volatilization, and estimation of potential evapotranspiration using the Haude equation (Klein, 1994). It has components for hydrology and is based on the USDA Soil Conservation Curve Number technique and the Modified Universal Soil Loss Equation (runoff and erosion) and chemical transport. The calculation of evapotranspiration is estimated using the Haude equation. PELMO was recently modified to include the prediction of volatilization from the soil, which the U.S. EPA does not account for (Wolters et al., 2004). Future modifications will require more detailed information at the soil–air–water interface.

PRZM3 is the most recent version (version 3.12.3 was released in May 2005) of a modeling system developed by the EPA that includes PRZM and VADOFT, to predict pesticide transport and transformation down through the crop root and unsaturated zone. PRZM is a one-dimensional, finite-difference model that predicts the fate of pesticides and nitrogen in the root zone of crops. It can also predict pesticide concentration in runoff water and solid particles. The latest version (PRZM3) includes the capability to simulate soil temperature, volatilization and vapor phase transport in soils, irrigation simulation, and microbial transformation, in addition to the transport and transformation of the parent compound and up to two daughter species. The input parameters include the characteristics of the pesticide, the application of the pesticide, and crop, climatic, and site information (soil and hydrological properties, agricultural practices, topography, etc.). VADOFT is a one-dimensional, finite-element code that solves Richard's equation for flow in the unsaturated zone. Several versions of the PRZM system are available on the EPA Center for Exposure Assessment Modeling (CEAM) Web site (http://www.epa.gov/ceampubl/gwater/przm3/index.htm). PRZM can also be used with the EXAMS model to simulate the fate and transport of the pesticide in water. Volatilization, sorption, hydrolysis, biotransformation, and photolysis are processes that are included in the model structure.

LEACHM is a suite of models for simulating the leaching and fate of water and chemicals within the soil (Hutson and Wagenet, 1989). Input data are similar to the PRZM-EXAMS model. One or more growing seasons can be simulated. Output includes a profile of the pesticide concentration throughout the soil and water and pesticide fluxes in the groundwater. The model can be found on the following Web site: www.scieng.flinders.edu.au/cpes/people/hutson_j/leachweb.html.

For the modeling of transport and fate of heavy metals, a dynamic modeling approach is used for a newly developed model named Dynabox (Heijungs, 2000). It is built on the

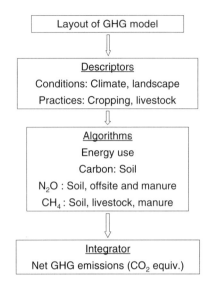

FIGURE 6.9
Layout of computer model for calculation of greenhouse gas (GHG) emissions from agriculture. (Adapted from Gibb, D.J. et al., A Computer Model for Estimating Greenhouse Gas Production from Agriculture, poster presented at BIOCAP Canada Foundation's First National Conference, 2005.)

Uses model, which incorporates transport to and from the compartments of the air, surface water, soils, sediment, etc. The regional and continental model included agricultural soils. Also, unlike organic compounds, heavy metals exhibit a negligible degradation period. Concentrations of the heavy metals, such as copper, lead, and zinc, can be calculated in air, surface water, and intakes by humans.

A computer model is under development by Agriculture and Agri-Food Canada to estimate the sources and sinks of greenhouse gas production from agriculture and to evaluate mitigation alternatives (Gibb et al., 2005). For example, greenhouse gas (GHG) emissions can be calculated from beef and dairy production. Some of the inputs would include animal type, weight and number, feed used, and manure handling. A layout of this model is shown in Figure 6.9.

6.4.3 Indicators of Agro-Ecosystem Sustainability

Although crop yields are a measure of sustainability, they do not provide any indication of the impact on the ecosystem and the geoenvironment or land environment in particular. Monitoring of ecosystem damage must be determined, particularly to differentiate natural changes from those due to human activity. GIS, remote sensing, and landscape ecology are new measuring approaches to indicate land use changes. Indicator organisms can be monitored for changes in the ecosystem. Various organizations, such as the Organization for Economic Cooperation and Development, the Food and Agriculture Organization, the World Bank, and the Commission of Sustainable Development, have published various indicators for agriculture related to economic, social, and environmental processes, farming practices, and environmental impacts.

For example, the U.K. Sustainable Development Strategy by the U.K. Local Government Management Board and Touche Ross Management Consultants (1994) provides a list of indicators to determine if development is improving in sustainability. The indicators were grouped into 21 families. Those related to agriculture are shown in Table 6.4. Soil quality is included, as it was deemed vital for food production and an ecosystem for vital organisms.

TABLE 6.4

Sustainability Indicators Developed by the United Kingdom

Area	Indicator
Agriculture, Rural Economy, Society	
Structure of the agriculture industry	Agricultural assets and liabilities
	Age of farmers
	Percentages of holdings that are tenanted
Farm financial resources	EU Producer Support Estimate (PSE)
	Agri-environment payments to farmers
	Total income from farming
	Average earnings of agricultural workers
Agricultural productivity	Agricultural productivity
Agricultural employment	Agricultural employment
Farm Management Systems	
Management	Adoption of farm management systems
Organic farming	Area converted to organic farming
Codes of practice	Knowledge of codes of agricultural practices
Inputs	
Pesticide use	Pesticides in rivers and groundwater
	Quantity of pesticide active ingredients used
	Area treated with pesticides
	Pesticide residues in food
Nutrients	N and P losses from agriculture
	P levels due to agriculture in soil
	Manure management
	Ammonia emissions
Greenhouse gas emissions	Emissions of methane and NO_x from agriculture
Resource Use	
Energy	Direct consumption by farms
	Trends in energy inputs to agriculture
Water	Use of water for irrigation
Soil	Organic matter in soil
	Heavy metals in topsoils
Agricultural land	Area of agricultural land
	Change in land use from agriculture to hard development
Nonfood crops	Planting of nonfood crops
Conservation Value of Agricultural Land	
Environmental conservation	Area of agricultural land committed to environmental conservation
	Characteristic features of farmland
Landscape	Areas of cereal field margins under environmental management
Habitats	Area of seminatural grassland
Biodiversity	Populations of key farmland birds

Source: Adapted from MAFF, *Towards Sustainable Agriculture: A Pilot Set of Indicators*, Ministry of Agriculture, Fisheries and Food, MAFF Publications, London, 2000.

TABLE 6.5

Canadian Agro-Environmental Indicators

Indicator Group	Subgroup	Brief Description
Farm Management	Soil cover by crops	Number of days per year soil is covered by crops
	Nutrient and pesticide management	Management practices for fertilizer, manure, pesticides
Soil quality	Water erosion	Soil loss due to water runoff
	Wind erosion	Soil loss due to wind and landscape conditions
	Soil organic carbon	Change in soil organic content
	Tillage erosion risk	Soil redistribution due to tilling and cropping
	Soil compaction risk	Degree of soil compaction due to cropping
	Soil salinization risk	Increases in soil salinity due to land use, hydrologic, climate, and soil conditions
Water quality	Nitrogen contamination risk	Increase in nitrogen levels in water leaving farm
	Phosphorus contamination risk	Increase in phosphorus levels in water leaving farm
Greenhouse gas emissions	Agricultural greenhouse gas budget	Estimates of N_2O, CH_4, CO_2 due to agriculture in CO_2 equivalents
Biodiversity	Wildlife on farmland	Change in number of habitat use units
Production intensity	Energy use	Energy content of agricultural input and output
	Residual nitrogen	Difference in N added to soils and amount in crops

Source: Adapted from McRae, T. et al., *Environmental Sustainability of Canadian Agriculture: Report of the Agri-Environmental Indicator Project*, Agriculture and Agri-Food Canada, Ottawa, Ontario, 2000.

Concentrations of organic matter, acidity, nutrient concentration (P and K), and heavy metals are the parameters included. This varies somewhat from the indicators in Canada (McRae et al., 2000), which are shown in Table 6.5. Other groups, such as Sustainable Measures (North Andover, MA), have developed a searchable database for evaluating sustainability in various sectors, including agriculture. Their indicators for soil include soil erosion per acre of cropland, average soil erosion, area affected by soil erosion, and soil organic matter. Surface water indicators included phosphorus concentration and biological oxygen demand (BOD) in county streams. Numerous other indicators were included in environmental parameters.

Pesticide indicators have also been developed by the OECD (2002). Pesticide use and pesticide risks are the two indicators. Most countries have decreased the use of pesticides. Reduction in risks to human health and the environment can be achieved by reducing particular pesticides. For water use, the three indicators developed include (1) intensity of water use, (2) water volume consumed, and (3) economic value of water use. Water use is very high for many OECD countries. Technical and economic efficiency information is difficult to obtain, as well as water stress caused by diversion of water from rivers for agricultural use. The environmental impacts from agricultural practices consider impacts on (1) soil and water quality, (2) land conservation, (3) production of greenhouse gases, (4) biodiversity, (5) wildlife habitats, and (6) landscape. Soil quality was evaluated based on risk of water and wind erosion. These were considered of higher concern than soil compaction or salinization. Overall, about 10% of agricultural land is at risk for erosion. Conservation or no tillage of land, less intense crop production, and retiring lands can reduce the effects of soil degradation.

6.5 Changes in Agricultural Practices To Reduce Impact

The reader is reminded that the discussion in this section, and in the previous sections concerning agricultural practices, their impacts, and the means to mitigate impacts, is structured from the geoenvironmental perspective. It can be argued that many of the measures for mitigating, minimizing, and even eliminating impacts from contaminant loading and soil quality impairment run counter to intensive agricultural practices. From the perspective of the agro industry, there is validity to this set of arguments. However, as has been realized for countless years, conflicts between agricultural productivity and protection of the geoenvironment have always existed. The hope is that these conflicts will continue to lessen as one strives toward structuring sustainable practices for both agricultural production and the geoenvironment.

6.5.1 Practices To Reduce Impact

Changes in agricultural practices can lead to reduction in soil degradation rates. Rotation of crops protects soils from erosion, increases organic matter contents and nutrient contents, and restructures the soil. As is well known in agriculture, it is common to alternate between high-yield cereal and tuber plant with leguminous or fallow land planting. Adding mulches, composts, or manures can also increase the organic content. Mechanical measures, including contouring, terrace cultivation, and contour hedges, can reduce erosion and increase yields. In the following sections, we will look at some measures to reduce impact. They are by no means an exhaustive list, but serve as examples of more sustainable practices, many of which are being practiced by farmers today.

Appropriate water management techniques are required to avoid subsidence and groundwater depletion. Cattle herds and flocks should be limited to avoid excessive trampling and overpasturing. Waterlogging should be avoided to prevent loss of the land to salinization. Adequate drainage is required to leach out the salts and remove the excess water from the soil.

Soil erosion must also be minimized. In New Zealand, the Franklin Sustainability Project evaluated different methods, including raised access ways, benched headlands, silt fins and traps, and contour drains for erosion minimization (Ministry of Environment, New Zealand, July 2001, www.mfe.gpvt.nz). It has been estimated that 98 mm of topsoil has been lost since 1952. Grasses and sedges are effective as sediment filters and, in combination with benched headlands, can be very effective for trapping sediment, but do not prevent erosion.

The British Agrochemical Association has developed an approach called integrated crop management to "avoid waste, enhance energy efficiency and minimize pollution" (www.ecifm.rdg.ac.uk/integrated_crop_management.htm). This approach advocates (1) crop rotation, (2) selection of appropriate cultivation techniques and seed varieties, (3) minimization of fertilizers, pesticides, and fossil fuels, (4) landscape maintenance, and (5) encouragement of wildlife habitats. Use of farm-produced inputs for fuels, pesticides, and fertilizers is encouraged. Soils are protected to (1) minimize energy use, (2) reduce erosion, and (3) reverse adverse effects on beetles, spiders, and earthworms. Crop production that is appropriate for the climate, soil type, and topography is maintained. Trials have shown that with such practices, (1) costs are reduced by 20 to 30%, (2) pesticide use is reduced by 30 to 70%, and (3) there is a reduction in requirement for nitrogen by 16 to 25%. The results also show that biodiversity was increased and nitrate leaching and soil erosion were reduced.

Livestock generate significant impacts on the geoenvironment, such as overgrazing, erosion of the soil, river and lake pollution, desertification, and deforestation (Regenstein, 1991). Their numbers have increased substantially to the point where they now outnumber humans 3:1 (Goodland, 1998). Incentives and taxes could be used to promote good environmental practice in food and agriculture. Reduction in water use and recycling of manure can be practiced. Cattle feedlots provide the most significant impact. Sheep generate a lesser impact since they graze on more natural grassland.

The Farm Waste Management Plan of the U.K. (Saha, 2001) has listed the following items for pollution control:

- Delaying ploughing of crop residues
- Reducing the use of fertilizers, manure, and sewage sludge
- Sowing autumn crops early
- Managing farm waste carefully

With regards to water quality, Sagardoy (1993) listed the following items to avoid water pollution:

- Development of water quality monitoring schemes
- Optimization of the use of farm inputs and other agricultural activities that affect wetlands
- Establishment of water quality criteria for agriculture
- Prevention of soil runoff and sedimentation
- Proper disposal of animal and human wastes
- Ionization of agricultural chemicals for pest management
- Education of the community to minimize impact on water quality and ensure food safety

Reduction in use of various mineral and organic fertilizer and pesticide discharges through legislation is particularly effective. Stricter standards to maintain ground and surface water quality are required. Pollutant treatment and storage systems commonly used in industrial processes may be required to reduce the pollutants at the source. Other practices that can reduce pollutants and their impacts can include (1) optimization of natural pollinators and predators to conserve species and ecosystems by maintaining natural vegetation near agricultural lands (Thies and Tscharntke, 1999), (2) optimization of the water supply and quantity by correct use of water and chemicals to protect human health and ecosystems (Matson et al., 1997), and (3) use of technologies that reduce erosion, salinization, water consumption, chemical pollution, and other environmental effects.

To ensure sufficient food and reduce hunger, further advances in food production are required. In particular, declining agriculture with increasing populations in the lesser developed countries is of concern. Technologies to increase yields while decreasing the effects on the environment will be required to work toward sustainability.

Biotechnology has been employed to increase food quality and production through the availability of transgenic plants. The main objectives are to develop plants that are resistant to bacteria, fungi, viruses, and environmental stress. At this point, conventional crop breeding is commanding more attention, due in part to the need to broaden capabilities and alternatives to the production of genetically altered plants.

Optimized uses of both fertile and fragile soils, together with measures to rejuvenate production on degraded lands, are procedures that can be pursued. Lower-quality soils could support more species than high-yield agricultural lands (Dobson et al., 1997). Planting trees for shelter can also help to (1) reduce evaporation and transpiration by 13 to 25% (Mari et al., 1985), (2) reduce wind erosion, and (3) increase crop yields such as corn by 10 to 74% (Gregerson et al., 1989).

In addition to minimization of tillage to prevent runoff, other methods, such as the use of intercropping and ground cover, can also be used. For example, Wall et al. (1991) found that interplanting with red clover was able to reduce water runoff by 45 to 87% and soil loss by 46 to 78%. Reducing water and soil loss conserves water, decreases nonpoint sources of pollution, and increases water availability for the plants (NGS, 1995). Water use can be minimized through precise applications. Night application, use of surge flow irrigation, low-pressure sprinklers, and drip irrigation can reduce water use substantially (Verplancke, 1992; Goldhamer and Snyder, 1989).

Other means can be used to enhance the soil's capability for retention of contaminants. A case in point could be the cereal cultivation in northeastern France that has led to pollution of the Rhine Valley groundwater by nitrates (Bernard et al., 1992). These agricultural soils contained much lower organic contents than forest soils. Considering that forest soils with their higher organic contents will show greater capability in retaining the pollutants and converting the nitrates to N_2O through denitrification, it would appear that conversion of these agricultural lands through reforestation can significantly reduce groundwater pollution.

Whereas attention is normally given to soil permeability in consideration of transport of fertilizers in the ground, the attention received by the underlying geology has not been as significant. This oversight can lead to serious consequences. The example of Brittany, France, with underlying densely fractured and weathered granite rocks, is a good demonstration. These densely fractured weathered rocks allow infiltration of contaminated water — in contrast to metaphoric schists that prevent seepage (Chamley, 2003). Nitrates and phosphates from farms are retained by the granite, which can then contaminate areas downstream. In contrast, in schistose regions, these fertilizers do not penetrate into the underlying rock. Geologic maps and corresponding laboratory data on the rock properties are useful tools in predicting the impact of pollution from farm fertilizers. Réunion, a sloped territory with increasing agricultural activities and urban and tourist areas, produced a map in the year 2000 for such purposes.

The U.S. Soil Conservation Service (1993) developed a wetland process for treatment of agricultural runoff consisting of a wet meadow, followed by a marsh and pond with an optional vegetated polishing area. It is applicable for the removal of sediments and nutrients such as phosphorus. The wet meadow with a slope of 0.5 to 5% consists of permeable soils with cool-season grasses. The depth of the marsh with cattails varies from zero at the surface of the meadow to 0.46 m at the deep pond. The deep pond performs as a biological filter for the removal of nutrients and sediments. Fish, such as common or golden shiners, should be included in the pond to feed on the plankton. Average sediment and phosphorus removal in a system for potato growing in northern Maine over two seasons were determined to be 96 and 87%, respectively (Higgens et al., 1993).

Losses to the environment of pesticides and herbicides through volatilization and runoff must also be minimized. Large quantities of agro-chemicals have been found in various water bodies, such as the Great Lakes in North America. Application of these chemicals during calm conditions can minimize losses due to drift. Biological methods can be used to control weeds and insects. It has been proposed that an integrated pest management (Janzen, 1998) strategy be utilized through chemical, biological, and cultural methods to optimize the use of pesticides.

6.5.2 Impact of Soil Additives

Various agricultural and industrial wastes are utilized as amendments for agricultural soils. Irrigation with reclaimed sewage water is practiced in arid and semiarid areas. In Israel, irrigation for 28 years with sewage effluents has led to the accumulation of cadmium, copper, nickel, and lead in the topsoil layer in coastal plain soils (Banin et al., 1981).

Significant amounts of these metals accumulated in the oxide fraction of the soils. The amounts recorded show that (1) for Cd the percentages retained varied from 9 to 20%; (2) for Cr, from 5 to 11%; (3) for Cu, from 16 to 23%; (4) for Ni, from 10 to 14%; and (5) for Pb, from 1 to 3.9%. Approximately 20 to 45% of produced sewage sludge is added for agricultural use in the United Kingdom, Germany, and the United States, among other countries (Mullins, 1990). These sludges can contain high levels of heavy metals, in addition to N, P, K, and other micronutrients. While extensive data are available on the distribution of copper and zinc in sludge-amended soils, the same cannot be said for such heavy metals as cadmium, chromium, nickel, and lead. After more than 10 years of application, copper, chromium, cadmium, and zinc have moved to the soil layer below the layer of application, and zinc has been found below the plough layer (Han et al., 2001). Zinc has been shown to be bioavailable and has been taken up by wheat, rice, soybean, and maize plants. The degree of uptake depends on the conditions of pH, soil type, and Eh. Heavy metals added via wastes can run off or seep into the soil, potentially contaminating groundwater or impacting the quality of the food and animals within the food chain. The heavy metals may also impact the soil microorganisms.

Heavy metals are a major concern in poultry and swine manure used as amendment to soils. Heavy metals in poultry litter, in particular, include As, Co, Cu, Fe, Mn, Se, and Zn (Sims and Wolf, 1994). The accumulation of heavy metals from manure amendment can occur over the long term. Extractable Cu and Zn concentrations increase over time, in addition to the heavy metal concentrations in runoff after amendment with poultry waste. Cu concentrations have also increased in soil and plants such as grass and corn after the addition of swine waste amendment (Kornegay et al., 1976; Sutton et al., 1983; Mullins et al., 1982; Payne et al., 1988).

Although beneficial, the practice of spreading manure can negatively impact air, soil, and water quality. Besides providing nutrients, manure addition can reduce soil erosion and improve soil water holding capacity (USDA, 1992). However, excessive levels of nitrogen, phosphorus, and organic matter can accumulate in the soil if the manure is not spread properly (Figure 6.10). Heavy metals such as copper and zinc may also accumulate in the manure, and subsequently in the soil due to their use as food additives. In countries

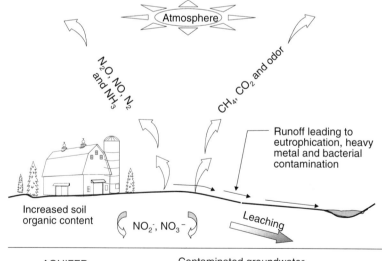

FIGURE 6.10
Illustration of the emissions and leachates due to manure spreading practices.

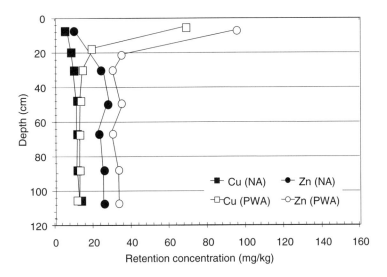

FIGURE 6.11
Retention profiles for total Cu and Zn for nonamended (NA) and poultry waste amended (PWA). (Data from Han, F.X. et al., in *Trace Elements in Soil, Bioavailability, Flux and Transfer*, Iskandar, I.K. and Kirkham, M.B., Eds., Lewis Publishers, Boca Raton, FL, 2001, pp. 145–174.)

such as the Netherlands, Belgium, and Germany, 25% of the groundwater has nitrate levels higher than 50 mg/l, most likely due to manure loading on the land (Eurostat, 1991). Large facilities in particular do not have sufficient land to apply the manure as a fertilizer. In the United States, it has been reported that 350 million tonnes of manure is produced each year, and that only 18% of hog farms and 23% of dairy farms spread the manure on enough land surface so that nitrogen levels in the drinking water are maintained within water quality guidelines. Approximately 50% of lake water and 27% of river water in the United States are also contaminated with nutrients (Gleick, 1993). Runoff of the manure during rain events causes high BOD levels in surface waters. Nonpoint source pollution of surface waters (rivers, lakes, and oceans) and groundwater is considered to be the major source of water pollution.

Coarse-textured soils are particularly susceptible to (1) increased movement of nitrate below the surface and (2) increased salinity levels and groundwater nitrate concentrations as manure application rates increase. More effective means of spreading fertilizers by correlating with plant needs are required. Plant uptake capacities should not be exceeded.

Han et al. (2001) reviewed the accumulation and distribution of heavy metals in soils amended with animal wastes, sewage sludges, and other wastes (Figure 6.11). For poultry litter-amended soil, copper and zinc concentrations increased at a rate of 2 mg/kg-yr. Most of the copper was present in the organic (46%) and residual fractions (52%), whereas most of the zinc was found in the easily oxidizable (48%) and organic (23%) fractions. Mobility studies indicated that there was slight movement of zinc downward in the soil, and copper moved to the 40-cm depth. Zinc was particularly mobile. At depths of 60 cm, iron oxide and residual fractions were twice those of the nonamended soils.

6.5.3 Alternatives for Manure Treatment

6.5.3.1 Aerobic Composting

Composting can be used to stabilize manure, particularly storage. This method is preferable over other aerobic treatments that are subject to high operation costs and sludge

TABLE 6.6

Comparison of Anaerobic Digestion Residue and Compost for Nitrogen Emissions

Treatment for Waste	Global Warming (tonne CO_2-equiv/year)	Acidification (tonne SO_2-equiv/year)	Eutrophication (tonne O_2-equiv/year)
Anaerobic digestion	654	43	688
Composting by reactor	2618	20	991

Source: Adapted from Dalemo, M. et al., *Resour. Conserv. Recycling*, 24, 363–381, 1998.

production. It is, however, labor intensive and costly due to the requirement for aeration. The thermophilic conditions of 54 to 71°C destroy most pathogens. Most of the ammonia is volatilized early in the process. Losses of nitrogen during the composting process, however, can be as high as 50% (Thomsen, 2000). Mixing with other substrates, such as the bedding, is usually required for bulking. Normally, straw or wood shavings are used as bulking agents, with adjustments for nitrogen content to be about 1:20 or 1:40. Compared with untreated manure, volatilization, and leaching of nitrogen, the risk of pathogen spreading and odor release is reduced. Yang et al. (2003) reported that (1) greenhouse gas (GHG) emissions of compost were 1.9 times less than those of slurry manure and 1.5 times less than those of stockpiled manure for dairy manure (Table 6.6); (2) composted liquid pig manure, in comparison to liquid pig manure, was more stabilized in terms of manure carbon; and (3) composted liquid pig manure produced reduced emissions of carbon dioxide and N_2O. Vervoort et al. (1998) determined that composting of poultry litter stabilizes phosphorus and reduces losses by runoff. Others, such as Delschen (1999), however, found that although the addition of 2.6 tonnes of composted manure led to higher accumulation rates of soil organic matter (SOM) than untreated manure, the amounts were approximately the same if the 40 to 60% carbon loss during composting was taken into account.

Various composting systems have been described in Mulligan (2002). Production of compost for soil conditioning leads to a more stable product with practically no odor. The more simple processes may also lead to increased ammonia emissions. The selected bulking material can influence ammonia emissions. Emissions of methane and nitrous oxide depend on the aeration rates. The benefits of compost for soil conditioning are well known.

6.5.3.2 Anaerobic Digestion

The organic content of manures can be treated by anaerobic digestion. The methane produced can be used for fuel or electricity production. Treatment of the manure by anaerobic digestion can significantly reduce the impact on water resources. Greenhouse gas (carbon dioxide in particular) emissions are reduced, and the products have improved fertilizer capability. Nitrogen and phosphorus availability for crops is increased, reducing chemical fertilizer requirements. A comparison of the mineralization of N of anaerobically stored manure with composted ruminant manure showed that anaerobic manure loses less nitrogen than during composting, and therefore is a better source of inorganic N for fertilizer (Thomsen and Olesen, 2000). Only some of the nitrogen in anaerobic residues is organically bound, whereas most of the nitrogen in compost is in this form. It has been estimated that digestion can reduce greenhouse gas emissions by 1.4 kg of carbon dioxide per kg of volatile solids (VS) in manure. Anaerobic treatment reduced N_2O emissions by more than 50% due to VS reduction after spring application onto soil — in comparison with untreated manure (Sommer et al., 2004). Although pathogens are reduced, the inactivation may only be about 1 to 2 log at 30°C (Burton and Turner, 2003).

In 2000 in Germany, there were more than 1000 anaerobic plants using agricultural substrates (Langhans, 2002). The combination of government and environmental factors has led to the substantial growth in the numbers used on animal farms. Thirty of these had capacities of greater than 1000 m^3. Although all types of manure are digestible, cow manure is more difficult to digest because of the higher fiber content, as opposed to pig and poultry manures. Typically, yields of methane are in the order of 290 l/kg of volatile solids (VS) for pig manure and 210 l of methane per kg of cattle manure (Burton and Turner, 2003). Co-substrates are often used to enhance the carbon and nutrient contents. These substrates include fodder beet and other green wastes. Cow manure in particular is well suited to co-digestion, as it already has a high fiber content. The nitrogen content of the manure serves as pH buffering and as a continuous inoculum. The co-substrates also increase methane yields (Mulligan, 2002).

Lagoons have lower capital and operating costs than digesters. Organic matter is reduced, while nitrogen and phosphorus remain in the end product. The use of lagoons is more frequent in warmer climates. Proper sizing and management of lagoons is required to ensure odor control. Loading rates of 60 to 90 g of VS per m^3 of lagoon per day have been suggested by the Natural Resources Conservation Service in the Northeast of the United States. Since methane is a greenhouse gas, gas collection should be practiced to avoid release into the atmosphere. Odors may also be a problem. Floating covers are becoming more popular. Temperatures of 35°C are required to maintain optimal biogas production. Heating systems may be required, particularly in the winter. Pathogen reduction in lagoons is also minimal (Burton and Turner, 2003).

6.5.3.3 *Wetlands*

The liquid portion of effluents from farm operations must usually be treated because of their high nutrient and organic contents. Since the wastewater contains a higher concentration of soluble ammonia than the initial feed, land spreading is a viable option because this allows for easier uptake by plants. Algal ponds or constructed wetlands may also be used to improve water quality to allow water reuse. Constructed wetlands are operated as subsurface flow or free surface flow (Mulligan, 2002). Nutrients are removed by the plants growing in the wetlands. Subsurface systems are not as efficient for nutrient removal. Hammer et al. (1993) have reported on a two-cell surface system used for a 500-swine operation to reduce 90 kg of BOD per day to 36 kg of BOD per day in a 3600 m^2 wetland. Nitrogen loading rates should be from 3 to 10 kg/ha-day, and ammonia concentrations in the influents should not be higher than 100 to 200 mg/l. In general, N, P, and solids reductions should be greater than 50%, and BOD greater than 60% if the wetlands are not overloaded. Wetlands offer a method of enhancing biodiversity while treating liquid effluent discharges.

6.5.3.4 *An Integrated Manure Treatment Method*

A comparison of manure management methods is shown in Table 6.7. The choice of the most appropriate method for management of manure must include minimization of emissions and other impacts on the environment, such as decreased water, groundwater, and soil quality. Dalemo et al. (1998) utilized a simulation model (ORWARE) for the calculation of energy and nutrient flows from soil and liquid organic wastes from restaurants. Composting and anaerobic digestion were compared. A life cycle approach was used that included the process and the soil emissions from the product. Soil emissions were in the form of N$_2$O production. Since the content of ammonia is lower in the compost, emissions of ammonia in the soil are also lower, in comparison to anaerobic digestion residues, where

TABLE 6.7

Comparison of the Sustainability of Manure Treatment Methods

Treatment	Energy	Emissions	Products	Costs
Land spreading	None produced, energy for spreading	Ammonia, methane, N_2O, odor, leaching of N, P, bacteria	Fertilizer	Low
Composting	Energy for mixing and mechanical separation, heat generation	Carbon dioxide, ammonia losses higher than anaerobic	Compost for soil improvement	Depends on sophistication
Anaerobic digestor	Energy produced in form of methane, heat required for the digester	High BOD liquid effluent	Methane for electricity or energy	Depends on sophistication
Anaerobic lagoon	Energy produced in form of methane	Odors, N_2O, and methane, if not covered well	Methane for electricity or energy	Low

the organic nitrogen is mineralized. In respect to concerns for eutrophication, emissions are mainly in the form of nitrate leaching from the soil, ammonia release during composting, and NO_x release during combustion of biogas for electricity generation. Composting produces ammonia emissions during the composting process, in addition to NO_x and SO_x.

The integrated manure management system shown in Figure 6.12 takes advantage of the less severe impact of the anaerobic digestion process for primary treatment and subsequent composting for final treatment and disposal. As shown in the diagram, combined wastes (green and manure) are sent to the anaerobic digestion system for treatment. This reduces the VS content. The digestate is then removed from the digesters for solid/liquid separation by screw press or other means. The solids are sent for composting, while the liquid is sent for wastewater treatment. After biological treatment (wetlands or a reactor if space is limited), the water can be stored and reused as required or spread on the land surface. The remaining solids content is ideal for composting. The requirements for oxygen will be decreased due to the previous anaerobic digestion step. This step also requires energy that can be produced by the anaerobic digestion process. The resulting compost would be high-grade compost full of nutrients with adequate pathogen reduction.

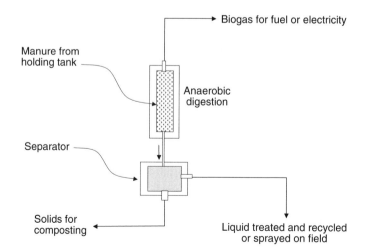

FIGURE 6.12

Schematic diagram showing flow sequence for sustainable management of manure.

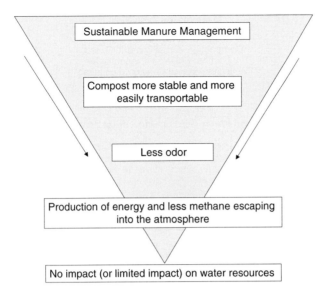

FIGURE 6.13
Impact of a sustainable manure management process on the geoenvironment.

The application rates to the soil would need to be based on N and P crop requirements to avoid excess soil nutrients (Cooperbrand et al., 2002).

Land spreading of manure creates numerous environmental problems in the agro-ecosystem. Manure treatment by anaerobic digestion is a step in the direction toward sustainable agricultural practice since it is a renewable source of energy. The main benefits are shown in Figure 6.13. Manure treatment enables farmers to reduce (1) pathogens, odor, N_2O, and carbon dioxide emissions, and (2) air, soil, and groundwater pollution caused by manure spreading. As anaerobic digestion is not a complete solution, an integrated one, including composting to produce a soil conditioner, will be required to ensure complete management of all aspects of manure treatment, as shown in Figure 6.12. Modeling of the emissions based on C, N, and P mass balances is an effective method to compare manure management methods. However, soil conditions, climate, and other factors can significantly influence the results of the comparison. This approach is a clear example of minimization of the impact of agricultural practices shown in Figure 6.14.

6.6 Concluding Remarks

The geoenvironment associated with agriculture plays a very important role in carbon and nutrient cycling, climate change, maintaining ecosystem biodiversity, and pollution management. As has been stressed previously, the discussion concerning the impacts from agricultural practices in production of food is motivated primarily in respect to concerns for protection of the geoenvironment. Accordingly, the impacts and means to mitigate impacts are all viewed from the geoenvironmental perspective. We acknowledge that many of these impacts and the means to alleviate and eliminate them are well known to the agricultural community, and that they share the same concern as those involved with geoenvironmental management. As has been mentioned, there is validity to the thesis that many of the measures for mitigating, minimizing, and even eliminating impacts from

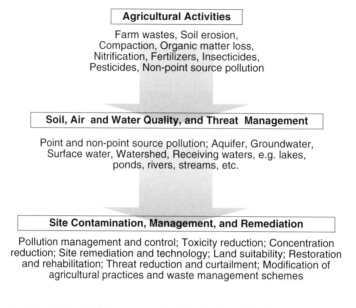

FIGURE 6.14
Summary of agricultural activities, their impacts, and minimization of the impacts.

contaminant loading and soil quality impairment run counter to intensive agricultural practices. Conflicts between agricultural productivity and protection of the geoenvironment have always existed. However, if sustainability is a goal that has merits, it is necessary to seek sustainable practices for both agricultural production and the geoenvironment.

While soils have been exploited for agricultural and livestock purposes, they are very vulnerable to environmental stresses and are renewable at very slow rates (over centuries). They are at the interface of the lithosphere, hydrosphere, atmosphere, and biosphere. Aquifers are subjected to overpumping, subsidence, and contamination. Excessive extraction leads to subsidence, erosion, aridification, and salination. Contamination from farm wastes and other activities reduces the use of the water and can cause health problems. Lack of water will have a significant impact on agriculture. Conservation of water, energy, and soil resources is required, and new technologies for agricultural practices such as irrigation are needed. Recycling of crop residues and other wastes is another area where considerable development is required.

Agricultural reshaping of the land leads to erosion and other soil displacements. It changes the landscape and causes disruptions to the ecosystems and the environment overall. More than 99% of the world's food comes from the land ecosystems, and more than 87% of the freshwater is used for agriculture. Lack of available and good-quality water and soil limit food production today and will be even more strained as the population increases.

Many contaminants are discharged by diffuse means from farming activities and are difficult to control. Water runoff and infiltration transport the contaminants to groundwater, lakes, rivers, and marine areas. The sediments are subsequently responsible for trapping many of these contaminants. Further investigations are required to better understand the retention of the contaminants by plants and soil, and the chemical and microbial reactions governing the fate of the contaminants. Long-term monitoring is needed to develop databases for predictive modeling and to evaluate the sustainability of the various agricultural practices. An integrated approach of mitigation methods and reduction of the contaminants at the source are required to reduce the impact on the surface and groundwater. Multidisciplinary efforts and those involving the public are required to increase

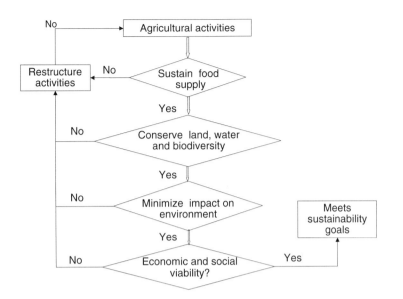

FIGURE 6.15
Evaluation of the sustainability of agricultural activities.

public knowledge and responsibility. To evaluate the sustainability of agricultural activities, a process as illustrated in Figure 6.15 will need to be followed.

If total costs involved in the production of food were to be determined, one would need to include not only the cost of physical operations in tilling and reaping the harvest, but also the cost of all the items needed to mount the effort in production. This would include the costs for production of the fuel needed to operate the machinery, production costs for the fertilizer and pesticides, etc. Expressing these all in terms of calories of energy, Pimental and Wen (1990) calculated that more than 10 million kcal of energy is required for the operation of agricultural machines, production of fuel, fertilizer, and pesticide, and for irrigation and other inputs for 1 ha of U.S. corn. This is a useful basis for determining efficiency of operation and production of food. Government and nongovernmental organizations, the scientific community, and farmers must all work together, as outputs in the order of two or three times the present levels must be achieved by 2050 (NRC, 1999).

References

Addiscott, T.M., Falloon, P.D., and Tuck, G., (2002), Using the PSALM Model to Interpret the Change Point and Other Aspects of Phosphate Behaviour, paper presented at the BSSS Meeting, Plymouth, September.

Banin, A., Navrot, J., Noi, Y., and Yoles, D., (1981), Accumulation of heavy metals in arid-zone soils irrigated with treated sewage effluents and their uptake by Rhodes grass, *J. Environ. Qual.*, 10:536.

Barbash, J.E., (2005), The geochemistry of pesticides, in *Environmental Geochemistry*, Sherwood Lollar, B. (Ed.), Elsevier, Amsterdam, pp. 541–612.

Bernard, C.C., Carbiener, R., Cloots, A.R., Groelicher, R., Schenck, Ch., and Zilliox, L., (1992), Nitrate pollution of groundwater in the Alsatian plain (France): a multidisciplinary study of an agricultural area, *Environ. Geol. Water Sci.*, 20:125–137.

Burton, C.H. and Turner, C. (Eds.), (2003), *Manure Management: Treatment Strategies for Sustainable Agriculture*, 2nd ed., Silsoe Research Institute, Wrest Park, U.K.

Canadian Biodiversity Information Network (CBIN), (1998), http://www.cbin.ec.gc.ca.

Chamley, H., (2003), *Geosciences, Environment and Man: Development in Earth and Environmental Sciences*, Elsevier, Amsterdam, 450 pp.

Christen, P., (1996), Konzept des "Sustainable Development" bzw. Für den landwirtschaftlichen Bereich "Sustainable agriculture" (gemäß dem erichtöur common future'bzw dem Brundtland-Report), *Berichte Landwirtschaft*, 74:66–86.

Commoner, B., (1971), *The Closing Circle, Nature, Man and Technology*, Alfred A. Knopf, New York, 326 pp.

Cooperbrand, L., Bollero, G., and Coale, F., (2002), Effect of poultry litter and composts on soil nitrogen and phosphorus availability and corn production, *Nutr. Cycling Agroecosyst.*, 62:185–194.

Daily, G.C., Dasgupta, P., Bolin, B., Crosson, P., du Guernay, J., Ehrlirch, P., Folke, C., Jansson, A.M., Jansson, B.O., Kautsky, N., Kinzig, A., Levin, S., Maler, K.G., Pinstrup-Anderson, P., Siniscalco, D., and Walker, B., (1998), Food production, population growth and the environment, *Science*, 281:1291–1292.

Dalemo, M., Sonesson, U., Jonsson, H., and Bjorklund, A., (1998), Effects of including nitrogen emissions from soil in environmental systems analysis of waste management strategies, *Resour. Conserv. Recycling*, 24: 363–381.

Delschen, T., (1999), Impacts of long-term application of organic fertilizers on soil quality parameters in reclaimed loess soils of the Rhineland lignite mining area, *Plant Soil*, 213:43–54.

Dobson, A.P., Bradshaw, A.D., and Baker, A.J.M., (1997), Hopes for the future: restoration ecology and conservation biology, *Science* 277:515–522.

Donaldson, D., Kiely, T., and Grube, A., (2002), *Pesticide Industry Sales and Usage: 1998 and 1999 Market Estimates*, U.S. Environmental Protection Agency, Office of Pesticide Programs, Washington, DC, 33 pp., http://www.epa.gov/oppbead1/estsales/99pestsales/market_estimates 1999.pdf.

El Bassam, N., (1998), Fundamentals of sustainability in agricultural production systems, and global food security, in *Proceedings of the International Conference on Sustainable Agriculture for Food, Energy and Industry* (Joint Conference of the Food and Agriculture Organization (FAO), Society of Sustainable Agriculture and Resource Management (SSARM), and Federal Agriculture Research Center (FAC)), Braunschweig, Germany, June 1997.

Eurostat, (1991), *Environment Statistics*, Eurostat Office for Official Publications of the European Communities, Luxembourg.

Falkenmark, M., (1989), Water scarcity and food production, in *Food and Natural Resources*, Pimentel, D. and Hall, C.W. (Eds.), Academic Press, San Diego, pp. 16–191.

FAO, (1995), *Livestock: A Driving Force for Food Security and Sustainable Development*, WAR/RMZ 84/ 85, FAO, Feed Resources Group, Rome, pp. 5–17.

FAO, (1996), *Integration of Sustainable Agriculture and Rural Development Issues in Agricultural Policy*, Winrock, Morrilton, AR.

FAO, (1997), *Review of the State of the World's Forest*, United Nations, Rome.

FAO, (1998), *The State of Food and Agriculture 1998*, FAO, Rome, 373 pp.

Fischer Taschenbuch, (1996), *Der Fischer Weltalmanach 1997*, Fischer Taschenbuch, Frankfurt.

Fischer Taschenbuch, (1999), *Der Fischer Weltalmanach 2000*, Fischer Taschenbuch, Frankfurt.

Galloway, J.N., Levy, H., II, and Kasibhatla, P.S., (1994), Year 2020: consequences of population growth and development on deposition of oxidized nitrogen, *Ambio* 23:120–123.

Giampietro, M. and Pimentel, D., (1995), *Food, Land, Population, and the U.S. Economy*, College of Agriculture and Life Sciences, Cornell University, Ithaca, NY.

Gibb, D.J., Helgason, B.L, Janzen, H.H., and McAllister, T.A., (2005), A Computer Model for Estimating Greenhouse Gas Production from Agriculture, poster presented at BIOCAP Canada Foundation's First National Conference, New York.

Gleick, P.H., (1993), *Water in Crisis*, Oxford University Press, New York, 473 pp.

Goldhamer, D.A. and Snyder, R.L. (Eds.), (1989), *Irrigation Scheduling: A Guide for Efficient On-Farm Water Management*, Publication 21454, Division of Agricultural and Natural Resources, University of California, Oakland.

Goodland, R., (1998), Environmental sustainability in agriculture: bioethical and religious arguments against carnivory, in *Ecological Sustainability and Integrity: Concepts and Approaches*, Kluwer Academic Publishers, Dordrecht, the Netherlands, pp. 235–265.

Gregerson, H.M., Draper, S., and Elz, D., (1989), *People and Trees: The Roles of Social Forestry in Sustainable Development*, World Bank, Washington, DC.

Hammer, D.A., Pullen, B.P., McCaskey, T.A., Eason, J., and Payne, V.W.E., (1993), Treating livestock, wastewaters with constructed wetlands, in *Constructed Wetlands for Water Quality Improvement*, Moshiri, G.A. (Ed.), Lewis Publishers, Chelsea, MI, pp. 343–348.

Han, F.X., Kingery, W.L., and Selim, H.M., (2001), Accumulation, redistribution, transport and bioavailability of heavy metals in waste-amended soils, in *Trace Elements in Soil, Bioavailability, Flux and Transfer*, Iskandar, I.K. and Kirkham, M.B. (Eds.), Lewis Publishers, Boca Raton, FL, pp. 145–174.

Heijungs, R., (2000), Dynabox: a multi-media fate model, in *Heavy Metals: A Problem Solved? Methods and Models to Evaluate Policy Strategies for Heavy Metals*, van der Voet, E., Guinée, J.B., and Udo de Haes, H.A. (Eds.), Kluwer Academic Publishers, Dordrecht, the Netherlands, pp. 65–76.

Hesketh, N. and Brookes, P.C., (2000), Towards development of an indicator of risk of phosphorus loss from soil to water, *J. Environ. Qual.*, 29:105–110.

Higgens, M.J., Rock, C.A., Bouchard, R., and Wnegrezynek, B., (1993), Controlling agricultural runoff by use of constructed wetlands, in *Constructed Wetlands for Water Quality Improvement*, Moshiri, G.A. (Ed.), Lewis Publishers, Chelsea, MI, pp. 357–367.

Houben, G.J., (2002), Flow and transport modeling: natural attenuation of common agricultural and atmospheric pollutants: reactive transport of nitrate, potassium and aluminium, Vol. 275, IAHS, pp. 519–524.

Hutson, J.L. and Wagenet, R.J., (1989), LEACHM: Leaching Estimation and Chemistry Model, version 2, Center for Environmental Research, Cornell University, Ithaca, NY.

Janzen, D., (1998), The gardenification of the wildland nature and the human footprint, *Science*, 279:1312–1313.

Kattelmann, R., (1990), Conflicts and cooperation over floods in the Himalaya-Ganges region, *Water Int.*, 15:1–5.

Klein, M., (1994), Evaluation and comparison of pesticide leaching models for registration purposes: results of simulations performed with the Pesticide Leaching Model, *J. Environ. Sci. Health*, A29:1197–1209.

Knisel, W.G., (1980), *CREAMS: A Field Scale Model for C*hemicals R*unoff and E*rosion for A*gricultural M*anagement S*ystems*, Department of Agricultural Science and Education Administration, Washington, DC.

Knisel, W.G. and Davis, F.M., (1999), *GLEAMS (Groundwater Loading Effects of Agricultural Management Systems), Version 3.0, User Manual*, SEWRL-WGK/FMD-050199, USDA-ARS, Tifton, GA, 167 pp.

Kornegay, E.T., Hedges, J.D., Martens, D.C., and Kramer, C.Y., (1976), Effect on soil mineral storage and plant leaves, following application of manures of different copper contents, *Plant Soil*, 45:151–160.

Langhans, G., (2002), Manure and Biowaste Digestion in Germany: History, Trends and Practical Verification, Linde Technical Document, VIII/P9e/0111, http://62.27.58.13/en/p0091/p0094/pdf/Biowaste-Digestion.pdf.

Millennium Ecosystem Assessment, (2005), *Ecosystem and Human Well-Being Synthesis Report*, Island Press, Washington, DC, 155 pp.

MAFF, (2000), *Towards Sustainable Agriculture: A Pilot Set of Indicators*, Ministry of Agriculture, Fisheries and Food, MAFF Publications, London, 74 pp.

Malthus, T., (1798), *An Essay on the Principle of Population, as It Affects the Future Improvement of Society with Remarks on the Speculations of Mr. Godwin, M. Condorcet, and Other Writers*, printed for J. Johnson, in St. Paul's churchyard, London.

Mari, H.S., Rama-Krishna, R.N., and Lall, S.D., (1985), Improving field microclimate and crop yield with low cost shelter belts in Punjab, *Int. J. Ecol. Environ. Sci.*, 11:111–117.

Matson, P.A., Parton, W.J., Power, A.G., and Swift, M.J., (1997), Agricultural intensification and ecosystem properties, *Science*, 277:505–509.

McCaffrey, S.C., (1993), Water, politics, and international law, in *Water in Crisis: A Guide to the World's Freshwater Resources*, Gleick, P.H. (Ed.), Oxford University Press, New York, pp. 92–104.

McRae, T., Smith, C.A.S., and Gregorich, L.J. (Eds.), (2000), *Environmental Sustainability of Canadian Agriculture: Report of the Agri-Environmental Indicator Project*, Agriculture and Agri-Food Canada, Ottawa, Ontario.

Meadows, D.H., Meadows, D.L., and Randers, J., (1992), *Beyond the Limits*, Chelsea Green Publishing Co., White River Junction, VT, 299 pp.

Micklin, P., (1988), Dessication of the Aral Sea. A water management disaster in the USSR, *Science*, 241:1170–1176.

Mulholland, P.J. and Lenat, D.R., (1992), Streams of the southeastern Piedmont, Atlantic drainage, in *Biodiversity of the Southeastern United States: Aquatic Communities*, Hackney, C.T., Adams, S.M., and Martin, W.H. (Eds.), John Wiley & Sons, New York, pp. 193–231.

Mulligan, C.N., (2002), *Environmental Biotreatment*, Government Institutes, Rockville, MD, 395 pp.

Mullins, C.L., (1990), Physical properties of soils in urban areas, in *Soils in the Urban Environment*, Bullick, P. and Gregory, P.J. (Eds.), Blackwell Scientific Publications, Oxford, pp. 87–118.

Mullins, C.L., Martens, D.C., Miller, W.P., Kornegay, E.T., and Hallock, D.L., (1982), Copper availability, form and mobility in soils from three annual copper-enriched hog manure applications, *J. Environ. Qual.*, 11:316–320.

Nabhan, G.P. and Buchmann, S.L., (1997), Services provided by pollinators, in *Nature's Services: Societal Dependence on Natural Ecosystems*, Daily, G.C. (Ed.), Island Press, Washington, DC, chap. 8, pp. 133–150.

Nash, L., (1993), Water quality and health, in *Water in Crisis: A Guide to the World Fresh Water Resources*, Gleick, P. (Ed.), Oxford University Press, New York, pp. 29–35.

Nature Conservancy Council, (1986), *Worcestershire Inventory of Ancient Woodland* (Provisional), NCC, Peterborough, U.K.

NGS, (1995), *Water: A Story of Hope*, National Geographic Society, Washington, DC.

NRC (National Research Council), (1991), *Toward Sustainability: A Plan for Collaborative Research on Agriculture and Natural Resources Management*, Panel for Collaborative Research Support for AID's Sustainable and Natural Resources Management, Academic Press, Washington, DC, 164 pp.

NRC (National Research Council), (1999), *Our Common Journey: A Transition toward Sustainability*, Board on Sustainable Development Policy Division, National Academy Press, Washington, DC, 384 pp.

OECD, (1996), *Selected Key Agrienvironmental Issues of Relevance to OECD Policy Makers*, Organization for Economic Cooperation and Development, Paris.

OECD, (2002), *Evaluating Progress in Pesticide Risk Reduction*, summary report on the OECD project on pesticide aquatic risk indication, Organization for Economic Cooperation and Development, Paris.

Ogaji, J., (2005), Sustainable agriculture in the UK, *Environ. Dev. Sustainability*, 7:253–270.

Payne, G.G., Martens, D.C., Kornegay, E.T., and Lindermann, M.D., (1988), Availability and form of copper in three soils following eight annual applications of copper-enriched swine manure, *J. Environ. Qual.*, 17:740–746.

PCAST, (1998), *Teaming with Life: Investing in Science to Understand and Use America's Living Capital*, President's Committee on Advisers on Science and Technology, Washington, DC.

Pimentel, D., (1989), Impacts of pesticides and fertilizers on the environment and public health, in *Toxic Substances in Agricultural Water Supply and Drainage*, Summers, J.B. and Anderson, S.S. (Eds.), U.S. Committee on Irrigation and Drainage, Denver, CO, pp. 95–108.

Pimentel, D., Acquay, H., Biltonen, M., Rice, P., Silva, M., Nelson, J., Lipner, V., Giordano, S., Horowitz, A., and D'Amore, M., (1992), Environmental and economic costs of pesticides use, *BioScience* 42:750–760.

Pimentel, D., Houser, J., Preiss, E., White, O., Fang, H., Mesnick, L., Barsky, T., Tariche, S., Schreck, J., and Alpert, S., (1998), Water resources: agriculture, the environment, and ethics, in *Ecological Sustainability and Integrity: Concepts and Approaches*, Kluwer Academic Publishers, Dordrecht, the Netherlands, pp. 104–134.

Pimentel, D. and Pimentel, M., (1996), *Food, Energy and Society*, 2nd ed., John Wiley & Sons, New York, 297 pp.

Pimentel, D. and Wen, D., (1990), Technological changes in energy use in U.S. agricultural production, in *Agroecology*, Carrol, C.R., Vandermeer, J.H., and Rosset, P.M. (Eds.), McGraw-Hill, New York, pp. 147–164.

Pinstrup-Anderson, P., Pandya-Lorch, R., and Rosegrant, M.W., (1997), *The World Food Situation: Recent Developments, Emerging Issues, and Long-Term Prospects*, International Food Policy Research Institute, Washington, DC.

Postel, S., (1992), *Last Oasis: Facing Water Scarcity*, W.W. Norton and Co., New York, 239 pp.

Postel, S.L., Daily, G.C., and Ehrlich, P.R., (1996), Human appropriation of renewable fresh water, *Science*, 271:785–788.

Quendler, T. and Reichert, T., (2002), Environmental issues and their significance for agriculture and the food industry, in *WTO, Agriculture and Sustainable Development*, Wohlmeyer, H. and Quendler, T. (Eds.), Greenleaf Publishing, Sheffield, U.K., pp. 234–251.

Quendler, T. and Schuh, T., (2002), Sustainability, a challenge for future economic and social policy, in *WTO, Agriculture and Sustainable Development*, Wohlmeyer, H. and Quendler, T. (Eds.), Greenleaf Publishing, Sheffield, U.K., pp. 193–205.

Quendler, T., Weiß, F., and Wohlmeyer, H., (2002), Conclusions and proposals for solutions, in *WTO, Agriculture and Sustainable Development*, Wohlmeyer, H. and Quendler, T. (Eds.), Greenleaf Publishing, Sheffield, U.K., pp. 234–251.

Redding, M., (2005), *Case Studies to Assess Piggery Effluent and Solids Application*, Department of Primary Industries and Fisheries, Toowooba, Queensland, Australia, February 2004, www.dpi.qld.gov.au/ilem.

Regenstein, L.G., (1991), *Replenish the Earth*, Crossroad Press, New York, 305 pp.

Rekolainen, S., Guoy, V., Francaviglia, R., Eklo, O.M., and Barlund, I., (2000), Simulation of soil water, bromide and pesticide behaviour in soil with the GLEAMS model, *Agric. Water Manage.*, 44:201–224.

Richard, S., Arnoux, A., Cerdan, P., Reynouard, C., and Horeau, V., (2000), Mercury levels of soils, sediments and fish in French Guiana, South America, *Water Air Soil Pollut.*, 124:221–244.

Richards, J.F., (1990), Land transformation, in *The Earth Is Transformed by Human Action: Global Warming and Regional Changes in the Biosphere over the Past 300 Years*, Clark, W.C., Turner, B.L., Kates, R.W., Richards, J., Mathews, J.T., and Meyer, W. (Eds.), Cambridge University Press, Cambridge, UK, chap. 10.

Rosanov, B.G., Targulian, V., and Orlov, D.S., (1990), Soils, in *The Earth Is Transformed by Human Action: Global Warming and Regional Changes in the Biosphere over the Past 300 Years*, Clark, W.C., Turner, B.L., Kates, R.W., Richards, J., Mathews, J.T., and Meyer, W. (Eds.), Cambridge University Press, Cambridge, U.K., chap. 12.

Sagardoy, J.A., (1993), An overview of pollution of water by agriculture, in *Prevention of Water Pollution by Agriculture and Related Activities, Proceedings of the FAO Expert Consultation*, Santiago, Chile, October 20–23, 1992, pp. 19–26.

Saha, A., (2001), Agricultural Pollution Control, *Envirospace Pollution*, http://www.envirospace.com/print.asp?article_id=461.

Said-Pullicino, D., Gigliotti, G., and Vella, A.J., (2004), Enviromental fate of triasulfuron in soils amended with municipal waste compost, *J. Environ. Qual.*, 33:1743–1751.

Sims, J.T. and Wolf, D.C., (1994), Poultry manure management: agricultural and environmental issues, *Adv. Agron.*, 52:2–84.

Sloggett, G. and Dickason, C., (1986), *Ground-Water Mining in the United States*, Economic Research Service, U.S. Department of Agriculture, Washington, DC.

Solley, W.B., Pierce, R.B., and Pearlman, H.A., (1993), *Estimated Use of Water in the United States, 1990*, U.S. Geological Survey, Washington, DC.

Sommer, S.G., Petersen, S.O., and Moller, H.B., (2004), Algorithms from calculating methane and nitrous oxide emissions from manure management, *Nutr. Cycling Agroecosyst.*, 69:143–154.

Statistics Canada, (1994), *Human Activity and the Environment*, Catalog 11-509E, Minister of Industry, Science and Technology, Ottawa, ON, 300 pp.

Sutton, A.L., Melvin, S.W., and Vanderholm, D.H., (1983), Fertilizer value of swine manure, Purdue University Co-operative Extension, *Pork Industry Handbook,* Factsheet, PIH-25.

Thies, C. and Tscharntke, T., (1999), Landscape structure and biological control in agroecosystems, *Science*, 285:893–895.

Thomas, G.W., (1987), Water: critical and evasive resource on semiarid lands, in *Water and Water Policy in World Food Supplies,* Jordon, W.R. (Ed.), Texas A&M University Press, College Station, TX, pp. 83–90.

Thomsen, I.K., (2000), C and N transformations in ^{15}N-cross-labelled solid ruminant manure during anaerobic and aerobic storage, *Bioresource Technol.*, 72:267–274.

Thomsen, I.K. and Olesen, J.E., (2000), C and N mineralization of composted and anaerobically stored ruminant manure in differently textured soils, *J. Agric. Sci.*, 135:151–159.

U.K. Local Government Management Board and Touche Ross Management Consultants, (1994), *Local Agenda 21 Sustainability Indicators Research Project: Report of Phase One*, The Local Government Management Board, Luton, UK, 85 pp.

United Nations Food and Agriculture Organization, (2003), Database on Pesticide Consumption, United Nations Food and Agriculture Organization, Statistics Division, http://www.fao.org/waicent/FAOINFO/economic/pestici.htm.

U.S. Soil Conservation Service, (1993), *Nutrient and Sediment Control System*, Technical Note 4, U.S. Department of Agriculture, Washington, DC, March.

USDA, Soil Conservation Service, (1992), *Agriculture Waste Management Field Handbook*, U.S. Department of Agriculture, Washington, DC.

U.S. EPA, (1987), *Quality Criteria for Water 1986*, EPA 440/5086-001, U.S. Environmental Protection Agency, Office of Water Regulations and Standards, Washington, DC.

U.S. EPA, (1997), *Monitoring Guidance for Determining the Effectiveness of Nonpoint Source Controls*, U.S. Environmental Protection Agency, Washington, DC.

USGS (U.S. Geological Survey), (1999), *The Quality of Our Nation's Waters: Nutrients and Pesticide*, U.S. Geological Circular 1225, water.usgs.gov.s

Verplancke, H., (Ed.), (1992), *Water Saving Techniques for Plant Growth*, NATO ASI Series, Series E: Applied Sciences, Kluwer Academic Publishers, Amsterdam, 241 pp.

Vervoort, R.V., Radcliffe, D.E., Cabrera, M.L., and Latimore, M., Jr., (1998), Nutrient losses in surface and subsurface flow from pasture applied poultry litter and composted poultry litter, *Nutr. Cycling Agroecosyst.*, 50:287–290.

Vitousek, P.M. and Mooney, H.A., (1997), Human domination of earth's ecosystems, *Science*, 277:494–499.

Wall, G.L., Pringle, W.A., and Sheard, R.W., (1991), Intercropping, red clover with sillage for soil erosion control, *Can. J. Soil Sci.*, 71:137–145.

Weyant, J., Davidson, O., Dowlabathi, H., Edmonds, J., Grubb, M., Parson, E.A.., Richels, R., Rotmans, J., Shukla, P.R., Tol, R.S.J., Cline, W., and Frankhauser, S., (1996), Integrated assessment of climate change; an overview and comparison of approaches and results, in *Climate Change 1995; Economic and Social Dimensions of Climate Change,* Bruce, J.P., Lee, H., and Haites, E.F., (Eds.), Cambridge University Press, Cambridge, UK and New York, NY, pp. 367–396.

WFS (World Food Summit), (1996), *Rome Declaration on World Food Security and World Food Summit Plan of Action*, FAO, Rome, http://www.fao.org/wfs/final/rd-e-.htm. Accessed on 7/8/99.

WHO, (1996), *The World Health Report 1996. Fighting Disease and Fostering Development*, World Health Organization, Geneva.

Wolters, A., Klein, M., and Vereecken, H., (2004), Atmospheric pollutants and trace gases: an improved description of pesticide volatilization: refinement of the pesticide leaching model (PELMO), *J. Environ. Qual.*, 33:1629–1637.

WRI, (1992), *World Resources 1991–1992*, Oxford University Press, New York.

Xu, J.M., Gan, J., Papiernik, S.K., Becker, J.O., and Yates, S.R., (2003), Incorporation of fumigants into soil organic matter, *Environ. Sci. Technol.*, 37:1288–1291.

Yagi, K. and Minami, K., (1990), Effect of organic matter application on methane emissions from paddy soils, *Soil Sci. Plant Nutr.*, 36:599–610.

Yang, X.M., Drury, C.F., Reynolds, W.D., Tan, C.S., and McKenney, D.J., (2003), Interactive effects of composts and liquid pig manure with added nitrate on soil carbon dioxide and nitrous oxide emissions from soil under aerobic and anaerobic conditions, *Can. J. Soil Sci.*, 83:343–352.

7

Industrial Ecology and the Geoenvironment

7.1 Concept of Industrial Ecology

The basic concept of *industrial ecology* is one that seeks to apply a systems approach to environment protection and environmental resource conservation as integral components of industrial production and development. By treating the industry–environment as an intertwined system, industrial ecology focuses its attention on the total picture of (1) renewable and nonrenewable environmental resource consumption and conservation at the one end (front end) of industry activities, (2) efficient industrial production through technology and resource conservation and adherence to the 4Rs (recycle, recovery, reduction, and reuse of waste products), and (3) environmentally sensitive management of emissions and disposal of waste products from industrial activities at the other end. In an ideal world, industrial ecology is a holistic approach to industrial production of goods; i.e., it takes into account the goals of environment and resources sustainability while meeting its goals of production of goods and other life support systems to the benefit of consumers.

This chapter directs its attention to the geoenvironment compartment of industrial ecology. The particular concerns are in respect to item 3 above and many aspects of the 4Rs in item 2. We will consider the interactions on the geoenvironment by activities associated with manufacturing and service industries. As we will see, as far as geoenvironmental resources are concerned, and in respect to sustainability goals, the primary concerns are (1) use of natural resources as both raw materials and energy supply, and (2) emissions and waste discharges. Since the purview of this book addresses resource use from the geoenvironment framework and not from the industry perspective, we will concentrate our attention on the smokestack emissions and discharge of liquid and solid wastes and waste products.

By definition, *manufacturing* means the transformation of raw or source materials into finished goods and products. The manufacturing industries considered in this chapter include those midstream and downstream industries that utilize the outputs (raw or source materials) from their upstream counterparts (e.g., those industries discussed in Chapters 5 and 6) to produce their goods as finished products or as inputs or source materials for further downstream industry use. The petrochemical and agro-processing industries are good examples of downstream industries. The source or raw materials for the petrochemical industries include oil and gas produced or obtained from oil and gas production upstream industries. Similarly, the source (raw) materials for the agro-processing industries include the products from agricultural and livestock production generated by the upstream agro-industries.

Industrialization, as commonly defined, is the process whereby *industry* becomes a dominant component in a socioeconomic order. The term *industry* is used here in the broadest sense to mean *the work undertaken (in a plant or facility) to produce goods using machines or technological aids.* A vital component of industrialization is manufacturing industries. These industries are essential to the health of the economy and to the support of a vibrant society. It is often said that activities of industries in general will inevitably come in conflict with the goals of a sustainable environment. To a very large extent, this argument or reasoning is based on the implicit assumptions that (1) nonrenewable natural resources (materials and energy) are required to fuel the engine of industry, (2) nonrenewable source materials are used in the production or manufacture of goods, and (3) the smokestack emissions and discharge of liquid and solid wastes from these industries are harmful to both human health and the environment.

In this chapter, we will be examining the dilemma of the apparent antagonistic actions from the activities of manufacturing industries established to meet societal needs in respect to the environmental sustainability indicators and goals. The problem reduces to a simple resolution of the required actions needed to accommodate the goals of societal and environmental sustainability. Within the context of the geoenvironment, this means the coexistence of societal and geoenvironmental sustainability aims and requirements. A clearer understanding of the interactions between the manufacturing industries and the land environment will allow for management strategies and ameliorative and preventative actions to be implemented. These should serve to promote harmony between societal and geoenvironmental sustainability goals.

Section 1.1.1 in Chapter 1 has pointed out that almost any external physical/chemical input or external sets of activity in the ecosphere will more than likely have an indirect or even direct impact on the geoenvironment and its ecosystems. This is particularly true when it comes to the various activities associated with the midstream and downstream industries dealing with the manufacture of goods. Included in the group of midstream–downstream industries are (1) life support industries such as the agro-, forest, mining, fisheries, and energy-producing industries and (2) production industries producing such varied goods as automobiles, pharmaceuticals, urban infrastructures, etc.

Although strictly speaking, service industries are not considered downstream industries because they do not produce goods; we include them in the discussions in this chapter because these industries provide services (as goods) to the consumer. Two simple examples of sources of potential geoenvironmental problems from service industries are (1) medical services (e.g., hospitals) and (2) military services (e.g., munitions handling and storage).

The industry concerns addressed in this chapter center around *land use in an industrial context*, i.e., land use in the context of activities in development of and in support of the various kinds of life support, manufacturing–production, and service industries. The upstream industries discussed in Chapters 4 to 6 have laid the basis for the production of consumer goods and facilities provided by downstream industries. Their interactions with the geoenvironment are direct, and most of their impacts on the geoenvironment can be readily recognized. Since downstream industries are the consumers of the raw goods and products from upstream industries, their interactions with the geoenvironment are more in terms of *what comes out from the processing and transformation end.* Figure 7.1 shows a simple illustration of the possible kinds of geoenvironmental interactions from downstream industries. The nature and composition of the outputs (airborne noxious particulates, liquid and solid wastes) will be direct functions of (1) the nature of the product being produced, (2) the nature and composition of the source material used for the production of the product, (3) the process technology and the kinds of treatments and additives used in the process technology, and (4) the various controls on emissions, wastes and wastewaters, treatments, and management of the production technology and system.

Given the wide extent of the various kinds of industries, it is clear that within the context of *geoenvironmental sustainability,* a discussion of all of them would not be possible or desired. Instead, the examination of the effects of land use and geoenvironmental man-

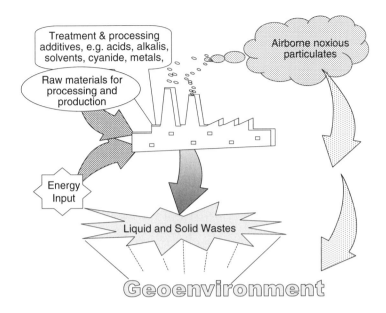

FIGURE 7.1
Interactions between manufacturing (midstream and downstream) industries and geoenvironment. Primary issues are energy usage, deposition of smokestack and other airborne noxious particulates, and discharge of liquid and solid wastes. The nature and composition of the liquid and solid wastes and noxious particulates will be a function of the source materials, process additives, process technology, and emission control.

agement requirements will confine itself to a few representative industries in the three groups of downstream industries. It is important to stress that the terms *effects* and *impacts* are not used in a negative sense. The nature of impacts ascribed to activities and events runs the gamut from beneficial through benign (neutral) to negative, depending on (1) the activity or event and (2) the indicators, markers, and criteria used to evaluate or assess the results of the impacts. Whether impacts from a specific set of activities or events will add or subtract value from the particular ecosystem in the geoenvironment is the question that needs to be answered. It is important to recognize that a comprehensive listing of all the impacts on the ecosystems of the geoenvironment accruing from a particular set of activities is not possible — to a very large extent because of the lack of knowledge of all the various items, activities, and interactions that comprise a functioning ecosystem. Chapter 2 has discussed many of these concerns.

7.2 Industry Activities and Geoenvironmental Interactions

The previous chapters have examined the impacts to the geoenvironment arising from urbanization (Chapter 4), resource exploitation (Chapter 5), and agriculture and food production (Chapter 6). The various kinds of industries that exist run the gamut from those that produce raw goods to those that provide finished products to service industries providing a variety of services. Most of the activities and industries discussed previously are upstream; i.e., they are industries devoted to production of the raw materials that need processing and transformation before reaching the consumer. Agricultural activities in aid of food production, for example, constitute an upstream industrial activity, whereas food preparation and food processing using materials from agricultural activities can legitimately be considered downstream industries. Downstream industries run the gamut from industries devoted to preparation of source materials as items and parts for manufacturers to transformation into final consumer products. For example, the production of automobiles requires countless numbers of parts, such as tires, engines, electronic parts, chassis, side panels, etc. Many of these parts are produced as downstream products by industries devoted to production of vital elements and parts for more intricate products. The following definitions apply:

- Upstream industries are those industries that produce the raw goods and source materials for downstream industries. Examples of these have been given in Chapters 5 and 6. Production of food, i.e., agricultural production of food (wheat, corn, barley, livestock, etc.), and mining and processing of metal ores are good examples of upstream activities and industries. The raw products can be used by the individual consumer and can also be used as resource material for midstream or even downstream industries.

- Downstream industries include those industries that use the raw goods produced by the upstream industries and prepare them as resource material for other downstream industries. Technically speaking, these can be called midstream industries. However, this term is not popular. A good example of this is the metal fabrication and processing industry discussed in the next section.

- Also included are production and assembly plants and industries that produce consumer goods and products that are directly utilized by the individual and collective consumers. These are the industries that transform source materials into consumer goods. Good examples of these are buildings, bridges, automobiles, leather goods, newsprint, electronic products, etc.

The subject of interest in the following sections relates directly to the net effect of the activities associated with these varied types of industries on the geoenvironment. The material covered in these sections will focus on some industries to highlight or demonstrate the particular geoenvironmental land use in question. At that time we will want to examine some of the major consequences and impacts of these interactions on the geoenvironment, with the aim of seeking solutions that would permit us to satisfy many of the geoenvironmental sustainability requirements.

Chapter 1, Section 1.4.1, shows four broad categories of life support and manufacturing–production industries and activities essential to the fulfillment of the WEHAB (water and sanitation, energy, health, agricultural, and biodiversity) goals articulated in the UN conferences on sustainability and growth. Taking the first goal (*water and sanitation*), as an example, the three recent declarations from world bodies demonstrate the significance of the need for adequate water and sanitation facilities. Both the United Nations Millennium Declaration (2000), adopted by all UN member states, and the World Summit on Sustainable Development (WSSD in 2002) supported the objectives of the Year 2000 and the Second World Water Forum. The principal objectives are stated as follows: (1) to reduce by one half the proportion of people without sustainable access to safe drinking water and without access to proper sanitation facilities by the year 2015, and (2) to reach full sustainable access to safe drinking water for all by the year 2025, including proper sanitation and hygiene. The figures, as stated in the year 2002, gave estimates of 1.2 billion people without proper access to safe drinking water, and 2.4 billion people without access to adequate sanitation facilities. The direct link to the third goal (*health*) and the associated issues is obvious. The questions to be asked are: (1) What technologies or strategies are available to allow one to harvest more safe drinking water in regions where water supply is scarce? (2) How do these technologies and strategies affect on the geoenvironment? (3) Will they lead to sustainable water supply?

Figure 1.6 and Figure 1.11 in Chapter 1 show that many of the activities and industries required to meet the WEHAB goals will have direct interaction with the geoenvironment and will incur major impacts on the geoenvironment. For the purpose of examination of geoenvironmental interaction by downstream industries, we will group them into groups of industries or activities associated with the following:

- *Mineral mining and processing industries*: The downstream mining processing industries are the metallurgical finishing and manufacturing–production types of industries. Also included are the nonmetal mineral processing industries, such as cement production, phosphoric acid production, etc.

- *Agro-processing industries*: The various interactions and actions of the upstream agro-industry have been discussed in Chapter 5. Associated downstream industries are generally classed as agro-processing-type industries. These are the industries that transform the products from the agricultural, forestry, and fisheries industries. The two general categories are (1) food industries and (2) nonfood industries.

- *Hydrocarbon, hydro, and other energy resources*: Other energy sources include biomass, hydrogen, solar, geothermal, wind, and tidal. Electric power generation and distribution for the nonhydro power systems fall somewhere in between upstream and downstream, depending on the type of energy resource being harvested. The major downstream industries are the various kinds of petrochemical industries. It is probably safe to say that downstream utilization of the products from this group of upstream industries is perhaps the largest of any of the categories of geoenvironmental resource usage.

- *Production of goods and facilities*: These are the downstream industries that transform the products issuing from upstream industries, such as hydrocarbon extraction, agricultural products, and metal ore production. The products from these downstream industries are either further transformed by other downstream industries or used directly by individual consumers.

- *Public and private services*: We classify these kinds of services as industries and downstream industries, even though they do not necessarily deliver hard goods to the consumer. The delivery of services as goods requires facilities and use of technology and goods that in one way or another will interact with the geoenvironment. The effects or results of these interactions need to be examined.

7.3 Mineral Mining and Processing Downstream Industries

For convenience in discussion, since this chapter is concerned with downstream industries and their interaction with the geoenvironment, it is understood that when we use the term *industry* this will mean *downstream industry.*

7.3.1 Metallurgical Industries

The detailed treatment of resource extraction and processing, covered in Chapter 5, dealt primarily with the various activities and processes associated with mineral mining and

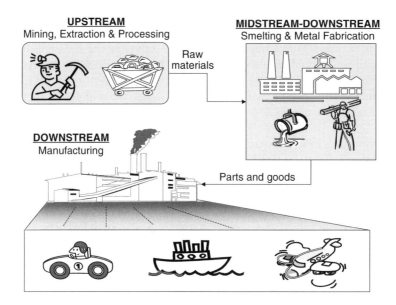

FIGURE 7.2
Illustration of upstream and downstream activities and industries. The example shown by the diagrams is for mining and extraction of metal for production–assembly of finished goods, e.g., automobiles, planes, and ships.

processing of the raw earth and rock materials. For this section, we want to outline the essential items for those downstream industries established to process the metal ores that directly or indirectly have an impact on the geoenvironment. Figure 7.2 gives a very simple example from the metal mining industry that shows the industries associated with mining extraction and processing as upstream and downstream types of industries. We show midstream–downstream industries in the upper right corner of Figure 7.2 because these industries can be classified as either midstream or downstream industries, depending on whether the products from these industries serve as source materials for other downstream industries or directly serve the individual consumer. A good example of this is the *parts* industry, which produces products such as parts and elements for other industries that will assemble the parts and elements into consumer goods (assembly plants or industries). The schematics shown in Figure 7.2 give a good illustration of the two kinds of industries. Note that midstream industries become downstream industries when their goods or products serve the individual consumer directly. An example of this from the diagram shown in Figure 7.2 is the production of metal consumer goods. Most reports on these industries are satisfied with the distinction between upstream and downstream types, and in general, these would be sufficient to encompass all the activities engaged in the exploitation and use of the products obtained from mining. However, the nature of the impacts on the geoenvironment can be vastly different, depending upon whether it is a midstream industry that produces or prepares parts for use by other downstream industries or a downstream industry that concentrates on consumer production goods. For example, the pollution impacts from the metal fabrication and processing midstream industries can be more significant than in the production and assembly of their counterpart downstream industries.

Metallurgical industries cover a vast variety of processes, depending on the final product issuing from the industry in question. The common practice of classifying these industries into three kinds of industries, in relation to the material source used in the process and the product produced, allows one to conveniently group the kinds of interaction of these industries with the geoenvironment. The midstream–downstream industries represented in the top right-hand corner of Figure 7.2, for example, are intensive users of energy. The foundries produce smokestack emissions that are sources of (1) acid rain generation and (2) land environment pollution when noxious airborne particulates find their way onto the solid land environment. The use of wet scrubbers or dry collectors can reduce harmful discharge of noxious gases and airborne particulates. Some of the discharges for foundries producing steel and aluminum include the various heavy metals, such as lead, zinc, manganese, chromium, arsenic, iron, nickel, and copper. For many other industries, in addition to the heavy metals, we will have a variety of organic chemicals included in the hazardous air pollutants (HAPs). A listing of many of these can be found at http://www.epa.gov/ttn/atw/188polls.html.

7.3.1.1 Metal Fabrication and Processing

Typical metal fabrication and processing midstream–downstream industries are those that produce value-added metals and metal goods. The types of metals include iron, aluminum, copper, lead, zinc, gold, tungsten, tin, silver, and cadmium, and they are produced in a variety of forms. These serve as downstream products or as resource material for other downstream industries. Some of the downstream industries in this present category include (1) metals finishing industries, e.g., electroplating, anodizing, and coatings, and (2) industries and assembly plants utilizing metals for production of goods, e.g., manufacture of automobiles, planes, trains, ships, ovens, refrigerators, and tin cans, to name a few.

Interactions between downstream industries and the geoenvironment are primarily in respect to energy resources needed to satisfy the energy requirements of the industries, and the handling and disposal of the waste discharges, including the inadvertent spills and overflows during processing, manufacture, and production. The demands for energy in downstream industries, especially those in the category of material preparation and finishing, are of considerable concern in the overall strategy to reach the sustainability indicators that define the sustainability goals in the energy resource field. Chapter 5 has discussed many of these concerns and strategies. Increased efficiency in manufacturing and production technology can alleviate some of the demands on energy use in these kinds of industries.

The metal finishing industries use a significantly large proportion of toxic chemicals in their various processes to produce, for example, corrosion resistance, wear resistance, electrical resistance, hardness, chemical resistance, and tarnish resistance metals. Figure 7.3 gives an example of the processes associated with metal fabrication. The sources of solid and liquid waste discharges are seen in the diagram. Most of the chemicals used in the metals finishing business end up as wastes. Considering that the various inputs to the processes include acids, solvents, alkalis, cyanide, loose metals, and complexing and emulsifying agents, it will not come as a surprise to note that the discharge and wastewaters can contain these chemicals and various other residues, especially since many of the processes include rinsing and bathing operations. These wastes predominantly result from the use of (1) organic halogenated solvents, ketones, aromatic hydrocarbons, and acids during the surface preparation stage of the overall finishing process, and (2) cyanide and metals in the form of dissolved salts in the plating baths during the surface treatment stage. These will all be found in the discharge streams shown in the bottom right of the diagram in Figure 7.3, as sludge, solid waste, and wastewater.

In the pig iron production process, the flue dust generated is captured by wet dust cleaners. In common with most of the metal finishing processes shown typically in Figure 7.3, steel finishing involves a number of necessary processes in the production of the desired surface and mechanical characteristics for the steel. The sulfuric acid used in most pickling processes creates hazardous by-products. Pickling solutions contain free acids, ferrous sulfate, undissolved scale and dust, and the various inhibitors and wetting agents, as well as dissolved trace elements. The concentration of acids and their types largely

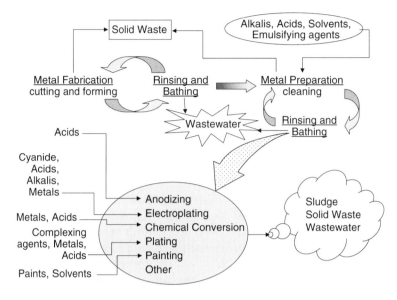

FIGURE 7.3

Processes and waste products generated in metal fabrication and preparation. (Adapted from U.S. EPA, *Profile of Metal Products Industry*, U.S. EPA, Office of Enforcement and Compliance Assurance, Washington, DC, 1995.)

depends on the type of iron being produced. Finishing operations generate effluents containing rolling oils, lubricants, and hydraulic oils that are in free and emulsified states. Other oils found in the effluents can also originate from cold reduction mills, electrolytic tin lines, and a variety of machine shop operations.

Water leaving the wet dust cleaners usually contains anywhere from 1000 to 10,000 mg/l of suspended solids, depending upon the furnace burden, furnace size, operating methods employed, and type of gas washing equipment. Disposal of these onto the land environment will require treatment and containment to minimize ground, groundwater, and receiving water pollution. Some of these aspects will be discussed in Section 7.3.3, when sustainability targets are considered. The more detailed treatment will be found in Chapter 9.

7.3.2 Nonmetal Mineral Resources Processing

Nonmetal mineral resources include clays and clay minerals, crushed stone, sand and gravel, dimensional rock slabs such as granite and marble slabs, phosphate, potash, gypsum, peat, sulfur, diamond, vermiculite, and natural alkali. Examples of utilization of the minerals supplied by upstream mining and processing industries include:

- Clays: For pottery and ceramics industries; construction industries for construction of roadway fills and clay-engineered barriers.

- Clay minerals: For paper coatings in paper industries: as catalysts for chemical industries; as expandable slurry materials (bentonites) in oil exploration industries and also construction industries.

- Crushed stone, silica, sand, and gravel: Glass industries; smelting industries (silica used as flux material); concrete production industries and bituminous concrete industries; cement industries; construction industries.

- Dimensional slabs: Construction industry.

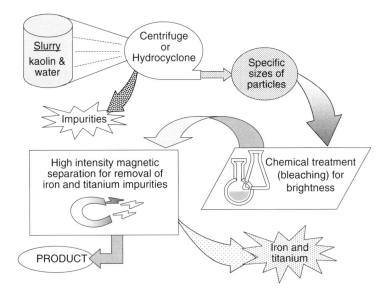

FIGURE 7.4
Production of coatings-grade kaolin using the hydrous procedure. Note that the discharge of impurities needs to be managed to ameliorate the impact on the geoenvironment.

The two minerals chosen as examples to illustrate the geoenvironmental interactions of activities of downstream industries transforming nonmetal minerals obtained from upstream mining extraction industries are the kaolin clay mineral and limestone. Figure 7.4 shows the principal steps taken to produce coatings-grade kaolin for use as kaolin-based coating pigment for the paper industry or in other industries, such as latex and alkyl paints and primers. The process is not energy intensive in comparison to the metal industries and cement production industries, and the discharges from the processing technique can be controlled. These discharges are the clay impurities and the iron and titanium impurities that are generally associated with kaolinites. The iron and titanium can be captured and reused for other applications. As far as geoenvironmental interactions are concerned, this downstream industrial activity is relatively benign.

The use of limestone for production of cement is shown in a simplified form in Figure 7.5. In respect to geoenvironmental interactions, two significant factors are evident: (1) intensive energy consumption attending almost every step of the production procedure, especially in the cyclones preheating stage and in the rotary kiln, and (2) discharge of fugitive cement kiln dust (CKN) as airborne particulates and as land discharge from electrostatic precipitators (ESPs) and other kinds of scrubbers. The significant human health issues attending the emissions of CO_2, NO_x, SO_2, dioxins, and furans cannot be ignored. For example, the release of metals into the atmosphere by copper–Ni refining, fossil fuel combustion and iron manufacture in the northern former USSR has contributed up to 90% of the metal loading in the European Arctic — a very sensitive environment (Pacyna, 1995). Arsenic (As), Cd, Pb, and Zn emissions in this region contribute 4.5, 2.4, 3, and 2.4%, respectively, of the global emissions. Wet and dry deposition of the metals with sulfuric acid has resulted in the accumulation of the metals in the soils, surface waters, and sediments. Acid rain enhances the environmental mobility of the metals and the bioavailability of these heavy metals. These metals may also bioaccumulate in the foods in the food chain, thus endangering the health of the consumers. Seabirds, seals, and polar bears have shown elevated levels of As, Cd, and Hg from ingesting fish and shellfish that eat contaminated algae, and plankton that accumulate heavy metals from sediments.

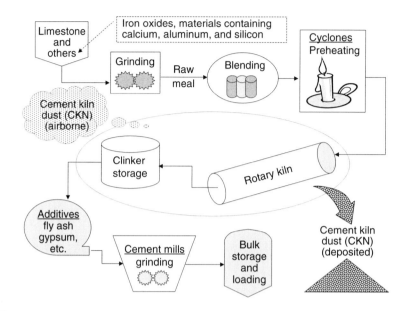

FIGURE 7.5

Basic elements in the manufacture of cement. Fugitive cement kiln dust (CKN) from the rotary kiln and clinker storage elements of the process will be airborne, and captured CKN from scrubber systems will be deposited on land.

TABLE 7.1

Release of Heavy Metals through Effluent Discharge, Emissions, or Waste Disposal from Selected Industries

Industry	Heavy Metals Released
Fossil fuel combustion (electricity)	As, Cd, Hg, Pb, Sb, Se
Mining, smelting, metallurgy	As, Be, Cd, Cr, Cu, Hg, Mn, Ni, Pb, Sb, Se, Ti, Tl, V, Zn
Petroleum refining	As, Co, Cr, Cu, Ni, Pb, V, Zn
Pulp and paper	Co, Cr, Hg, Ni, Pb

Kansanen and Venetvaara (1991) have shown that lichens and mosses are effective bioaccumulators of heavy metals.

In 1995, more than 2 million tonnes of materials with heavy metals were emitted from a Ni-Cu smelter in Siberia. Reports indicate that the population suffers from respiratory illnesses. The precipitation of the materials in the areas has killed the lichens, which affects the grazing reindeer (Klein and Vlasova, 1992). The affected area has extended up to 70 km in the SSE direction. Heavy metals may also be discharged from industrial effluents into rivers, ponds, lakes, lagoons, wetlands, and oceans. Metals generated by various industries are shown in Table 7.1. Mercury has been particularly problematic. Fish such as swordfish are known as hyperaccumulators of Hg. Fish samples in the Smithsonian Institute contain up to 500 ppb Hg, the limit permissible by the World Health Organization (WHO). In Lake Ontario, Hg exceeding this level was found in fish in the early 1970s. Hg reached the lake as a result of industrial discharges.

7.3.3 Land Environment Impacts and Sustainability Indicators

The indicators that define or establish the path toward sustainability goals are specific to the activity or industry under consideration. One needs to define or identify the various processes or activities that impact directly or indirectly on the geoenvironment. This can be a very detailed accounting of all the activities and their outcome, or it could be a broad sweep of the major categories, elements, or issues. The basic elements that contribute directly as land environmental impacts for most types of downstream industries considered in this section, shown in Figure 7.1, include (1) deposition of airborne noxious particulates, (2) acid precipitation provoked by smokestack emissions of SO_2 and NO_x, (3) wastewater and other liquid waste discharges, and (4) solid wastes and other disposable solids. As stated previously, the composition, distribution of the various components in the wastes and particulates, and nature of the discharges are all functions of the type of process technology, technological efficiency, smokestack emission control, housekeeping efficiency, and waste management capabilities and strategies.

Acid precipitation's impact on the geoenvironment (including the receiving waters contained in the land environment) is felt in several ways:

1. Soil quality: The increased soil acidity will release the metal ions and other positive ions bound to the soil particles, and also the structural ions, such as aluminum. The mechanism for such release is found in the ionic bonds formed between the charged soil particle surfaces and the positive ions (metals and salts such as Ca^{2+}, K^+, Na^+, and Mg^{2+}). The sulfate and nitrate ions from acid precipitation act as counterions and have the effect of releasing the sorbed cations. Weathering of the silicate minerals will release the structural metals, such as aluminum, manganese, and iron. The release of the salts Ca^{2+}, K^+, Na^+, and Mg^{2+} will result in nutrient depletion for plant growth, and the release of aluminum especially will be harmful to aquatic life and growth of plants.

2. Biology of the forest: Reduction in rates of decomposition of the forest floor, damage to roots and foliage, and changes in respiration rates of soil microorganisms.

3. Water quality and aquatic habitats: Acidification of lakes and rivers, deposition of soil-released aluminum, species destruction, and alteration of food supply for higher fauna.

Other not-so-evident land environment impacts from the mining processing downstream industries are (1) use of nonrenewable mineral resources as source material for the downstream industries and (2) excessive use of nonrenewable energy resources to drive the various processes in production of the final product. Except for the aggregate and slab production industries, metal mining and processing downstream industries and the cement-producing industries are heavy users of energy. Until alternative renewable energy resources become more available, these industries have a direct impact on the land environment when land energy resources are used to fuel their many process requirements. Cement kilns are perhaps the best potential users of alternative energy sources, using recycled or recoverable materials. Because of the high burn temperatures required in the cement kilns, municipal biowaste and various kinds of combustible solid and liquid wastes can be used as burn energy sources.

Land environment sustainability objectives of direct relevance to the metal mining and processing downstream industries are (1) preservation or minimal use of nonrenewable energy resources, (2) preservation or minimal use of metal mineral resources, (3) elimination of smokestack emissions and airborne noxious particulates, and (4) the 4Rs (reduce, recover, reuse, and recycle) and nontoxic and nonhazardous discharge of liquid and solid wastes. Figure 7.6 shows some of the elements that can serve to drive downstream industries toward sustainability goals. Industry initiatives are needed for many of the elements shown, e.g., (1) better control of smokestack emissions and airborne particulate discharge

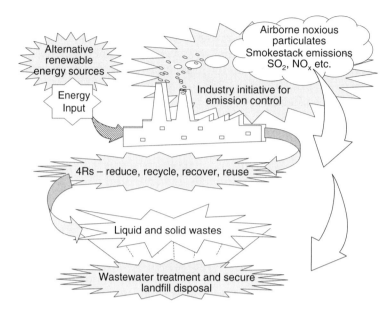

FIGURE 7.6
Industry initiatives for amelioration of geoenvironmental impacts.

to eliminate deposition of particulates and generation of acid precipitation, (2) more efficient use of metal mineral resources to produce more yield, (3) use of the 4Rs strategy as part of the process efficiency technology, and (4) use of alternative renewable energy sources to aid in reduction of consumption of nonrenewable energy resources. Chapter 9 gives a brief discussion of these initiatives. A full treatment of these industry initiatives is not within the purview of this book. There exists much concern and interest in the development of these initiatives, and without a doubt, much is being done by industry to resolve these issues.

The direct connections between the impacts shown in Figure 7.6 and the geoenvironment are seen in terms of contamination or pollution of the land and water elements of the geoenvironment. Amelioration of the pollutant loads and protection of the land environment are necessary requirements to ensure and maintain the quality of the land and water elements of the geoenvironment. The sustainability goals for these land environment elements are (1) maintenance of the quality of the land environment and the receiving waters, (2) protection of biodiversity and the natural habitats, and (3) protection of the natural (e.g., biotic and organic) and geoenvironmental (e.g., mineral and energy) resources in the region of consideration. The procedures and protocols in respect to impact minimization, avoidance, and amelioration are developed in Chapter 10, together with considerations and requirements for remediation and management.

7.4   Agro-Processing Industries

Agro-processing industries are the group of industries that form part of the larger group identified broadly as *agro-industry*. There are essentially two groups of industries that constitute the agro-industry: (1) agricultural production industries and (2) agro-processing industries. Details of the agricultural production industries and their relationship to the

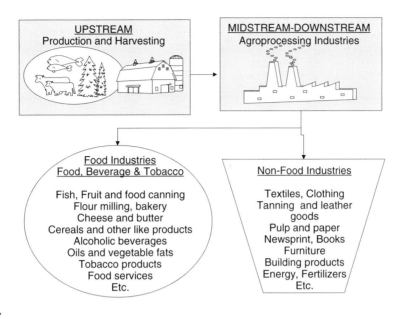

FIGURE 7.7
Various types of midstream–downstream agro-processing industries.

geoenvironment have been covered in the previous chapter. The agro-industry covers many sectors, ranging from production of the raw farm goods (e.g., crops and livestock, as noted in the previous chapter) to the processing and production (i.e., transformation) of end-use products, including such consumables as cheese and bread, cotton fabrics, leather goods, etc. (Figure 7.7). The range of agro-industries covers agricultural production to transformation, manufacture, and processing of the agricultural products. We can conveniently divide the agro-industry into two categories: (1) agricultural production, including crops and livestock (upstream industries), and (2) agro-processing, including midstream (initial processing prior to manufacture and production) and downstream (manufacture and production of final or finished goods) processing. As shown in Figure 7.7, the agro-processing industries include both food-producing and nonfood-producing midstream and downstream industries. We include forestry and forest products in the agro-industry category. The downstream industries associated with forest products include all the wood products, such as pulp and paper, furniture, etc. Although the use of agents such as pesticides, fungicides, herbicides, and fertilizers is a major part of regular and intensive agricultural practices, the industries involved in the manufacture of these agents are not included in the agro-processing or general-agro industry classification. Instead, these will be considered in the grouping of *petroleum, petrochemical, and chemical industries.*

7.4.1 Leather Tanning Industry

The leather tanning industry, in a sense, has a unique position in the food and nonfood agro-processing industries in that it has to rely on the availability of an animal source for its raw material. The different kinds of leather produced from raw hides and skins are a function not only of the production process, but also of the source material. This refers to the animal species. The variety of source materials includes (1) cattle, sheep, pigs, and horses for hides and (2) reptiles, crocodiles, snakes, etc., for the more exotic leathers. The size of the total industry, together with the range in sizes of the industries and variety of raw materials (animal species) devoted to production of hides and leathers, requires one

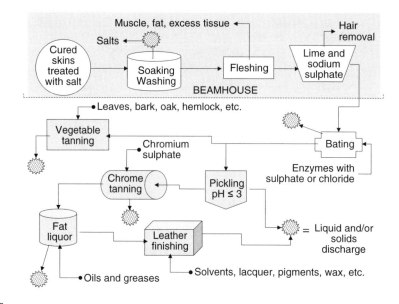

FIGURE 7.8
Major processes in leather tanning. Note the additives and aids used in the various processes and also the many discharge points.

to pay attention to the impacts on the geoenvironment in the processing of the raw material into product form. The estimates are that about 2 billion m² of leather is produced a year, with a market value of about $40 billion. When some of the leather material is processed into consumer goods, as for example leather footwear, where it is estimated that about 65% of the 2 billion m² is utilized, the market value of the produced footwear is in the order of about $150 billion. There is an interesting counterargument that states that production of leather is in effect an environmental conservation effort since it removes the burden of disposal of the hides and skins of slaughtered animals.

Figure 7.8 shows some of the main elements of the leather tanning process. There is intensive use of water mixed with various additives during the beam house process, and also in the many other processes accompanying tanning and fat liquor and finishing. The additives and aids used in the diverse processes include lime, sodium sulfate, enzymes, tannin, sulfuric acid, ammonium salts, sodium bicarbonate, linosulfate, trivalent chromium salts, pigments, dyes, resins, formaldehyde, glutaraldehyde, and heavy oils. The liquid discharge, i.e., wastewater, from the various processes has been reported to be from 20 to 80 m³/t (cubic meters per metric tonne) of hide treated. The composition of the wastewater contains many of the additives and aids described above, with high levels of chromium, sulfide, chloride, biological oxygen demand (BOD), and chemical oxygen demand (COD). It has been reported that residues of pesticides from hides have been found in the wastewaters, and because of the removed fats, unwanted tissue, muscle, hair, skin, and other body meat items, the solid wastes containing the decaying and decayed body parts pose severe human health threats. The treatment, handling, and disposal of the liquid and solid wastes are problems that must be addressed satisfactorily if impact on the geoenvironment is to be avoided. These will be addressed in Section 7.4.3 and in Chapter 10.

7.4.2 Pulp and Paper Industry

Wood industries supply building materials, cellulose fiber, resins and other pulp chemicals, and pulp and paper products. By far, the supply of building materials and pulp

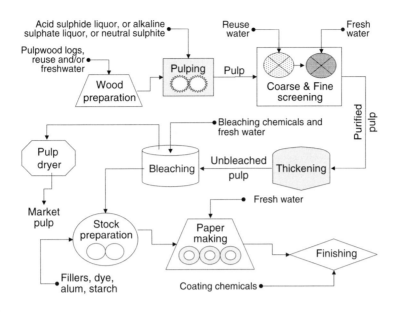

FIGURE 7.9

Simple flow diagram showing typical pulp and paper processes. Note the inputs into the various processes and the extensive use of freshwater for many of the processes.

and paper products constitutes the major portion of the overall wood industries. There is general agreement that the many thousands of pulp and paper industries in the world, when taken as a whole, constitute one of the largest contributors to water, land, and air pollution problems in the locations of those industries. Figure 7.9 shows a simplified flow diagram for the typical pulp and paper process. Debarking and chipping are the major operations in the *wood preparation* process. Recovery of the removed bark for subsequent use as fuel has reduced the waste discharge while decreasing energy costs. The pulping and bleaching processes contribute the more significant impacts to the geoenvironment. The various means for pulping range from mechanical and thermomechanical to chemical. The most frequently used process is the kraft process, i.e., the sulfate process. This chemical process (sulfate process) is not to be confused with the sulfite pulping process that is becoming less popular because the source wood species (spruce, balsam, fir, and western hemlock) are not as plentiful as the other species used for the kraft process. This kraft process generates sludges high in chromium (Cr), lead (Pb), and sodium (Na).

The various processes require considerable amounts of energy input and extensive use of freshwater. Recycled water, identified as whitewater, is used to augment the water input to the coarse screening, bleaching, and wood preparation processes. For the fine-screening–washing and paper machine processes, freshwater is required. Other inputs into the different processes include acid sulfite liquor, alkaline sulfate liquor, and neutral liquor; chlorine-type bleaching chemicals, such as hydrogen peroxide, ozone, peracetic acid, sodium hypochloride, chlorine dioxide, and fillers; dyes; alum; starch; and paper-coating chemicals. The wastewater discharges from the many processes include pulping liquor, mill washings and acid plant wastes, solvents and chlorine-based organic bleach compounds, and general wastewaters. The organochlorines discharged in the wastewater and solid wastes present some difficult toxic issues to the biotic receptors in the immediate geoenvironment. Discharged mill wastewater in the rivers causes considerable breeding problems to several types of aquatic species. Solid wastes discharged include sludges from secondary treatment plants containing inks, dyes, pigment, etc.; boiler ash; and chemical processing and fiber wastes, including organochlorines (if chlorine has been used in the bleaching process).

7.4.3   Palm Oil Industries

While palm oil industries may be considered as belonging in the larger group of vegetable and oilseed processing industries, its variety of transformed products have greater application. The grouping of vegetable and oilseeds includes palm, soybean, rapeseed (canola), sunflower, and cottonseed. As agro-processing industries, palm oil industries fit into both the food- and nonfood-producing agro-industries — with application in food industries as production of nonhydrogenated and refined oils and shortening and margarine, and broad application in nonfood industries such as chemical, cosmetic (personal care), and pharmaceutical. Food industries are grouped into two kinds: (1) production of nonhydrogenated and refined oils and solid fat products (e.g., margarine, shortening, nondairy ice cream, specialty fats, chocolate) for human consumption and (2) production of animal feed (palm kernel cake and palm kernel meal) from the by-products. Nonfood industries are wide ranging, from industries devoted to production of personal care products (cosmetics), such as creams, soaps, and moisturizers, to chemical products using epoxidized palm to produce polyurethanes and polyacrylates, and palm methyl esters, glycerol, and fatty acids as source materials to produce a variety of products, ranging from a diesel substitute to plasticizers. Harvesting of the palm fruit bunches and milling belong to the upstream industry category. The output from the mill is sent to the downstream refining plant, and further on to the manufacturing or production plants (e.g., chemical, cosmetic, and pharmaceutical).

The two different kinds of oils or fats extracted from the palm fruit come from two separate sources, i.e., palm oil from the outer portion of the fruit and palm kernel oil from the kernel. Refinement of crude palm oil to refined oil is required to remove the moisture and impurities and other items, such as the oxidation products, free fatty acids, and caretenoids. The two methods for refinement include physical (steam) and chemical (alkali) refining processes. Fatty acid distillate and refined bleached and deodorized palm oil are obtained from the physical process, and soap stock and acid oil, together with the refined, bleached, and deodorized palm oil, are obtained with the chemical process. Separation or fractionation into liquid and solid phases is performed by thermomechanical means.

The major areas of concern in respect to geoenvironment impacts are (1) energy consumption in the refining processes, (2) excessive water consumption, and (3) effluent and solid waste discharge. It is estimated that effluent discharge (palm oil mill effluent, or POME) is from two to three times that of the amount of palm oil processed. Since the discharge contains organic compounds, these can have some severe consequences on the receptors in the area of discharge.

7.4.4   Land Environment Impact and Sustainability Indicators

Land environment impacts from the upstream phases of the various sectors comprising the agro-industry (production of raw goods and source materials) have been discussed in Chapter 6. The indicators and impacts discussed in Section 7.3.3 for the metallurgical industries apply equally well to the agro-processing industries. What has been omitted from a discussion on the impacts in the upstream agro-industries and from Section 7.3.3 is the effect of a monoculture on the land environment and the ecosystem. Survival of the natural indigenous species is often difficult, especially since it is common practice to use integrated pest management to enhance the growth of the monoculture species. Loss of soil quality and impact on biodiversity are some of the more prominent impacts.

7.5   Petrochemical and Chemical Industries

Petrochemical industries constitute the bulk of the chemical (inorganic and organic) industries in the world. The overwhelming portion of organic chemicals is derived from feedstock obtained from crude oil, natural gas liquids, and coal. The two major processes involved in obtaining organic chemical feedstock are (1) chemical reaction and (2) purification of reaction products (U.S. EPA, 2002). The simplest chemical reactions process is obtained in the batch reaction method where chemicals used to obtain the desired reactions and products are introduced into a reaction vessel. At completion of the reaction process, the reaction products obtained are removed from the vessel and extraneous by-products and unreacted inputs removed (U.S. EPA, 2002). This step is identified as product separation or purification. The techniques used involve filtration, distillation, and extraction either singly or in combination. The continuous reactions method, as the procedure implies, is a continuous reaction process technique and is suited more for greater production of reaction products — in comparison with the batch reaction method, which is used for production of smaller quantities of reaction products. As in the case of the batch method, the reaction products of the continuous reaction technique require product separation.

The feedstock obtained includes alkanes, benzene, butane, butadiene, butylenes, ethane, ethylene, methane, propane, propylene, toluene, and xylene. The feedstock organic chemicals are both end-use products and intermediate chemicals; i.e., they serve as feedstock for production (generally by conversion processes) into other end-use organic chemicals or products. Typical end-use products include the various pesticides and fertilizers used in agriculture, the various forms and types of plastics, textiles, solvents, detergents, pharmaceuticals, appliances, synthetic lubricants, nylon, plumbing, and even chewing gum. By and large, the various kinds of plastics are perhaps the largest and most important product group emanating from petrochemical industries.

By definition, inorganic chemical industries manufacture chemicals that do not contain the carbon molecule. To a large extent, the feedstock or raw materials for the inorganic chemicals are mineral in origin, and the products manufactured are acids, alkalis, salts, and chemicals that are used as aids in producing other products, especially fertilizers. Ammonium nitrate (NH_4NO_3), ammonium sulfate (($NH_4)_2SO_4$), urea ($CO(NH_2)_2$), and superphosphates are some of the fertilizers manufactured. The processes involved for production or manufacture of the fertilizers, for example, vary depending on the type of fertilizer. Neutralization of nitric acid (HNO_3) with ammonia (NH_3) will yield ammonium nitrate (NH_4NO_3) through the simple reaction

$$NH_3 + HNO_3 \rightarrow NH_4NO_3$$

On the other hand, there are at least three different ways or sources to obtain ammonium sulfate (U.S. EPA, 1979). These include (1) combining anhydrous ammonia and sulfuric acid in a reactor to obtain synthetic ammonium sulfate, (2) as a by-product from production of caprolactam (($CH_2)_5COHN$), and (3) as a coke oven by-product obtained by reacting ammonia from coke oven off-gas with sulfuric acid.

The acids produced are to a large extent mainly utilized as intermediates in industrial and manufacturing activities. The fertilizer industry is a big beneficiary. Nitric acid (HNO_3) and sulfuric acid (H_2SO_4) contribute significantly to the production of ammonium nitrate and phosphate, respectively. Hydrochloric acid (HCl) is used in steel pickling, etching and metal cleaning, and hydrometallurgical production.

Perhaps one of the largest groupings of inorganic chemical industries is the chlorine–alkali group. This group produces chlorine, sodium hydroxide, sodium carbonate and bicarbonate, and potassium hydroxide. Their products are greatly utilized as intermediates in the organic chemical manufacturing industries (U.S. EPA, 1995), ranging from (1) for chlorine, vinyl chloride monomer, ethylene dichloride, glycols, chlorinated solvents, and methanes, and (2) for caustic soda, propylene oxide, polycarbonate resin, epoxies, synthetic fibers, soaps, detergents, and rayon. The raw or source material for the industry is both natural salt deposits and seawater. Removal of the impurities, such as calcium, iron, aluminum, sulfate, magnesium, and trace metals, is required before the electrolysis process used to obtain the end product (chlorine, caustic soda, and hydrogen). The three types of cells used in the electrolysis processes for the manufacture of the products are mercury, diaphragm, and membrane.

The major areas of concern for land environment protection are similar to the other industries described in the previous sections. These areas are (1) deposition of smokestack (point) and fugitive (from process equipment, leaks, and spills) emissions of particulates and noxious substances, including SO_x and NO_x; (2) discharge of wastewater and fugitive process waters; and (3) solid wastes. The nature of the chemicals, especially the organic chemicals, makes it critical to monitor both the acidity and chemistry of precipitations. For example, ammonia (NH_3) and nitric acid (HNO_3) have been recorded as emissions from processing for ammonium nitrate. NO, NO_2, and SO_2 have been detected as emissions from plants producing HNO_3 and H_2SO_4, respectively.

Wastewater from petrochemical industries manufacturing organic chemicals will contain excess chemicals (spills?), hydrocarbons, and other dissolved solids in suspended form. Also included are the wastewaters from maintenance procedures and washing of equipment. These would likely contain solvents, lubricants, and detergents. In the case of the inorganic chemical manufacturers, corresponding wastewater discharges will be obtained, except that the surplus chemicals will be inorganic chemicals. Maintenance and cleaning procedures will supply wastewater that would also contain lubricants and detergents and perhaps some solvents.

Land disposal of waste materials would generally be in a sludge form since raw materials and manufacturing processes generally do not involve solid materials. Brine muds are perhaps the greatest solid wastes derived from the chemical industry. These are obtained from the chlorine–alkali industry and are regularly disposed in brine mud ponds in a manner similar to that of the holding ponds of the mining–metal industries.

7.5.1 Land Environment Impacts and Sustainability Indicators

The impacts to the land environment from the emissions, discharges, and land disposal of the waste items are in general similar to those for the other industries described in the previous sections. The only significant difference in the case of petrochemical and chemical industries is the added chemical nature of the various emissions and discharges. The record shows that utilization of proper scrubber systems, together with stringent wastewater treatment procedures, has served to reduce the levels of impact to the land environment. Indicators of proper sustainability of the land environment are obviously zero-threat emissions and discharges.

7.6 Service Industries

Service industries do not create tangible goods or products. By definition, they are industries that create or provide services to the consumer. The major service industries include (1) health — all aspects of medical, dental, and social services, (2) military, (3) educational, (4) government, (5) technical, and (6) financial. Because they do not create goods, the interactions between these industries and the geoenvironment are limited to the discard and fugitive liquid and solid items associated with the service and the "tools of the trade." Military services, for example, present an added dimension to the composition of waste streams. This relates to waste products associated with munitions and the storage and use of ammunitions. A major feature of the decommissioning of military sites is the decontamination of sites contaminated by all of these. The bulk of the discards and wastes generated by the large service industries can be classified as institutional wastes. This category includes the paper discards and general housekeeping items. The exception to the preceding will be the health service industry. The problem arises from services associated with the care of patients in hospitals.

7.6.1 Hospital Wastes and the Geoenvironment

Outside of the regular housekeeping type of waste products, such as paper consumables and kitchen waste, hospitals generate wastes that are classified under the category of hazardous, toxic, and infectious wastes. The contributors to these are the biomedical, radioactive, and chemical-pharmaceutical wastes. The sources for biomedical wastes include biological, medical, and pathological. Contributors to these are services associated with surgery, pathology, biopsy, laboratories, and autopsy. For the radioactive wastes, the sources include x-ray discards, liquid scintillation vials, and all other treatment procedures and equipment utilizing radioactive materials. Sources of chemical-pharmaceutical wastes include research laboratories, pathology, and histology.

The record shows that among hospital wastes, infectious biomedical wastes pose the greatest threat to human health. Management and disposal of these wastes to the land environment are critical issues. Special regulations have recently been structured by most state regulatory agencies concerning hospital wastes in general and infectious wastes in particular. Sorting, storage, transportation, treatment, disinfection, and incineration are some of the principal steps in the handling and disposal of these infectious wastes. Liquid infectious wastes are required to be treated before discharge, and only noninfectious and nonanatomical wastes are permitted to be disposed of in landfills. Unhappily, technical and economic constraints may deny full and safe/secure disposal of these wastes.

7.7 Contaminating Discharges and Wastes

Table 7.2 gives a very short summary of many of the contaminating substances and chemicals found in the geoenvironment as a result of deliberate discharges, spills, leaks, emissions, and disposal of liquid waste and solid waste materials from the downstream service industries. For the processing and manufacturing industries, the discharges come during the processing and manufacturing stages. These include (1) inadvertent losses of

TABLE 7.2

Typical Composition of Discards, Spills, and Waste Streams from Some Representative Industrial Activities and Industries

Industry	Discards, Spills, and Waste Streams
Metal manufacturing and finishing	Acids, bases, cyanide, reactive wastes, heavy metals, ignitable wastes, solvents, spent platings, oil and grease, emulsifying agents, particulates, polishing sludges, scrubber residues, complexing agents, wastewater treatment sludges
Agro-industries — food and nonfood	Acids, bases, sulfides, chromium, hides, skins, suspended solids, pesticides, solvents, chlorine compounds, waste (fiber) sludge, dyes, pigments, kaolin clay sludge
Petrochemical and chemical industries	Acids, bases, ignitable wastes, heavy metals, inorganics, pesticides, reactive wastes, solvents, lubricants, spent catalysts, spent caustic and sweetening agents, organic waste sludges
Hospitals — medical facilities	Biomedical wastes, infectious wastes, acids, bases, radioactive materials, solvents, heavy metals, ignitable wastes

raw and intermediate products and materials utilized during processing and manufacturing, and (2) liquid and solid waste products associated with processing and manufacturing procedures and technology. Acids, bases, heavy metals, inorganics, organic chemicals, and solvents are common to the industries shown in the table. Many of the inorganics are composed of chemical compounds that do not contain carbon as the principal element. Most of the inorganic compounds are stable and soluble in water. They tend to have rapid chemical reactions and large numbers of elements. They are generally less complex than the organic chemical compounds. Other, more specialized wastes include pesticides, herbicides and the like, solvents, cyanides, and reactive wastes.

Other than general wastewater, liquid waste streams emanating directly from downstream industries can be grouped into four categories: (1) aqueous–inorganic, including brines, electroplating wastes, metal etching, and caustic rinse solutions; (2) aqueous–organic, including wood preservatives, water-based dyes, rinse water from pesticide and herbicide containers, organic chemical production, etc.; (3) organic, including oil-based paint wastes, production of pesticides, herbicides, fungicides, etc., spent motor oil, cleaning agents, refining and reprocessing wastes, etc.; and (4) high solid content and high-molecular-weight hydrocarbon sludges.

In respect to managed liquid and solid wastes in landfills, there are two types of liquid plumes or streams that emanate from the waste pile in the landfill. The first type, which is identified as *primary leachate*, consists of the liquid waste originally submitted to the landfill, combined with dissolved constituents in the waste pile. The primary leachate may be aqueous–organic, aqueous–inorganic, or organic. Leachate generated from water entering into the waste pile is defined as *secondary leachate* and is generally composed of the percolating water and the solutes from dissolution products. This leachate may be aqueous–organic, aqueous–inorganic, or a combination of inorganic and inorganic solutes and compounds in the liquid phase. Since it is not really possible to distinguish between the two kinds of leachates when they exit from the bottom of a landfill, the general term *leachate* is used — with no attempt at categorization. The predominant liquid in a leachate may be water, an organic liquid, or a combination. The solutes and inorganic and organic chemicals in the leachate are the products of the dissolution of the materials in the waste pile. The relative abundance of a given dissolved component is a function of the composition of the principal liquid. Neutral nonpolar organic liquids will have large carrying capacities and will easily carry other neutral nonpolar organic chemicals. Aqueous liquids have very limited carrying capacity and will not be capable of carrying nonpolar organics in their dissolved phase. Water, on the other hand, has a relatively large carrying capacity for polar organic chemicals (they may be miscible in each other in all proportions) and for inorganic acids, bases, and salts.

7.7.1 Physicochemical Properties and Processes

Water is the carrier for contaminants in the subsoil. The movement and distribution of the various contaminants in the soil depend not only on the hydrogeological setting, but also on the interactions between the contaminants carried in the liquid phase and the soil fractions. Many of these interactions have been described in Chapter 2 as transport processes. In this section, we will look at the physicochemical properties and processes involved when the kinds of contaminants listed in Table 7.2 are introduced to the geoenvironment.

7.7.1.1 *Solubility*

The amount of solute needed to reach a saturated state in a given quantity of solvent at a specific temperature is defined as the solubility of the given solvent. For considerations of ionic equilibria in aqueous solutions, the solutes are those that fall into the class of sparingly or slightly soluble ionic solids. We define the *solubility product* k_{sp} as the equilibrium constant for the equilibrium that exists between the sparingly soluble ionic solid and its ions in a saturated solution. Because of the significant electrostatic attraction between ions, crystals composed of small ions packed closely together are generally harder to pull apart than crystals made up of large ions. For example, fluorides (F^-) and hydroxides (OH^-) are less soluble than nitrates (NO_3^-) and perchlorates (ClO_4^-).

Solubility equilibria are useful in predicting whether a precipitate will form under specified conditions, and in choosing conditions under which two chemical substances in solution can be separated by selective precipitation. Substances that are more soluble are more likely to desorb from soils and less likely to volatilize from water. On the other hand, substances with no hydrogen bonding groups or little polar character, such as hydrocarbons or halogenated hydrocarbons, usually have very low solubilities compared with compounds such as alcohols, which are capable of interaction with water. The solubility of an organic compound depends primarily upon the sorption–desorption characteristics of the sorbate (organic compound) in association with the sorbent (soil and sediment).

7.7.1.2 *Partition Coefficients*

Partition coefficients have been described in Chapter 2. They provide a measure of the distribution of a given inorganic contaminant — between sorption onto the soil solids and the pore water. For organic chemicals, the octanol–water partition coefficient is used to describe partitioning. The *octanol–water partitioning coefficient* (k_{ow}) has been defined in Chapter 2 as the ratio of the amount of a solute dissolved in octanol and water in octanol–water immiscible mix. k_{ow} is well correlated with the solubility of several organic chemicals. Log k_{ow} values normally range from –3 to 7. Highly water soluble compounds, such as ethanol, have values of log $k_{ow} < 1$, and hydrophobic compounds, such as certain polychlorinated biphenyls (PCBs) and chlorinated dioxin congeners, have values of 6 to 7.

7.7.1.3 *Vapor Pressure*

The vapor pressure of a liquid or solid is the pressure of the gas in equilibrium with the liquid or solid at a given temperature. Gasoline, for example, will evaporate rapidly since it has a high vapor pressure and is very volatile. Volatilization is a significant factor in disposal for compounds with vapor pressure greater than 10^{-3} mmHg at room temperature. Chemicals with relatively low vapor pressures and high solubility in water are less likely to vaporize and become airborne. The transport of a compound from the liquid to the vapor phase is called volatilization. This could be an important pathway for chemicals with high vapor pressures or low solubilities. Evaporation depends on the equilibrium vapor pressure, diffusion, dispersion of emulsions, solubility, and temperatures.

7.8   Concluding Remarks

To implement the basic concepts of industrial ecology from a geoenvironment perspective, one is required to determine the connections or interactions between downstream manufacturing industries and the geoenvironment. For manufacturing and other kinds of downstream industries, the major areas requiring detailed scrutiny include (1) use of nonrenewable resources as energy input and also as raw materials for the industries, (2) spills and debris, together with liquid and solid waste discharges, and (3) gaseous and noxious particulate airborne emissions. Figure 7.10 gives a schematic of what one might call *common denominator descriptors* for the industries and their interaction with the geoenvironment. The various input and output items shown in the schematic are common to most of the midstream and downstream industries. While the geoenvironment perspective developed in this book does not consider industry manufacturing and processing technology, the use of nonrenewable resources as raw materials and as energy sources has a direct impact on the mandate of industrial ecology, and must therefore be identified as

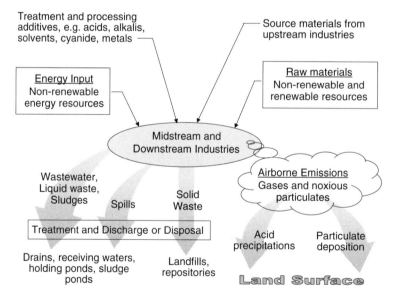

FIGURE 7.10
Common denominator descriptor identifying interactions between midstream–downstream industries and the geoenvironment. While all the descriptors shown are central to determination of industrial ecology and sustainability indicators, the mandate for this book does not cover industry technology for manufacturing and processing (top half of diagram). The geoenvironmental concerns are directed to the issues identified in the bottom half of the diagram.

issues that need resolution. The issues of direct concern in this chapter in respect to the land compartment of the geoenvironment are shown in the bottom half of Figure 7.10 — identified by the broad arrows leading to the bottom of the diagram. From a geoenvironment perspective, the main points that require attention in establishing geoenvironment sustainability indicators as a step toward assessment of capabilities to attain geoenvironmental sustainability objectives have been stated in the preceding paragraph. These are grouped into (1) resource utilization and (2) discharges.

The discussions and examples given in this chapter have focused principally on the discharges, since these are the direct concern of this book. In discussing the various processes in this chapter, and in showing the diagrams for some of the processes, some appreciation of the many sources of interactions with the geoenvironment can be gained. In the final analysis, these all fall into the *discharge* group of *points for study*. In the bottom half of Figure 7.10, the interactions with the geoenvironment are given as:

- Wastewater and liquid spills: These are treated before discharge into receiving waters. The concern is in regard to whether treatment is capable of removing all the noxious and toxic substances.

- Liquid wastes and sludges: As noted in Table 7.2, these consist of inorganic and organic chemicals and also inorganic and organic sludges. Disposal of these in the geoenvironment is generally performed by constructing holding ponds or various kinds of secure containment ponds. Escape of these contained liquids or sludges into the subsurface environment will pose problems to the environment and to human health.

- Solid wastes: These are contained in waste landfills and in underground repositories. Leachates generated can escape into the subsurface and can pose health and environmental threats.

- Airborne emissions: Gaseous and noxious particulates pose problems when they return to the land surface under gravitational forces and through rainfall and snowfall. The classic problems of acid rain are clear demonstrations of the *return* to land surface.

It is clear that the primary issues in respect to the discharges are the interactions of the inorganic and organic chemicals with the subsurface geologic material. Chapter 2 has given a brief overview of these interactions in the context of contaminant–soil interactions. These interactions are fundamental elements that govern the transport and fate of these inorganic and organic chemicals in the ground. We will examine these further in the context of the material developed in this chapter and the previous chapters. Chapters 9 and 10 consider most of these issues in terms of land environment impacts and geoenvironmental sustainability indicators.

References

Kansanen, P.H. and Venetvaara, J., (1991), Comparison of biological collectors of airborne heavy metals near ferrochrome and steel works, *Water Air Soil Pollut.*, 60:337–359.

Klein, D.R. and Vlasova, T.J., (1992), Lichens, a unique forage resource threatened by air pollution, *Rangifer*, 12:21–27.

Pacyna, J.M., (1995), The origin of arctic air pollutants: lessons learned and future research, *Sci. Total Environ.*, 160/161:39–53.

U.S. EPA, (1979), *Ammonium Sulfate Manufacture: Background Information for Proposed Emission Standards*, Report EPA-450/3-79-034a, U.S. EPA, Washington, DC.

U.S. EPA Office of Compliance, (1995), *Profile of the Inorganic Chemical Industry*, Report EPA/310-R-95-004, U.S. EPA, Washington, DC.

U.S. EPA Office of Compliance, (2002), *Profile of the Organic Chemical Industry*, 2nd ed., Report EPA/310-R-02-001, U.S. EPA, Washington, DC.

8

Coastal Marine Environment Sustainability

8.1 Introduction

For many different reasons, the coastal marine environment can be considered to be an important part of the geoenvironment. It is the recipient of (1) liquid discharges from surface runoffs, rivers, and groundwater, and (2) waste discharges from land-based industry, municipal, and other anthropogenic sources. It is also a vital element in the geoenvironment that provides the base for life support systems. The combination of these two large factors, with their direct link to human population, makes the coastal marine environment an integral part of the considerations on the sustainability of the geoenvironment and its natural resources.

8.2 Coastal Marine Environment and Impacts

The coastal marine environment is a significant resource. A healthy coastal marine ecosystem ensures that aquatic plants and animals are healthy and that these do not pose risks to human health when they form part of the food chain. In this chapter, we will discuss (1) the threats to the health of the coastal sediments resulting from discharge of pollutants and other hazardous substances from anthropogenic activities, (2) the impacts already observed, and (3) the necessary remediation techniques developed to restore the health of the coastal sediments.

8.2.1 Geosphere and Hydrosphere Coastal Marine Environment

The discussion in Section 1.2 in Chapter 1 on the inclusion of many elements of the hydrosphere in the geoenvironment points out that the coastal marine environment is included in the discussion on the receiving waters of the geoenvironment. This is because it is affected by the outcome of anthropogenic activities. The sea provides the habitat for many living organisms and is an important ecosystem in the geoenvironment. Seawater is one of the major supply sources for our water resources through evaporation of the seawater and subsequent deposition on land as rainfall. Evapotranspiration constitutes the other major source of water for rainfall. Circulation and recycling of water on land is a very important process for preservation of the global environment and life support for all living organisms. This is not only because no living organism can survive without water, but also because many substances required for the preservation of the sea environment are transported with the circulation of water.

Nutrients such as phosphorus, nitrogen, silicate, etc., are mainly produced during the decomposition of plants by the actions of organisms and are transported into the sea by water. Phosphorus and nitrogen are produced from the decomposition of withered leaves, while silicate originates primarily from inorganic soils. These nutrients are essential to the organisms in the sea and are basic elements of the ecosystems in the sea. In general, there are fewer nutrients in shallow coastal zones, in comparison with deep seawater. This makes deep seawater more attractive for fish farming and for creating fishing grounds. To increase the amount of nutrients in the fishing grounds, fishermen have begun to plant broad-leaved trees in mountainous areas as a means to increase production of phosphorus and nitrogen as decomposition products for eventual rainfall (land surface flow) transport into the sea. Recognition of the high mineral and nutrient values that can be obtained from deep seawater has led to harvesting of deep seawater for extraction of salt and other products.

8.2.2 Sedimentation

The sea bottom is the interface between the seawater as the hydrosphere and the sediments and rock as the geosphere. Sediments are formed from substances deposited in the hydrosphere or produced in the hydrosphere itself. Because of the concentration of cations in seawater, suspended clay particles can aggregate more easily and settle faster than in freshwater. In addition, most sediment solids have a specific gravity greater than that of seawater, which explains why most of the nonaggregated solids will finally settle to the bottom of the sea. The settlement or sedimentation of solids is probably one of the strongest agents responsible for the purification of seawater, because:

1. Various (harmful and noxious) substances sorbed or attached to the sedimenting substances (particles) will be sedimented with the particles, resulting in a measure of purification of the seawater.
2. Turbidity is reduced and transparency is promoted.

8.2.3 Eutrophication

In some closed sea areas, increased concentration of nutrients can be found. This phenomenon is called *eutrophication*. This can happen naturally. More often than not, this phenomenon is developed as a result of the input of additional nutrients due to anthropogenic sources. This is sometimes called anthropogenic eutrophication. The main sources of these nutrients are sewage effluents, nutrients washed out of farmland, golf courses, and lawns, and deposition of nitrogen from nitrous oxide emissions. Low to moderate eutrophication is beneficial because it enhances production of microscopic plants that live in the ocean, called phytoplankton. Because phytoplankton are bait for zooplankton, they are the basis for the marine food chain, and their increased presence means a better food supply for the fish that rely on them as their source of food. However, when eutrophication is high, an excess amount of phytoplankton will be produced, and the resultant phytoplankton bloom will contribute to the reduction in the amount of dissolved oxygen in the immediate region — creating problems for the fish population (red tide), as explained later in this section.

Resuspension of decayed algae and inorganic and organic particles will contribute to the turbidity of water, adding to the sunlight shading effect. Turbidity affects growth of sea plants that need sunlight. Decay of algae by bacteria removes large amounts of oxygen from the water and may kill living organisms. Oxygen deficiency will cause sediments to change from an aerobic to anaerobic state. When this occurs, sulfides such as pyrite can be formed (Fukue et al., 2003). Hydrogen sulfide produced is a hazardous substance for the fish. The combination of oxygen depletion with hydrogen sulfide makes the seawater look blue or greenish blue — a condition sometimes called *blue tide*.

Accelerated production of phytoplankton algae will lead to algae bloom — an overpopulation of certain types of algae that are readily distinguished on the water surface as patches of bloom because of their high population density. The red algae bloom, known as red tide, is a vivid demonstration of such an occurrence involving certain types of algae that contain red pigment. Some, but not all, of the red algae species are toxic. The prominent ones are the dinoflagellate *Alexandrium tamarense* and the diatom *Pseudo-nitzschia australis* (The Harmful Algae Page, 2005). Their production of neurotoxins makes them harmful to fish and other aquatic life-forms — and even humans. Remediation treatment is needed if one wishes to maintain a healthy environment in the coastal marine ecosystem to counter the effects of the preceding.

Many countries, states, and communities have set up programs for monitoring of surface water quality. The term *surface water* is used here in the overall defining sense to mean all water that is naturally exposed to the atmosphere. This includes rivers, lakes, reservoirs, ponds, streams, impoundments, seas, estuaries, and all springs, wells, or other collectors directly influenced by surface water. In general, the quality of surface water has a direct influence on its sediments, benthic organisms, and plankton. Data obtained on surface water quality are used to structure measures to counter eutrophication and contamination of surface water.

8.2.4 Food Chain and Biological Concentration

Some organic matter, such as phytoplankton, deposited onto the sea bottom or suspended in seawater will be decomposed by microorganisms or consumed by small benthic animals called zooplankton. Phytoplankton is the first level in the marine food chain, and zooplankton is the second trophic level. Decomposition of the phytoplankton and other organic matter will produce detritus. This is a good source of nutrients. The food chain starting from detritus is often called the *detritus food chain*, in contrast to the food chain starting from phytoplankton in seawater. These food chains are the most important process for the preservation of marine environment and living organisms, such as fish, shellfish, sea plants, etc. These are important sources of protein and minerals for humans. However, the food source for these marine living organisms can contain toxic and hazardous substances bioaccumulated through uptake, as for example through the gills of fish.

Bioaccumulation refers to the uptake and storage of a pollutant (toxic substance) by an organism in the food chain. In the case of fish, these can be considered the intermediate trophic level of the food chain for humans — with the phytoplankton being the first trophic level. While uptake of a toxic substance is an important measure, we need to account for possible excretion of the substance and metabolic transformation of the substance as factors that will reduce the uptake amount of the substance. Hence, storage becomes an important consideration. The literature shows that the term *bioconcentration* is quite often used in place of bioaccumulation. Bioconcentration is related to bioaccumulation, but is more specific in that it expressly refers to the uptake and storage of toxic substances from water. The term *bioconcentration* has also been used to indicate the condition when concentrations of a substance in a particular biota are higher than in the surrounding medium. For this discussion, we use the first meaning of the term *bioconcentration* — uptake and storage of pollutants from water. When uptake and storage occur

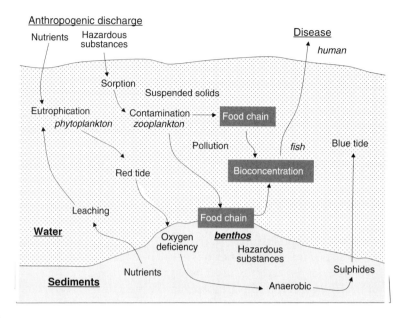

FIGURE 8.1
Various processes of dispersion and accumulation of substances in the sea.

from water and also from food, the more general term *bioaccumulation* is used. Available evidence (Chiou, 2002) indicates that there is a correlation between bioaccumulation and the octanol–water coefficient k_{ow} (see Section 2.5.3) for many of the organic chemical pollutants reported in Section 2.3.2.

8.2.5 Contamination of Sediments

Historic discharge of contaminants and pollutants into the environment and the receiving waters from industrial development is a matter of record. This discharge became more significant during and after the industrial revolution at around the eighteenth century. Many of the discharges included such substances as heavy metals, polychlorinated biphenyls (PCBs), dioxins, polycyclic aromatic hydrocarbons, tributyltin, triphenyltin, etc. — all of which are known toxicants. When these substances find their way into the ocean environment, some of them will dissolve. However, most of them will find their way onto the sea bottom through eventual sedimentation or attachment to suspended solids — a process that is considered to be dispersion of noxious substances in the sea. Their effects on the human food chain can be deduced from Figure 8.1. This figure shows the food chain beginning with phytoplankton anchoring the lowest trophic level, progressing upward through the zooplankton, fish, and other marine aquatic species, and finally to humans. Mineral particles and organic matter settling in seawater can adsorb toxic and hazardous elements and compounds. The record shows the bioaccumulation of such elements and compounds as PCBs, polychlorinated dibenzo-p-dioxins/dibenzo furans/dioxins, tributyltins (TBTs), heavy metals, etc., in living organisms such as seaweed, sea turtles, shellfish, fish, etc. (Jensen et al., 2004; Gardner et al., 2003; Green and Knutzen, 2003). In some coastal areas and bays near urban centers, studies show that the sediments are heavily polluted with heavy metals and other hazardous substances (Jones and Turki, 1997; Kan-Atireklap et al., 1997; Fukue et al., 1999; Ohtsubo, 1999; Cobelo-García and Prego, 2003; Romano et al., 2004; Selvaraj et al., 2004). Makiya (1997) reports an intake of dioxins in the range of 60 to 70% from a 1-day ingestion of fish and shellfish. To eliminate the root of the human food chain problem requires one to decontaminate the contaminated sediments. This will eliminate the food source for the lower trophic levels. Until such is achieved, the danger of ingesting fish and shellfish that have bioaccumulated toxic and hazardous substances will always be present.

8.2.5.1   Some Case Studies of Sediment Contamination

Evidence of sediment contamination from land-based anthropogenic activities can be found both in marine coastal regions and at the bottom of rivers, lakes, and other bodies of receiving waters. At a site in Germany, a lignite seam was found to accumulate aliphatic and aromatic chlorinated hydrocarbons downstream from a chemical plant (Dermietzel and Christoph, 2002). An initial fast desorption occurred from the outer surface of the sediment, followed by a slower-diffusion controlled release from the interior of the sediment.

Sediment samples from Lake Harwell, SC, were taken at five places in 1998, to determine if natural attenuation of polychlorinated biphenyls (PCBs) was occurring (Pakdeesusuk et al., 2005). From an analysis of the mole percentage of each congener of PCB or the total of meta, para, and ortho chlorines and total chlorines per biphenyl, it was determined that solubilization and desorption were negligible according to mass balances since 1987. *In situ* dechlorination was occurring, though, after an initial rapid rate, followed by a slow rate since 1987. Microcosm studies supported the findings. There was a lack of information on organic matter and electron acceptors such as nitrate, sulfate, iron, and manganese, which made it difficult to predict optimal dechlorination conditions. To reduce the risk of bioaccumulation in fish, capping with fresh sediment may need to be increased.

In 1982, trichloroethene contamination in the groundwater was first detected at a Michigan National Priorities List site (An et al., 2004). Since then, samples were taken in 1991, 1992, 1994, 1995, and 1998, 100 m from the shore and later 3 m from the shore. Anaerobic degradation was indicated as the products of dichloroethene (DCE), vinyl chloride (VC), ethene, and methane were found. Analysis of the water within the lake sediments indicated natural attenuation.

At the Columbus Air Force Base, the fate and transport of jet fuel contaminants were evaluated in 60 sediment samples (Stapleton and Sayler, 1998); 10^7 to 10^8 organisms per gram of sediment were found by DNA probes, compared to 10^4 to 10^6 organisms per gram by traditional methods. There was evidence of the degradation of benzene, toluene, ethyl benzene, and xylene (BTEX) and naphthalene, particularly after 5 to 7 days. Without nutrient addition, more than 40% of these ^{14}C-labeled compounds were mineralized in the sediments. Correlations of laboratory assay and field analyses are required, and thus further field tests will be performed.

At the Dover Air Force Base, which was contaminated with chlorinated ethenes, low biomass levels ($<10^7$ bacteria per gram of sediment) were found (Davis et al., 2002). However, mineralization of vinyl chloride and cis-DCE was found to be occurring. The 16 S rRNA gene sequence indicated the presence of anaerobic microorganisms capable of anaerobic halorespiration and iron reduction. It was concluded that microorganisms were the major mechanism for reductive and oxidative attenuation of the chlorinated ethenes.

A study of heavy metals in fish and shellfish was conducted in Port Philip Bay, Australia (Fabris et al., 1999), with the objective of determination of the partitioning of heavy metals in dissolved and particulate species in the bay waters. Concentrations in the near-shore and estuarine areas were not higher than in the coastal marine waters despite a flushing time of 10 to 16 months in the bay. The mechanisms for partitioning were related to co-precipitation of iron and manganese oxyhydroxides with dissolved heavy metals. A strong correlation of iron with chromium, nickel, and zinc was seen in the particulates. Contrary to the metals, arsenic concentrations (as As(III)) increased in depth in the sediments, and

TABLE 8.1

Outlines of Sediment Quality Guidelines

Country or Organization	Criterion or Standard for Guideline
U.S. EPA	Screening concentrations for inorganic and organic contamination
	SQCoc: Draft sediment quality criteria; oc = organic carbon
	SQALoc: Sediment quality advisory levels
	ERL: Effects: range — low
	ERM: Effects: range — median
	AET-L: Apparent effects threshold — low
	AET-H: Apparent effects threshold — high
	TELs: Threshold effects levels
	PELs: Probable effects levels
U.S. NOAA[a]	ERL: Effects: range — low
	ERM: Effects: range — median
Canada	ISQGs: Interim sediment quality guideline
	PELs: Probable effects levels
Australia	ISQG–low: Interim sediment quality guidelines — low
	ISQG–high: Interim sediment quality guidelines — high
The Netherlands	Target value
	Intervention value

[a] U.S. National Oceanic and Atmospheric Administration.

thus did not seem to be the result of anthropogenic activity. Near the surface layer of sediments, arsenic is oxidized to As(V) and leaves the sediments, while Fe(III) can co-precipitate some of the arsenic and become trapped in the sediments.

8.2.5.2   Sediment Quality Criteria

Table 8.1 gives an indication of the different criteria or definitions structured by some countries for characterization of sediment quality. Although many countries and jurisdictions have established guidelines for water quality, especially for drinking water, very few countries have set up sediment quality guidelines. The importance of available guidelines and criteria can be seen in the need to protect pollution of the sea from dumping and indiscriminate discharge of hazardous wastes — all of which will eventually find their way onto the sea bottom and ultimately into the human food chain. The 1954 International Convention on the Prevention of Pollution of the Sea by Oil and the 1958 Geneva Convention on the High Seas were the earliest formal attempts to regulate and control discharge of hazardous substances into the sea. These have been reinforced with more attention paid to discharge of various kinds of hazardous wastes, especially wastes originating from land sources.

8.3   London Convention and Procotol

The London Convention and Protocol consists of the original London Convention of 1972, which expanded the Oslo Convention for North-East Atlantic to cover marine waters worldwide, except for the inland waters of the various 80 states that were signatories to the convention. The Oslo Convention came into force in 1974, and the expanded Oslo Convention (to worldwide marine waters), which became the London Convention, came into force in 1975. With this convention, elimination of future marine pollution from deliberate discharge of industrial and other wastes is to be achieved through regulation of dumping of wastes at sea. The wastes of concern included the original oily wastes from the 1954 and 1958 conventions, dredging spoils and wastes, and industrial wastes (i.e., land-based generated wastes).

Adoption of the *1996 Protocol to the Convention on the Prevention of Marine Pollution by Dumping of Wastes and Other Matter, 1972* essentially strengthened the London Convention with the *precautionary approach* and the *polluter pays principle*. With the precautionary approach, the burden of responsibility for determining whether a waste designated for ocean dumping is potentially hazardous is now borne by the originator of the waste. The annex in the 1996 protocol, "assessment of wastes or other matter that may be considered for dumping," states that "acceptance of dumping under certain circumstances shall not remove the obligations under this Annex to make further attempts to reduce the necessity for dumping." The wastes or other matter that may be considered for dumping in accordance with the objectives of the protocol and the precautionary approach include

1. Dredged material
2. Sewage sludge
3. Fish waste or material resulting from industrial fish processing operations
4. Vessels and platforms or other man-made structures at sea
5. Inert, inorganic geological material
6. Organic material of natural origin
7. Bulky items, primarily comprising iron, steel, concrete, and similar nonharmful materials, for which the concern is physical impact, and limited to those circumstances where such wastes are generated at locations, such as small islands with isolated communities, having no practicable access to disposal options other than dumping

Of particular interest in the annex is the following:

> For dredged material and sewage sludge, the goal of waste management should be to identify and control the sources of contamination. This should be achieved through implementation of waste prevention strategies and requires collaboration between the relevant local and national agencies involved with the control of point and non-point sources of pollution. Until this objective is met, the problems of contaminated dredged material may be addressed by using disposal management techniques at sea or on land.

Apparently, dredged material from the sea bottom has always occupied a special position under the convention — to a large extent because dredging is an important requirement and a necessity for keeping navigation routes open, and also for ports and harbors. The volume of dredged materials is considerable. However, there is incontrovertible evidence to show that a significant portion of the dredged material ports, harbors, and coastal regions is highly polluted. This realization has now energized many countries to begin considering dredged sediments as contaminated-polluted, and to insist that proper disposal of dredged materials be obtained.

8.4 Quality of Marine Sediments

Changes in seawater quality occur quickly, over a very short time period, because of the effect of currents and dilutions. While these effects may lead to low concentrations of pollutants in seawater, the record shows that pollutants can be (and will be) adsorbed onto suspended and sedimenting particles. This is especially true when the settling particles are of biological origin — since they have the capability for sorbing heavy metals. This sorption process is known as biosorption, and the sorbent is called a biosorbent. Generally speaking, a biosorbent refers to the capability of a biomass to sorb heavy metals from solutions. Of particular importance is the fact that biosorbents have a combination of functional groups, such as those described in Section 2.5.2 and in Figure 2.12 in Chapter 2 for organic chemicals. The combination of functional groups endows the combined group with significantly enhanced biosorbent capability to sorb various kinds of heavy metals, as opposed to monofunctional groups. Microalgae, for example, are well-known biosorbents (Wilde and Benemann, 1993). The study reported by Inthorn et al. (2002) on 52 strains of microalgae and their capabilities for removal of Pb, Cd, and Hg from various solutions showed that both green algae and blue-green algae functioned well in removing the heavy metals. While deposition of the pollutant-associated settling particles onto the seabed may serve to remove a proportion of the waterborne pollutants from seawater itself, the accumulation of polluted particles in the sediment presents a significant problem for the benthic population.

Determination of the quality of marine sediments is required for at least two different purposes:

1. Preservation of the ecosystem: Knowledge of the types and nature of the pollutants in the sediments is essential, to enable one to structure the necessary measures for remediation and management of the quality of the sediments, in conformance with established guidelines for sediment quality.

2. Safety assurance for the lower trophic levels in the marine-derived human food chain: Since phytoplankton and zooplankton are the first and second trophic levels, respectively, in this marine-derived human food chain, preservation of the quality of these trophic levels means eliminating or reducing the concentrations of pollutants in the sediments — inasmuch as these can be bioaccumulated or bioconcentrated by the organisms that occupy these trophic levels. The record shows that organisms such as benthos, fish, and mammals have been more or less contaminated through the food chain and through bioconcentration (Kavun et al., 2002; Do Amaral et al., 2005; Moraga et al., 2002). While guidelines have yet to be established, it is evident that in the absence of detailed records and tests, specification of allowable concentrations will likely be severely conservative, in the interest of eliminating health threats to the human population.

8.4.1   Standards and Guidelines

8.4.1.1   *Guidelines*

Guidelines have been established in some countries for the purpose of evaluating sediments contaminated with toxic chemicals. The aim of these guidelines is to limit the concentration of the toxic chemicals through the use of various criteria, such as those shown in Table 8.1. As with all criteria based on observed effects on human health, differences exist between the guidelines used by the various countries. To a large extent, this is because of the different means for determination of the effective levels, and also in the perception of what constitutes an acceptable risk. For example, since apparent effects threshold (AET) values are essentially determined by a single result (i.e., the highest nontoxic sample), as opposed to the entire distribution of results (e.g., as with threshold effects level [TEL] or probable effects level [PEL]), the final AET values used by the regulatory agency may vary substantially, depending on the outcome of their analyses. A considerable amount of work remains to be done in this area. The use of interim values as preliminary values at the present time recognizes the fact that additional technical work on individual AET values, together with reliability analyses and discussions with other involved agencies, is required.

8.4.1.2 Chemicals

The guidelines issued by the various countries and agencies for environmental quality for sediments include trace and heavy metals and different types of organic chemical compounds. There is no definitive common listing of elements and chemicals between the agencies and countries, and no common agreement as to criteria used to evaluate and target the listed elements and chemicals. An example of sediment quality guidelines can be seen in the interim sediment quality guidelines (ISQG) and PEL values given in the 2003 Canadian Environmental Quality guidelines (Table 8.2). The U.S. National Oceanic and Atmospheric Administration (NOAA), on the other hand, has established sediment quality guidelines that contain 9 trace metals, 13 individual polycyclic aromatic hydrocarbons (PAHs), 3 classes of PAHs, and 3 classes of chlorinated organic hydrocarbons. As an example, for lead the ERL (effects: range — low) is 46.7 ppm dry wt, the ERM (effects: range — medium) is 218 ppm dry wt, and the incidence of effects is 90.2% when the concentration is higher than the ERM. When concentration exceeded the ERM values, the incidence of adverse effects increased from 60 to 90% for most trace metals, and 80 to 100% for most organics. However, the reliability of the ERMs for nickel, mercury, DDE, total DDTs, and total PCBs are much lower than those for other substances.

8.4.2 Background and Bioconcentration

8.4.2.1 Background Concentration

As discussed in Section 1.5 in Chapter 1, the elements shown in Table 1.3 are known to exist naturally in the environment, generally in the form of compounds and minerals such as sodium chloride, copper carbonate (azurite $Cu_3(CO_3)_2(OH)_2$ and malachite $Cu_2CO_3(OH)_2$), and magnesite (for magnesium), and in food sources such as spinach and nuts (for magnesium). Not all the elements shown are toxic or totally harmful to human health. In Table 1.3, both lead (Pb) and cadmium (Cd) are known as toxic elements and can be safely identified as pollutants. There are no acceptable daily intake values for these. The rest of the elements shown in Table 1.3 are known to be essential elements, and the lack of any of these can be harmful to human health. However, ingestion of concentrations of these essential elements in excess of the acceptable daily intake (ADI) can be harmful to human health. Table 8.3 shows some of the effects to human health for some of the essential elements when ingested concentrations are deficient (*lack of*) or in excess (*toxic*). The harmful effects shown in the last two columns are not meant to be totally definitive. They should be considered as *potential effects* since very few totally controlled studies have been conducted to fully isolate the noted *harmful effects*.

The record shows that background concentration of many of these elements and several other known toxicants exists in the environment, especially in the coastal marine environment and sediments — naturally derived and more likely due to anthropogenically derived discharges. Exposure to concentrations of these elements and toxicants that are higher than the PEL (probable effects level) or ERM (effects: range — median) raises questions relating to safety and risks to human health. Since the PEL- and ERM-type of criteria and guidelines

TABLE 8.2

Environmental Quality Guidelines for Sediments

Substance	Fresh Sediments		Marine Sediments	
	ISQG (µg/kg)	PEL (µg/kg)	ISQG (µg/kg)	PEL (µg/kg)
Arsenic	5900	17,000	7240	41,600
Cadmium	600	3500	700	4200
Chlordane	4.5	8.87	2.26	4.79
Chromium	37,300	90,000	52,300	160,000
Copper	35,700	197,000	18,700	108,000
DDD (2,2-bis(p-chlorophenyl)-1,1-dichloroethane; dichloro diphenyl dichloroethane)	3.54	8.51	1.22	7.81
DDE (1,1-dichloro-2,2-bis(p-chlorophenyl)-ethene diphenyl dichloroethylene)	1.42	6.75	2.07	3.74
DDT (2,2-bis(p-chlorophenyl)-1,1,1-trichloroethane; dicholoro diphenyl trichloroethane)	1.19	4.77	1.19	4.77
Dieldrin	2.85	6.67	0.71	4.3
Endrin	2.67	62.4	2.67	62.4
Heptachlor (heptachlor epoxide)	0.6	2.74	0.6	2.74
Lead	35,000	91,300	30,200	112,000
Lindane (hexachlorocyclohexane)	0.94	1.38	0.32	0.99
Mercury	170	486	130	700
Polychlorinated biphenyls (PCBs)	34.1	277	21.5	189
Arochlor 1254	60[V]	340[V]	63.3[V]	709[V]
Polychlorinated dibenzo-*p*-dioxins/dibenzo furans (PCDD/Fs)	0.85 ng-TEQ/kg dry wt	21.5 ng-TEQ/kg dry wt	0.85 ng-TEQ/kg dry wt	21.5 ng-TEQ/kg dry wt
Acenaphthene	6.71	88.9	6.71	88.9
Acenaphthylene	5.87	128	5.87	128
Anthracene	46.9	245	46.9	245
Benzo(a)anthracene	31.7	385	74.8	693
Benzo(a)pyrene	31.9	782	88.8	763
Chrysene	57.1	862	108	846
Dibenzo(a,h)anthracene	6.22	135	6.22	135
Fluoranthene	111	2355	113	1494
Fluorene	21.2	144	21.2	144
2-Methylnaphthalene	20.2	201	20.2	201
Naphthalene	34.6	391	34.6	391
Phenanthrene	41.9	515	86.7	544
Pyrene[V]	53	875	153	1398
Toxaphene	0.1[x]		0.1[x]	
Zinc	123,000	315,000	124,000	271,000

Note: [V] = sediment quality guideline for Aroclor 1254: provisional; 1% total organic carbon (TOC), adoption of severe effect level of 34 µg g^{-1} TOC (Persaud et al., 1993); [x] = sediment quality guidelines for toxaphene: 1% TOC, adoption of the chronic sediment quality criterion of 0.01 µg g^{-1} TOC (NYSDEC, 1994); ISQG = interim sediment quality guideline; PEL = probable effect level.

Source: Adapted from Canadian Environmental Quality Guidelines, 2003, http://www.ccme.ca/assets/pdf/e1 062.pdf.

have been provided for determination of the direct toxic effects on aquatic life, the potential risk to human health resulting from bioconcentration (of toxicants) has yet to be fully evaluated. Excessive bioconcentration in marine species constituting seafood for humans will pose problems to human health and may well result in chronic or acute poisoning. The Minamata disease was an unhappy event that originated from methylmercury pollution of fish. This is discussed in greater detail in a later section.

Target concentrations reflecting maximum allowable background values for sediments are necessary to provide the necessary protection for both aquatic species and human

TABLE 8.3

Average Daily Intake of Some Inorganics in Typical North American Adults Compared with Typical Dosages in a Common Dietary Supplement

Element	Daily Intake (mg/day)[a] *(Typical average value for adult)*	Typical Dosage in Dietary Pills (mg/day)	Possible Effects from Deficiency[b]	Possible Toxic Effects[b]
Potassium	3750	32	Hypokalemia, muscle weakness, abnormal heart rhythms	Diarrhea, nephrotoxicity, hyperkalemia, muscle fatigue, cardia arrhythmia
Calcium	420	530	Loss of calcium from bone, muscle spasms, leg cramps	Calcium deposition in soft tissue, kidney stones
Sodium	5660		Muscle cramps	High blood pressure
Phosphorus	1500	400	Weakness, rickets, bone pain	Kidney/liver damage
Magnesium	375	100	Electrolyte imbalance of Ca and K	Muscle weakness
Zinc	13	22.5	Reduced appetite and growth	Irritability, nausea
Iron	19.5	12	Anemia	Gastrointestinal irritation
Chromium	0.115	0.027	Atherosclerosis	Tubular necrosis of the kidney
Fluoride	3		Possible osteoporosis	Dental fluorosis, possible osteoscicrosis
Copper	1.7	2	Anemia, loss of pigment, reduced growth, loss of arterial elasticity	Disorder of copper metabolism, hepatic cirrhosis

[a] Information from Lappenbusch (1988).

[b] Deficiency and toxicity effects are *probable effects* and are very much dependent on initial health, diet, local environment, cultural attitudes, body features, physiology, etc.

Source: Adapted from Yong, R.N., *Engineering Geology of Waste Disposal*, Special Publication 11, Bentley, S.P., Ed., Geological Society of London, 1996, pp. 325–340.

health. These have yet to be set because (1) available data on present backgrounds are not available, (2) the links between these and bioconcentration in aquatic species and human health are also not available, and (3) the nature of background concentrations of a substance is highly variable. The variability and changing nature of background concentrations are functions of many factors, such as sediment characteristics, local geology, mineralogical aspects of soils near the deposited place, and discharges from land-based industries. Isolating discharges from land-based industries, it is possible to consider the deviation of the background concentrations with sediment type as an indication of a range of tolerable intake.

8.4.3   Heavy Metals

Many heavy metals are trace metals. These are metals that exist in extremely small quantities and are almost at the molecular level. They reside in or are present in animal and plant cells and tissue, and are a necessary part of good nutrition, as shown, for example, in Table 8.3. Because excess intake of heavy metals may cause damage to human health, it is necessary to take into consideration the background concentrations in sediments in structuring safe limits for ingestion of aquatic species. We define the *background concentrations* (values) as the concentrations of substances under natural conditions without any significant input or effect from human activities.

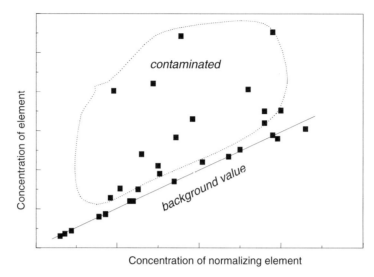

FIGURE 8.2
Concept of a normalization technique for obtaining a background value.

Background values of metals were found to agree well with average values for gneisses rocks (Carral et al., 1995). However, the background values for copper, zinc, lead, and cobalt do not necessarily agree with previously obtained values. Various techniques have been used to reduce data scattering and to allow for a more accurate statement of background values. Fukue et al. (1999) used carbonates as the normalizing substance in their measurements. Cobelo-García and Prego (2003) obtained the baseline relationships between the concentration of iron and pollutant, while Din (1992), Cortesão and Vale (1995), and Santschi et al. (2001) used aluminum to normalize heavy metals. Titan has also been used as a normalizing element. Fukue et al. (2006) calculated the specific surface area of sediment particles, assuming them to be spheres and that the specific surface areas will be related to the background values of heavy metals. This is consistent with the thesis that the amount of metals sorbed onto particles is to a large extent dependent on the particle size. The results reported by Fukue et al. (2006) using these normalization characteristics were found to be reasonable. Santschi et al. (2001) reported that concentrations and fluxes of most trace metals found in sediment cores recovered from the Mississippi River delta, Galveston Bay, and Tampa Bay in the United States, when normalized to Al, were typical for uncontaminated Gulf Coast sediments. Similar results can be cited with other normalizing elements. The concept for the normalization technique is shown in Figure 8.2. The drawback of these methods is that noncontaminated sediment samples are required in order that relationships between normalized substances and the objective concentrations can be constructed. If the depth and extent of sampling are insufficient, normalization characteristics cannot be obtained.

Many pollutants in water are sorbed on sorbates that may be inorganic and organic particulate materials. These particulates, together with the sorbed pollutants, eventually settle to form sediments on the seabed. The properties and surface characteristics of the soil solids that comprise the particulate materials have been discussed briefly in Sections 2.4 and 2.5 in Chapter 2 in respect to pollutant–particle interactions. Some of the more pertinent surface activities include such parameters as charge density, cation exchange capacity (CEC), specific surface area, the equilibrium (natural) concentration of contaminants, etc. A more detailed explanation of all of these surface properties and their interactions with pollutants can be found in Yong (2001). The background concentration values of sediments will be directly related to the inherent properties and characteristics of the constituents.

There are two concepts of background concentration: (1) concentration in the mother rocks and (2) mother rock concentrations together with sorbed substances under natural conditions during the transport process and deposition. The second concept is more appropriate when background concentrations are used to evaluate the risk to human health, especially when bioconcentration is factored into risk evaluation.

8.4.3.1 Profile of Heavy Metal Concentration

Most profiles of metal concentration for sediments show trends similar to those portrayed in Figure 8.3. The total concentration consists of the background and contributions from the discharges originating from land-based industries. Beginning from the bottom of the profile shown in Figure 8.3, the total concentration consists only of the background values. As one progresses upward, toward the top of the sediment, the total concentration begins to include pollutant discharge contributions from land-based industries. The beginning point shown in the diagram assumes that this can be tied into the start of the industrial revolution in about 1750. The total concentration increases as one progresses upward in the sediment, indicating increased discharge of pollutants into the marine environment. As an example, Figure 8.4 shows zinc (Zn) concentrations in various sediments obtained from coastal regions in Japan. Because the sediment samples were obtained from various depths in the total sediment layer, the concentration levels were not uniform, and some of the shallower samples were contaminated. The background concentrations can be obtained using a normalizing factor. The procedure adopted for the zinc concentrations shown in Figure 8.4 was to use the fines content as the normalizing factor. The fines content refers to the fine particle sizes in the sediment, generally clay and silt, and is defined as the content of clay and silt fractions that is <0.075 mm. The line representing the approximate lower limit shown in Figure 8.4 can be defined as the background zinc concentrations for the sediments. The detailed theoretical approach to obtain the background line has been reported by Fukue et al. (2006).

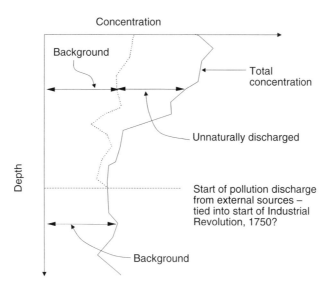

FIGURE 8.3

Concept of background value of metal concentrations in sediments. The *unnaturally* discharged concentrations are likely due to discharges from land-based industries and land-based nonpoint pollution sources.

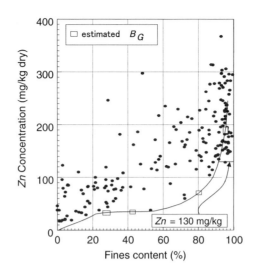

FIGURE 8.4
Deviation of zinc (Zn) concentration in various sediments. B_G = background concentration.

Similar relationships with lead (Pb) and copper (Cu) have been obtained, as shown in Figure 8.5 and Figure 8.6. Fukue et al. (2006) have used relationships between calculated specific surface area and fines content, and also limitation of sorption of fine particles under a relatively low equilibrium concentration. This is due to the limited concentrations of the substance in nature. The background values for Zn, Pb, and Cu shown in the figures are lower than the values presented as the interim sediment quality guidelines (ISQGs) in Table 8.2. When the background concentration, B_G, is known, the degree of pollution, P_d, will be obtained as

$$P_d = (C\ B_G)/B_G \qquad\qquad (8.1)$$

where C is the current concentration of specific substance or element (Fukue et al., 1999).

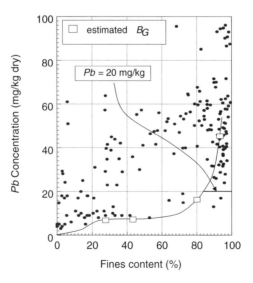

FIGURE 8.5
Deviation of lead (Pb) concentration in various sediments.

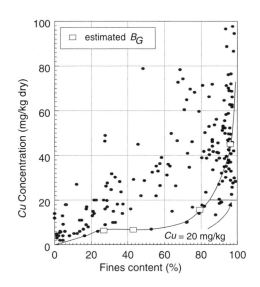

FIGURE 8.6
Deviation of copper (Cu) concentration in various sediments.

As was stated in the previous section and in Chapter 1, and shown in Table 8.3, there is a risk that the intake of hazardous substances beyond the PEL values may be harmful to human health.

8.4.3.2 Minamata Disease

In the 1950s, a significant number of people succumbed to a disease, later identified as Minamata disease, that was traced to ingestion of poisoned fish and shellfish in Minamata Bay in Kyushu Island (Japan). Measurements of methylmercury chloride showed very high concentrations, up to approximately 50 ppm in fish and 85 ppm in shellfish, from the contaminated areas in Minamata Bay. The more than 100 people that ingested the contaminated fish and shellfish showed initial symptoms of numbness of the limbs and areas around the mouth, sensory disturbance, lack of coordination, weakness and tremor, and speech and audiovisual difficulties. Progressive worsening of all of these with time led to general paralysis, convulsions, brain damage, and death. Traces of mercury poisoning were also found in animals living around the bay. Many of these animals also died. More details of this terrible demonstration of the impact of polluted sediments can be found in http://www.env.go.jp/en/topic/minamata2002/ch2.html and http://www.unu.edu/unupress/unupbooks/uu35ie/uu35ie0c.htm.

It is useful to note that while Minamata disease is well documented and serves as a dramatic demonstration of the chain of effects originating from polluted sediments, discovery and diagnosis of the health problems and source of the health problem took some considerable time, effort, and tracking. There are undoubtedly countless cases of poisoned aquatic animals in the first level of the food chain serving the human population, and unreported or undiagnosed (underdiagnosed) poisoning of the affected population.

8.4.4 Organic Chemical Pollutants

8.4.4.1 Organotins

Other than from direct and indirect discharge of land-based industrial waste pollutants and leachates from waste piles, some hazardous substances enter the marine environment through direct contact and use, such as organotins used as antifoulants for ships, quays, bouys, etc. Organotins are compounds of tributyltin (TBT). They are highly toxic chemicals comprised of tin combined with organic molecules. They are used not only as antifoulants, but also as wood preservatives, slimicides, and biocides. In the context of a coastal marine environment, organotins are used essentially as biocides to prevent the buildup of barnacles and algae. They are self-polishing co-polymers and generally have a service life between 3 and 5 years — meaning that they have to be reapplied to the marine structures at the end of their service lives. They are poisonous to marine life, including whales, dolphins, seals, fish, and seabirds. Linley-Adams (1999) reports that concentrations in bottlenose dolphin liver from the U.S. Atlantic and Gulf Coasts, found in the period between 1989 and 1994, ranged from 110 to 11,340 ng/g wet wt. He further reports on the presence of organotins in sea otters in California waters and harbor porpoises in Turkish coastal waters in the Black Sea. TBT has been reported to result in the development of imposex, a pseudo-hermaphroditic condition in female gastropods (snails) at ng/l levels of concentration (Horiguchi et al., 1994). The U.K. guideline provides 0.002 µg/l of TBT for seawater quality and 0.008 µg/l for seawater quality for triphenyltin (TPT) guideline. The Canadian environmental guideline for seawater quality is 0.001 µg/l for TBT.

TBT can enter the marine environment (1) through leaching of the antifoulant paints, (2) during application of TBT as an antifoulant, and (3) when the paint is removed from the pieces of equipment painted with the antifoulant. The degradation of TBT is relatively slow in sediments. The half-life of TBT in sediments has been reported to be approximately 2.5 years (de Mora et al., 1995), while it is only one week or so in marine waters (Seligman et al., 1988). Since 1990, their use has been banned for all vessels in certain countries (Japan, Australia, and New Zealand), and in some other countries, their use has been restricted to vessels with lengths greater than 25 m. In 2001, the International Maritime Organization (IMO) adopted the International Convention on the Control of Harmful Anti-Fouling Systems — a convention that prohibits the use of harmful organotins as antifoulants. The convention came into effect at the beginning of 2003 for all ships and will be expanded to include all floating platforms, floating units, floating production, and storage units at the beginning of 2008.

8.4.4.2   *Chlorinated Organic Micropollutants*

Chlorinated organic micropollutants have been accumulating in sediments and sea animals for many years. These pollutants are highly toxic, and direct or indirect ingestion of these will be life threatening. The particular group of chemicals known as chlorinated dibenzo-p-dioxins (CDDs) are perhaps the ones that have gained the most attention and concern, since they are known to enter the geoenvironment and into the marine environment from many sources. Of the more than 70 chemicals that make up the CDDs, the one known as 2,3,7,8-TCDD, or more simply as TCDD (2,3,7,8-tetrachlorodibenzo-p-dioxin), is considered to be one of the most toxic. The World Health Organization (WHO) has designated this as a known human carcinogen. In general, CDDs find their way into the land and marine environment as waste discharges from some processes that employ chlorine, such as paper mills that use chlorine bleaching processes (see Section 7.4 of Chapter 7), manufacturing of chlorinated organic intermediates, and water treatment plants. Generically, CDDs are known as dioxins. They also occur naturally and are discharged when substances containing such substances are degraded and combusted, as in forest fires. Dioxins released as atmospheric emissions from forest fires, combustion of fossil fuels, and incinerators serving industrial plants and municipal and hazardous waste facilities will subsequently be deposited onto land and water bodies at points distant from their sources. In water bodies, they will be attached to the suspended and sedimenting particles and will form part of the sediment. A limit of 3×10^{-5} μg of 2,3,7,8-TCDD per liter of drinking water (30 pg/l) has been set by the Environmental Protection Agency (EPA).

Dioxins generally occur in the environment with chlorinated dibenzofurans (popularly known as furans). They are ubiquitous and have been found in all media: air, water, soil, sediments, animals, and food (Johnson, 1992). They fall into the class of persistent organic pollutants (POPs) and are bioaccumulative. Since they have a strong affinity for soil and sediment particles, their presence in the media containing these POPs poses significant health threats. To determine the risks resulting from exposure to these POPs, many environmental regulatory bodies, such as the U.S. Environmental Protection Agency (EPA), have adopted an evaluation technique that uses *toxicity equivalency factors* (TEFs). This technique compares the toxicity of designated dioxins and furans to that of the most toxic dioxin, i.e., 2,3,7,8-TCDD. With this technique, the TCDD establishes the height of the toxicity bar and is given a TEF value of 1; all other organic chemical compounds being scrutinized are assigned TEF values according to how toxic they are perceived to be. Determination of toxicity intake (or uptake) of a particular organic chemical pollutant requires the use of *toxicity equivalents* (TEQs). A TEQ for a particular organic chemical pollutant ingested is determined by the product of the pollutant weight (in grams) and its assigned TEF, and the units are grams TEQ. As an example, Makiya (1997) stated that 60 to 70% of daily intake of dioxins (approximately 3.5 pg-TEQ/kg/day) is obtained from fish and shellfish. The tolerable daily intake (TDI) established in Japan is 4 pg-TEQ/day/kg of weight. The Japanese standard value for soils contaminated with dioxins is 1000 pg-TEQ/g or less, whereas the corresponding standard value for sediments is 150 pg-TEQ/g. This value drops to 1 pg-TEQ/l or less for water — in accordance with the Ministry of Environment (Japan) standards. Sediments containing dioxins that are more than the standard value are required to be treated. The classical remediation techniques used are dredging and disposal on land — procedures that are most often costly and sometimes prohibitive when dredged materials have to be cleaned before land disposal.

8.5 Rehabilitation of Coastal Marine Environment

Several techniques have been tried and used in the rehabilitation of coastal marine environments. Table 8.4 gives a short summary of some of the recent procedures used to treat contaminated seawater and sediments, and also measures taken to create coastal marine ecosystems. The methods used include (1) removal of the pollutants and hazardous substances by various techniques, (2) immobilization and isolation of these substances, and (3) neutralization and detoxification. For the treatment of seawater, the techniques include aeration, filtration, adsorption, accumulation, isolation, etc. In the case of contaminated sediments, dredging and capping are common techniques, while other techniques, such as cultivation, lime strewing, etc., are also used. The essence of the various techniques, present capabilities, economics of operation, and requirements are also briefly noted in the table.

The ultimate aim of sediment and seawater treatment procedures is to remove the threat posed by pollutants and hazardous substances in the water and sediments.

TABLE 8.4

Various Techniques Developed for Remediation of Contaminated Seawater and Sediments

Technology	Name	Feature	Example of Use	Economics	Durability	Maintenance	Comments
Seawater							
Aeration	Air bubble curtain	Aeration Aerobic condition Convection of seawater	Developing	Needs electricity	Durability of hardware	Clogging due to organisms	
	Microbubbling	Aeration Very small bubbles	Oyster farms				
	Aerator	Aeration Mixing of seawater	Developing	Needs electricity			
	Special seawall	Natural bubbling Mixing of seawater	Developing				
	Air mixing flow	Aerobic condition of deep seawater Mixing of seawater	Developing				Enclosed water area
Filtration	Filtration	Control and management of eutrophication Removal of suspended solids	Many examples	Exchange of filtered materials Removal of filtered materials	Depends on instruments	Exchange of filtered materials	
	Gravel oxidation	Control and management of eutrophication Removal of suspended solids Actions of microorganisms	Developing				
	Inclined wall	Control and management of eutrophication Removal of suspended solids Filtration	Used in rivers		Exchange of filters	Removal of sediments	

Category	Method	Purpose	Status/Example	Cost/Material	Effect	Countermeasure	Remarks
Adsorption	Artificial leaves	Control and management of eutrophication; Removal of suspended solids; Sedimentation by aggregation agents	Used in sediment ponds			Removal of sediments	
	Clays	Control and management of eutrophication	Many examples			Removal of sediments	Muddiness due to spraying agent; Impact on living organisms
	Aggregation agent	Control and management of eutrophication; Promotion of sedimentation	Many examples	Polymer		Removal of sediments	Impact on environment from aggregation agents
Sorption/accumulation	Lime strewing	Adsorption of sulfide, increase in pH	Many examples		Short-term effect	Strew often when no effects are observed	Fossil limestone
	Accumulation of plants	Reduction of eutrophication	Lakes			Recovery of plants	Management of aquatic plants
	Bioaccumulation of shellfish	Removal of suspended solids	Well known				
Isolation	Silt fence	Prevention of diffusion of suspended solids; Promotion of settlement of suspended solids	Many examples	$500–2000/m			
	Oil fence			$50/m	About 5 years	Treatment of adhesive organisms	
Sediments							
Cultivation	Oxidation		Established	Simple and easy			
In situ solidification	Solidification of seabed			Expensive			

Continued.

TABLE 8.4 *(Continued)*

Various Techniques Developed for Remediation of Contaminated Seawater and Sediments

Technology	Name	Feature	Example of Use	Economics	Durability	Maintenance	Comments
Dredging and disposal		Removal of contaminants	Many examples	$300/m^3$			Disposal using bags
Trench		Collection of mud in trench with about 2 m deep	Needs experience		Influenced by waves	Removal of sediments	Need disposal sites
Capping of sand		Capping of sand	Many examples	Easy task with a simple machine			Not for navigation route of ships
Artificial tidal land		Sand beach	Many examples	Lack of sand materials	Influenced by waves		
Lime strewing		Adsorption of sulfide, increase in pH	Many examples		Short-term effect	Strew often when no effects are observed	
Phytoremediation		Heavy metals, TBT, etc.	Developing				
			Need seawater purification				
Creation of Ecosystem							
Artificial wetlands		Biological diversity	Many examples	High cost			Difficulty in evaluation
Sea grass		Biological diversity, increase in haul of fish and shellfish					
Tidal pool		Recreation in seaside					Easy
Artificial coral reef		Biological diversity					
Artificial fish compartment		Biological diversity, increase in haul of fish and shellfish					

TABLE 8.5

Examples of Metal Concentrations in the Suspended Solids of a Coastal Sea
Area (mg/kg dry wt)

Element	Concentration	Element	Concentration	Element	Concentration
Fe	8000–30,000	Cu	240–500	P	2600–3000
Al	14,000–44,000	Pb	26–97	Mg	9600–14,000
Ti	1000–32,000	Cd	2.6–3.0	Ca	9000–90,000
V	40–53	Ni	2.8–36	K	9500–15,000
Zn	350–940				

8.5.1 Removal of Contaminated Suspended Solids

8.5.1.1 Confined Sea Areas

In confined sea areas, accumulation of hazardous substances and eutrophication are significant problems. A major factor in the constitution of seawater quality is the amount of suspended solids in the seawater. These solids include mineral particles, plankton, organic matter, and other kinds of particulate matter. The sorption characteristics and properties of suspended solids make them useful tools for sorption of hazardous substances, such as heavy metals, PAHs, chlorinated organic compounds, etc., in the seawater. Bacteria such as colon bacilli may also be found on the surfaces of suspended solids. Table 8.5 provides an example of the concentration of heavy metals adhering to some typical suspended solids. Removal of the contaminated suspended solids will be a step toward obtaining better seawater quality. Together with the process of bioaccumulation, these suspended solids can be removed from the seawater using pumping and filtration techniques; e.g., released phosphorus (into the sea) initially taken by phytoplankton and algae will be removed when the phytoplankton and algae are themselves subsequently removed. Figure 8.7 gives a schematic illustration of this simple concept. The ultimate aim is to ensure that the concentrations of the hazardous substances are lower than the allowed values in the guidelines. Improved seawater quality provides for (1) greater transmission

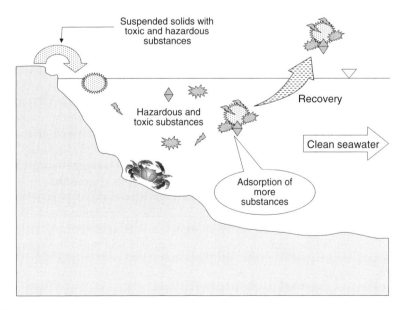

FIGURE 8.7
Concept of removal of suspended solids.

FIGURE 8.8
Purification vessel to remove suspended solids from seawater.

of light to seabed, (2) acceptable water quality for leisure use by the local coastal community, (3) reduction of potential for eutrophication and reduced capability for development
of red tide, and (4) better sediment quality, and also further reduction in potential for
eutrophication, benthic pollution, and development of blue tide. Technology and procedures for removal of suspended solids in closed sea areas using filtering techniques should
be determined by the targeted final requirements and results of site investigation.

8.5.1.2 Large Bodies of Water

To treat large bodies of seawater, filter units consisting of beach sand and steel slag have
been used successfully to remove the suspended solids. These units were installed on a
vessel (approximately 2500 tonnes), as shown in Figure 8.8. The case study conducted
with 38 filter units (1.5×3.6 m) for purification of seawater in a closed sea area achieved
a purification capacity of about 6000 m^3/day. The quality of the treated (purified) seawater
satisfied the regulatory requirements for allowable suspended solids, COD, pH, dissolved
oxygen (DO), etc. (Fukue et al., 2004). A comparison of the nature of the suspended solids
showed that these contained large amounts of substances — as much as the underlying
sediments. Figure 8.9 shows the results of removal of the suspended solids. The left-hand
lucite dish contains the seawater with suspended solids. The right-hand picture shows
that removal of almost 100% of suspended solids from the contaminated seawater produced seawater that to all intents and purposes was devoid of suspended solids and all
the substances adhering to the suspended solids. Studies show that this technique can be
used for purification treatment of water in various kinds of marine applications, including
dredging, and treatments to prevent development of red tide and blue tide.

8.5.1.3 Continuous Removal of Suspended Solids

In some closed sea areas, land-based hazardous substances are continuously discharged
and delivered by rivers, etc., to the coastal marine waters. Filter units similar to those

(a) Non-filtered   (b) Filtered

FIGURE 8.9
(See color insert following page 142.) Lucite dishes showing (a) nonfiltered and (b) filtered seawater obtained with purification units in the purification vessel.

used in the purification vessel, mounted on semipermanent pier fixtures, can function well to remove the hazardous substances adhering to the suspended solids.

8.5.2 Sand Capping

Containment and isolation of nutrients and hazardous substances in sediments will discourage eutrophication and prevent uptake of the hazardous substances. Two methods presently in use are (1) sand capping and (2) dredging removal and dump on land. Sand capping is a procedure that places a bed of clean sand on the contaminated sediment. It may be the easiest and one of the most economical ways to restore the health of the bottom environment. *In situ* capping refers to placement of a covering or cap over an *in situ* deposit of contaminated sediment (Palermo et al., 1998). The cap may be constructed of clean sediments, sand, or gravel, or may involve a more complex design using geo-textiles, liners, and multiple layers, as used for the capping of waste disposal sites in land. This capping isolation procedure not only prevents resuspension of polluted fines and other sediment particles, but also reduces the flux of contaminants into the water column above.

The sand cap can provide a new sand beach and tidal flats that will serve as a habitat for various kinds of organisms — from bacteria in pore spaces to macro-benthos such as bivalves, sea cucumbers, and seaweed. These organisms consume a large amount of nutrients. When sand is obtained from regions distant from the capping site, there may exist organisms and small animals in the transported sand that are not native to the capping site. This is especially true when the source of the sand is land based. Newly created sand cap beaches can function as physical purification units, i.e., filters, through the action of waves and tides. Removal of suspended solids from the seawater and decomposition of organic matter adhering to the suspended solids by microorganisms in the pores are additional benefits obtained in sand capping.

For sand capping to be robust and successful in providing the necessary isolation capability, the following must be accounted for in the design and construction of the cap:

TABLE 8.6

Purification Projects Using Sand Caps

Project Location	Contaminants	Site Conditions	Cap Design	Construction Methods	Reference
Kihama Inner Lake, Japan	Nutrients	3700 m²	Fine sand, 5 and 20 cm		
Akanoi Bay	Nutrients	20,000 m²	Fine sand, 20 cm		
Denny Way, Washington	PAHs, PCBs	1.2 ha near shore, with depths from 0.6 to 18.3 m	Average 80 cm of sandy sediment	Barge spreading	Sumeri et al., 1995
Simpson Tacoma, Tacoma, Washington	Creosote, PAHs, dioxins	6.8 ha near shore, with varying depths	1.2 to 6.1 m of sandy sediment	Hydraulic pipeline with sandbox	Sumeri et al., 1995
Hamilton Harbor, Ontario	PAHs, metals, nutrients	10,000 m² portion of large, industrial harbor	0.5 m sand	Tremie tube	Zeman and Patterson, 1996
Eitrheim Bay, Norway	Metals	100,000 m²	Geotextile and gabions	Deployed from barge	Instanes, 1994

1. Nature and type of currents, waves, flood, etc., that will result in drifting of the sand during and after emplacement. The record shows that hurricanes and typhoons have the capability of dramatically altering the near-shore bottom conditions.
2. Possible movement of some benthic organisms through the sand cap.
3. Disruption of the food chain.

Table 8.6 shows a sampling of some recent major sand capping projects. The benefits from sand capping are shown in Figure 8.10. Data reported by the Port of Kanda, Ministry of Land, Infrastructure and Transport, Japan, for COD, T-N, T-S, and T-P following sand capping indicate environmental benefits, such as less COD, less sulfide, and a lesser amount of nutrients in the sand cap, in comparison with that of the surrounding silty seabed. From the viewpoint of biological diversity, there is every indication that the sand cap is better than the surrounding silty seabed.

8.5.3   Removal of Contaminated Sediments by Dredging

8.5.3.1   *Dredging*

Dredging has historically been used (1) for beach reclamation, (2) to remove bottom sediments to deepen channels, waterways, and harbors and ports, (3) for maintenance of desired water depths for sea routes, and (4) to obtain materials for reclamation. Regardless of the type of equipment used — cutter wheels, augers, bucket wheels, pumping, suction hopper, self-propelled, automatic, etc. — dredging is a very invasive and destructive procedure. When used for removal of contaminated sediments, total disruption of the bottom ecosystem occurs. To return the benthic layer to full functionality, replacement of the food supply for benthic organisms is necessary. Removal of the first trophic level in the food chain will have severe consequences on the higher trophic levels. In that regard, planning for dredge removal of the contaminated bottom layer must include restoration of the functionality of the benthic layer.

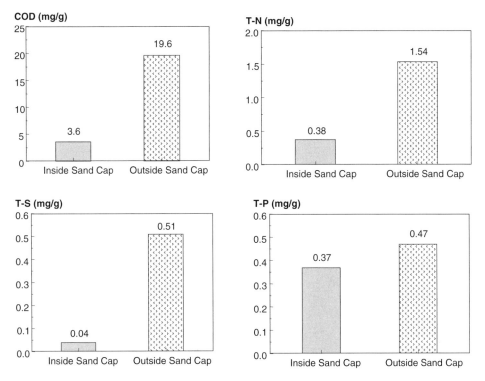

FIGURE 8.10
Effects of sand cap (Port of Kanda, the Ministry of Land, Infrastructure and Transport, Japan). COD = chemical oxygen demand, T-N, T-S, and T-P = total nitrogen, total sulfides, and total phosphorus, respectively.

8.5.3.2   Treatment of Dredged Sediments

Two options are available for disposal of dredged contaminated sediments: (1) disposal in a secure landfill and (2) treatment of the contaminated sediments and reuse of the treated sediments. Option 1 is not an option that has many proponents. Treatment of contaminated sediments (option 2) can be an expensive procedure, especially when the quantities are large. An expedient procedure is to perform gravity separation of the contaminated sediment and to remove the coarse fractions for treatment and reuse as construction material. A useful technique for sediments that do not contain much organic matter is to form larger particles by promoting aggregation of the fines with lime. Contaminated fine fractions can be treated or disposed in secure settling ponds. These settling ponds are not unlike those obtained in natural resource extraction processes (Chapter 5). Techniques for dewatering and hastening sedimentation of the suspended fines that constitute the fine fractions of the sediment have been discussed in Chapter 5. In the case of the fines in sediments, solidification and compression by filter pressing can be used (Yamasaki et al., 1995).

8.6   Creation of a Natural Purification System

8.6.1   Creation of Sand Beaches and Tidal Flats

Artificial sand beaches and tidal flat are created for one of the following purposes:

1. Formation of clean beach for resort areas and parks
2. Farming for shellfish
3. Recovery of beach following reclamation
4. Rehabilitation of coastal marine environment

Sand beaches and tidal flats possess natural capabilities for cleaning seawater under repeating waves and tides. This capability arises from a combination of their ability to filter a large amount of suspended solids (mostly organic matter) and the dissolution of the suspended solids by microorganisms. While the organic matter entrapped in the sand pores is food for microorganisms and benthic animals, there are no easy means to quantify the process and its benefits. Evaluation of the impacts arising from construction of the tidal flats and beaches cannot be readily performed. In part, this is due to the dynamic processes initiated by the actions of currents and waves. Stabilization of the new beaches and tidal flats will be a long-term process. The use of breakwaters on beaches brings with it problems of decrease in redox in the region due to the dead organisms and excrements. Figure 8.11 shows one of the three tidelands (Kasai Rinkai Park, 270,837 m^2) created artificially by the Tokyo Metropolitan Government in 1965, at a time when Tokyo Bay was losing its valuable natural environment. The area incorporates vast tidelands, which were once the breeding areas for birds and were also once abundant with fish and shellfish.

FIGURE 8.11
An example of creation of an artificial tideland.

FIGURE 8.12
Dense sward of eelgrass (*Zostera marina* L.) in a coastal region.

8.6.2 Creation of Seaweed Swards

Eelgrass (*Zostera marina*) is a water plant with long grasslike leaves. Figure 8.12 shows a dense sward of eelgrass. There are many different species of eelgrass. Shallow intertidal-water eelgrass has shorter and narrower leaves, whereas deeper subtidal-water eelgrass has longer and wider leaves. They tend to grow in tidal creeks, sandy bays, estuaries, and on silty–sandy sediments, and are a vital part of the food web chain for the coastal marine ecosystem. In dense swards of eelgrass, silt and clay particles tend to be deposited with organic matter. Decomposition of organic matter will render the seabed anaerobic, and the color of the sediments will become black because of the effect of sulfide.

The eelgrass family is one of the few flowering plants that lives in salt water, and the long grass blades are home to various kinds of small marine plants and animals. They are the breeding ground and habitat for all kinds of marine animals, including crabs, scallops, and other kinds of shellfish. They not only serve to foster and stabilize the benthic habitats, but they also have the potential for phytoremediation. Eelgrass can absorb trace metals and organotins (Brix and Lyngby, 1982; Fransois et al., 1989).

Table 8.7 gives a comparison of concentrations of various heavy metals in sediments with and without eelgrass. The sediments, which were taken from a small eelgrass sward at an estuary of Kasaoka Bay in Seto Inland Sea, Japan, consisted of a number of small communities — with bare parts between the communities. The total area of the eelgrass sward was 1491 m². The sampling points from A to H were located in the bare parts and communities. The results show that the sediments with eelgrass contain a lesser amount

TABLE 8.7

Comparison of Heavy Metal Concentrations in Sediments
with and without Eelgrass (Seto Inland Sea, Japan)

Sampling		Heavy Metal (mg/kg)		
Point		Cu	Pb	Zn
A	With eelgrass	11	20	76
B	Without eelgrass	13	10	83
C	With eelgrass	11	25	69
D	Without eelgrass	14	130	90
E	With eelgrass	15	17	77
F	With eelgrass	16	18	82
G	Without eelgrass	27	17	110
H	With eelgrass	17	18	83
	Average with eelgrass	14	19.6	77.4
	Average without eelgrass	18	52.3	94.3

of heavy metals — most likely attributed to heavy metal absorption (uptake) by the eelgrass. Since eelgrass grows from spring to summer, and their dead leaves drift to the sea surface at the end of their growing season, collection of the dead leaves can be simply implemented. This means that if eelgrass is used for phytoremediation, the absorbed heavy metals can be harvested with the dead sea grass leaves.

Reclamation and other near-shore industrial activities can negatively impact the coastal habitat, particularly the sea grass beds that form the seaweed fields. Reduction of seaweed fields not only decreases the habitat of marine living things, but will also result in a marked decrease in the haul of inshore fish. For example, in Japan, approximately 6000 ha of seaweed field has disappeared since 1978, and about one third of this was due to the impact of reclamation projects.

8.7   Sea Disposal of Waste

Section 8.3 has discussed the strict prohibitions articulated in the London Convention and Protocol. Dumping or discharging land-based industrial waste into the sea is essentially prohibited — with the burden of responsibility resting on the waste generator to ensure that any waste material entering the sea must be nontoxic and nonhazardous. Since many countries and jurisdictions with restricted land areas do not have sufficient land space for land disposal of waste, controlled and regulated discharge of municipal and industrial wastes into the sea remains the option of last resort. When such a need arises, waste disposal sites in the sea must be constructed to meet safety and health protection requirements. Isolation of the waste from contact and dispersion into the sea is a prime requirement. In some countries, artificial islands have been constructed for the principal purpose of emplacing secure disposal facilities. These island-based disposal sites must conform to all the regulations that attend land-based disposal sites — with the strict requirement for monitoring and control, to ensure no escape of leachate into the sea.

In some other cases, actual disposal sites have been constructed in the sea using seawalls as containment walls to prevent escape of waste and leachates into the surrounding sea. Typical examples are seen in Tokyo Bay and Osaka Bay. Table 8.8 gives a short summary of the waste disposal sites and materials disposed in four sites in Osaka Bay, constructed under the auspices of the Phoenix Project. The objectives of the project focused not only

TABLE 8.8

Examples of Sea Disposal of Various Kinds of Waste

| Site | Area (ha) | Volume of Waste Disposal (Million Cubic Meters) | | | | |
		Municipal Waste	Industrial Waste	Soils from Construction Site	Dredged Materials	Total
Off Amagasaki	113.0	2	3.9	5.8	4.1	16.0
Off Izumi-Otsu	203.0	4	9.4	12.7	4.8	31.0
Off Kobe	88.0	5	7.3	3.0		15.0
Osaka Port	95.0	5	6.3	2.8		14.0
Total	499.0	16	26.9	24.3	8.9	76.0

on the proper and safe disposal–discharge of the wastes in Osaka Bay, but also on preservation of the coastal marine ecosystem and development of the regional area through environmentally acceptable reclamation of shoreline.

8.8 Coastal Erosion

Coastal erosion is a problem that has confronted coastal communities for a long time. Some will argue that this is a natural phenomenon and that the problem arises when the human population occupies the coastal regions and imposes various requirements on the coast, such as building structures, infrastructures, wharfs, etc. Alteration of the coastline is a natural phenomenon, and erosion is a major factor in the alteration process. This alteration process becomes a problem because of the requirements and expectations imposed by the human population. Erosion arises not only from the aggressive action of waves and currents, but also from diminished sediment recharge from streams and rivers feeding into the sea. Flood control dams and other river restoration projects can curtail the flow of suspended particulates and sediments. Without this sediment recharge and with erosive forces acting on the coastal plains, erosion becomes a considerable issue. To counter the erosive forces and to prevent beach erosion, armoring of the beach with revetments and seawalls is a procedure that has been adopted by many coastal regions. Figure 8.13 shows an erosive beach on the coast of Japan facing the Pacific Ocean. A huge number of concrete blocks have been installed to protect the coastline and to gather sand. Figure 8.14 shows soil dumping as a countermeasure for erosion. The record shows that some period of time after leveling of the dumped soil, erosion of the slope occurred. To overcome this, geotextile tubes can be used to protect the coastline from waves and tides, as has been utilized in some countries. The tubes, which can be installed along the coastline, are a few meters in diameter and a few kilometers in length and can be filled with dredged soils (Figure 8.15). They can also serve as a breakwater for man-made islands and wetlands.

8.9 Concluding Remarks

The health of coastal marine environments is vital to the productivity of the marine aquatic resources in the ecosystems within this environment. The oceans and coastal marine environments are significant resource bases and are essential components of the life sup-

FIGURE 8.13
An erodible coast in Japan facing the Pacific Ocean.

FIGURE 8.14
(See color insert.) Countermeasure for coastal erosion by soil dumping and the action of erosion.

FIGURE 8.15
Geotextile tube used for protection and creation of coast.

port system for the human population. The two groups of events that pose significant threats to the health of this environment are distinguished on the basis of *natural* and *man-made*. In the first instance, the natural events include phenomena such as coastal erosion resulting from aggressive current, waves, and tidal action. Man-made events result in fouling of the coastal marine environment and the seas through discharge of wastes from ocean vessels and from land-based industries and activities. It has been argued that coastal erosion is also a victim of man-made events, according to the thesis that *man-generated variations* in sea level contribute to the aggressive actions of the currents, waves, and tidal actions. The thrust of this thesis is that global warming is in part responsible for the variations in sea level not only through ice melting, but also through changes in the seabed levels. While a discussion on global warming is not within the purview of this book, it is nevertheless important to point out that the many facets of geoenvironmental sustainability are affected by such an event. Some of these issues have been discussed in Section 1.5 in Chapter 1.

Fouling of the coastal marine environment and the seawater affects both seawater and sediments. Contaminants are accumulated in the sediments, while others are put into circulation through the food chain and bioconcentration — ultimately posing health threats to the human population. Two basic problems exist in respect to fouling of the coastal marine environment and the seawater: eutrophication and concentration of toxic and hazardous substances. Figure 8.16 gives an illustration of a simple strategy for rehabilitation of the coastal marine environment. Hazardous substances and nutrients have been discharged into the coastal marine environment for many decades. These have to be collected and removed. A balance in the amount of nutrients removed is needed. On the one hand, sufficient removal of excess nutrients is needed to avoid eutrophication; on the other hand, sufficient nutrients must be available for the aquatic animals that rely on nutrients for their food supply. A sustainable coastal marine environment requires a

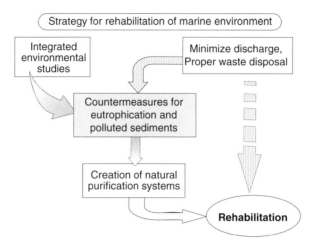

FIGURE 8.16
A strategy for rehabilitation of the marine environment.

natural purification system. Research and development into how this can be achieved would contribute substantially to the goals of a sustainable coastal marine environment. In the final analysis, since fouling of the marine coastal environment and the sea is in large measure attributable to the land-based activities and industries, proper control and management of the land pollution would go a long way toward mitigating fouling.

References

An, Y.-J., Kampbell, D.H., Weaver, J.W., Wilson, J.T., and Jeong, S.-W., (2004), Natural attenuation of trichloroethene and its degradation products at a lake-shore site, *Environ. Pollut.*, 130:325–335.

Brix, H. and Lyngby, J.E., (1982), The distribution of cadmium, copper, lead, and zinc in eelgrass (*Zostera marina* L.), *Sci. Total Environ.*, 24:51–63.

Canadian Environmental Quality Guidelines, (2003), http://www.ccme.ca/assets/pdf/e1 062.pdf.

Carral, E.,Villares, R., Puente, X., and Carballeira, A., (1995), Influence of watershed lithology on heavy metal levels in estuarine sediments and organism in Galicia (north-west Spain), *Mar. Pollut. Bull.*, 30:604–608.

Chiou, C.T., (2002), *Partition and Adsorption of Organic Contaminants in Environmental Systems*, John Wiley & Sons, New York, 257 pp.

Cortesäo, C. and Vale, C., (1995), Metals in sediments of the Sado Estuary, Portugal, *Mar. Pollut. Bull.*, 30:34–37.

Davis, J.W., Odom, J.M., DeWeerd, K.A., Stahl, D.A., Fishbain, S.S., West, R.J., Klecka, G.M., and DeCarolis, J.G., (2002), Natural attenuation of chlorinated solvents at Area 6, Dover Air Force Base: characterization of microbial structure, *J. Contaminant Hydrol.*, 57:41–59.

de Mora, S.J., Stewart, C., and Phillips, D., (1995), Sources and rate of degradation of tri(n-butyl)tin in marine sediments near Auckland, New Zealand, *Mar. Pollut. Bull.*, 30:50–57.

Dermietzel, J. and Christoph, G., (2002), *The Release of Pollutants from Aged Field Sediments*, IAHS Publication 275, IAHS, Oxfordshire, U.K., pp. 147–151.

Din, Z., (1992), Use of aluminium to normalize heavy metals data from the estuarine and coastal sediments of the strait of Melaka, *Mar. Pollut. Bull.*, 24:484–491.

Do Amaral, M.C.R., Rebelo, M. de F., Torres J.P.M., and Pfeiffer, W.C., (2005), Bioaccumulation and depuration of Zn and Cd in mangrove oysters (*Crassostrea rhizphorae*, Guilding, 1828) transplanted to and from a contaminated tropical coastal lagoon, *Mar. Environ. Res.*, 59:277–285.

Fabris, G.J., Monahan, C., and Batley, G.E., (1999), Heavy metals in waters and sediments of Port Philip, Australia, *Mar. Freshwater Res.*, 50:503–513.

Fransois, R., Short, F.T., and Weber, J.H., (1989), Accumulation and persistence of tributyltin in eel grass (*Zostera marina* L.) tissue, *Environ. Sci. Technol.*, 23:191–196.

Fukue, M., Nakamura, T., Kato, Y., and Yamasaki, S., (1999), Degree of pollution for marine sediments, *Eng. Geol.*, 53:131–137.

Fukue, M., Sato, Y., Inoue, Minato, T., Yamasaki, S., and Tani, S., (2004), Seawater purification with vessel-installed filter units, in *Geoenvironmental Engineering: Integrated Management of Groundwater and Contaminated Land*, Yong, R.N. and Thomas, H.R. (Eds.), Thomas Telford, London, pp. 510–515.

Fukue, M., Sato, Y., Yamashita, M., Yanai, M., and Fujimori, Y., (2003), Change in microstructure of soils due to natural mineralization, *Appl. Clay Sci.*, 23:169–177.

Fukue, M., Yanai, M., Sato, Y., Fujikawa, Y., Furukawa, Y., and Tani, S., (2006), Background values for evaluation of heavy metal contamination in sediments, *J. Hazard. Mater.*, (in press, on line 18 January 2006).

Gardner, S.C., Pier, M.D., Wesselman, R., and Arturo, J., (2003), Organochlorine contaminants in sea turtles from the Eastern Pacific, *Mar. Pollut. Bull.*, 46:1082–1089.

Green, N.W. and Knutzen, J., (2003), Organohalogens and metals in marine fish and mussels and some relationships to biological variables at reference localities in Norway, *Mar. Pollut. Bull.*, 46:362–374.

The Harmful Algae Page, www.whoi.edu/redtide/species/species.html, July 2005.

Horiguchi, T., Shiraishi, H., Shimizu, M., and Morita, M., (1994), Imposex and organotin compounds in *Tais clavigera* and *T. bronni* in Japan, *J. Mar. Biol. Assoc.*, 74:651–669.

Instanes, D., (1994), Pollution Control of a Norwegian Fjord by Use of Geotextiles, paper presented at the Fifth International Conference on Geotextiles, Geomembranes and Related Products, Singapore, September 5–9.

Inthorn, D., Sidtitoon, N., Silapanuntakul, S., and Incharoensakdi, A., (2002), Sorption of mercury, cadmium and lead by microalgae, *ScienceAsia*, 28:253–261.

Jensen, H.F., Holmer, M., and Dahllöf, I., (2004), Effects of tributyltin (TBT) on the seaweed *Ruppia maritima*, *Mar. Pollut. Bull.*, 49:564–573.

Jones, B. and Turki, A., (1997), Distribution and speciation of heavy metals in surficial sediments from the Tees Estuary, North-east England, *Mar. Pollut. Bull.*, 34:768–779.

Johnson, B.L., (1992), Public Health Implications of Dioxins, congressional testimony, Subcommittee on Human Resources and Intergovernmental Relations, Committee on Government Operations, U.S. House of Representatives, June 10.

Kan-Atireklap, S., Tanbe, S., and Sanguansin, J., (1997), Contamination by butyltin compounds in sediments from Thailand, *Mar. Pollut. Bull.*, 34:894–899.

Kavun, V.Ya, Shulkin, V.M., and Khristoforova, N.K., (2002), Metal accumulation in mussels of the Kuril Islands, north-west Pacific Ocean, *Mar. Environ. Res.*, 53:219–226.

Lappenbusch, W.L., (1988), *Contaminated Waste Sites, Property and Your Health*, Lappenbusch Environmental Health, Inc., Alexandria, VA, 360 pp.

Linley-Adams, G., (1999), *The Accumulation and Impact of Organotins on Marine Mammals, Seabirds and Fish for Human Consumption*, WWF-U.K. Report, Project 90854, WWF-U.K., Surrey, U.K., 26 pp.

Makiya, K., (1997), Current situation concerning government activities associated with dioxins, *Waste Manage. Res.*, 8:279–288 (in Japanese).

Moraga, D., Mdelgi-Lasram, E., Romdhane, M.S., El Abed, A., Boutet, I., Tanguy, A., and Auffret, M., (2002), Genetic response to metal contamination in two clams: *Ruditapes decussates* and *Ruditapes philippinarum*, *Mar. Environ. Res.*, 54:521–525.

NYSDEC (New York State Department of Environmental Conservation), (1994), *Technical Guidance for Screening Contaminated Sediments*, prepared by the Division of Fish and Wildlife and Division of Marine Resources.

Ohtsubo, M., (1999), Organotin compounds and their adsorption behaviour on sediments, *Clay Sci.*, 10:519–539.

Pakdeesusuk, E., Lee, C.M., Coates, J.T., and Freedman, D.L., (2005), Assessment of natural attenuation via *in situ* reductive dechlorination of polychlorinated biphenyls in sediments of the Twelve Mile Creek arm of Lake Hartwell, SC, *Environ. Sci. Technol.*, 39:945–952.

Palermo, M., Maynord, S., Miller, J., and Reible, D., (1998), *Guidance for In-Situ Subaqueous Capping of Contaminated Sediments*, EPA 905-B96-004, Great Lakes National Program Office, Chicago, http://www.epa.gov/glnpo/sediment/iscmain/one.html.

Persaud, D., Jaagumagi, R., and Hayton, A., (1993), *Guidelines for the Protection and Management of Aquatic Sediment Quality in Ontario*, Ministry of Environment, Toronto, Ontario, Canada, 30 pp.

Romano, E., Ausili, A., Zharova, N., Magno, M.C., Pavoni, B., and Gabellini, M., (2004), Marine sediment contamination of an industrial site at Port of Bagnoli, Gulf of Naples, southern Italy, *Mar. Pollut. Bull.*, 49:487–495.

Santschi, P.H., Presley, B.J., Wade, T.L., Garcia-Romero, B., and Baskaran, M., (2001), Historical contamination of PAHs, PCBs, DDTs, and heavy metals in Mississippi River delta, Galveston Bay and Tampa Bay sediment cores, *Mar. Environ. Res.*, 52:51–79.

Seligman, P.F., Valkirs, A.O., Stang, P.M., and Lee, R.F., (1988), Evidence for rapid degradation of tributyltin in a marina, *Mar. Pollut. Bull.*, 19:531–534.

Selvaraj, K., Ram Mohan, V., and Szefer, P., (2004), Evaluation of metal contamination in coastal sediments of the Bay of Bengal, India: geochemical and statistical approaches, *Mar. Pollut. Bull.*, 49:173–185.

Stapleton, R.D. and Sayler, G.S., (1998), Assessment of the microbiological potential for the natural attenuation of petroleum hydrocarbons in a shallow aquifer system, *Microb. Ecol.*, 36:349–361.

Sumeri, A., (1995), Dredged Material Is Not Spoil: A Status on the Use of Dredged Material in Puget Sound to Isolate Contaminated Sediments, paper presented at the 14th World Dredging Congress, Amsterdam, November 14–17.

Wilde, E.W. and Benemann, J.R., (1993), Bioremoval of heavy metals by the use of microalgae, *Biotechnol. Adv.*, 11:781–812.

Yamasaki, S., Yasui, S., and Fukue, M., (1995), Development of solidification technique for dredged sediments, dredging, remediation, in *Containment for Dredged Contaminated Sediments*, ASTM STP1293, ASTM, International, West Conshohocken, PA, pp. 136–144.

Yong, R.N., (1996), Waste disposal, regulatory policy and potential health threats, in *Engineering Geology of Waste Disposal*, Special Publication 11, Bentley, S.P. (Ed.), Geological Society of London, pp. 325–340.

Yong, R.N., (2001), *Geoenvironmental Engineering: Contaminated Soils, Pollutant Fate, and Mitigation*, CRC Press, Boca Raton, FL, 307 pp.

Zeman, A.J. and Patterson, T.S., (1996), *Preliminary Results of Demonstration Capping Project in Hamilton Harbour*, NWRI Contribution 96-53, National Water Research Institute, Burlington, Ontario.

9

Contaminants and Terrestrial Environment Sustainability Indicators

9.1 Introduction

We use *terrestrial environment* instead of the term *geoenvironment* because we do not include the coastal marine environment (considered in the previous chapter) in our discussions in this chapter. In terms of the terrestrial environment, other than depletion of natural resources and habitat destruction, ground contamination by all kinds of contaminants and pollutants poses one of the greatest threats to the sustainability of the natural capital of the geoenvironment. Within the context of the terrestrial environment, the term *natural capital* refers to the terrestrial ecosystem, which includes (1) the receiving waters in the land surface environment, such as rivers, lakes, ponds, and streams; (2) the solid land surface and the underlying soil–water system; (3) the natural resources, such as forests and mineral and carbon resources; and (4) the various biotic species that populate the ecosystem and contribute to the biodiversity of the system. The presence of contaminants in the ground affects not only soil and water quality, but also those living elements and biota that more or less depend on soil and water for their well-being. This would include forests, agricultural production, habitats, and the host of biotic species contributing to the biodiversity of the terrestrial ecosystem.

Anthropogenic activities associated with production of goods and services, such as resource exploitation, agricultural production, harvesting of forest and carbon resources, urbanization, industrial production, and manufacturing, are by far the greatest contributors to the generation of waste and contaminants that ultimately find their way onto and into the terrestrial environment. Sustainability of the terrestrial environment's natural capital cannot be easily or readily defined for many of the individual components that make up the natural capital of interest because of their dynamic nature. This is particularly true for the living components. It is easier to define sustainability goals for the natural capital. This task is facilitated if one can establish an acceptable natural capital baseline, and if actual numbers or quantities are not absolutely required.

9.2 Indicators

The general understanding of sustainability of a natural capital of the terrestrial environment can be stated as follows: *to ensure that each natural capital maintains its full and uncompromised functioning capability without loss of growth potential.* Clear baseline values for the various components that constitute the terrestrial environment natural capital are needed to define or establish sustainability requirements of the natural capital in quantitative terms. Since absolute sustainability is not always attainable, the use of indicators is a means to establish benchmarks or targets that point the way toward sustainability of specific components of the natural capital and of the natural capital itself.

9.2.1   Nature of Indicators

Indicators are essentially signs or markers used every day by many ordinary people and by professionals. Take, for example, the traffic lights at an intersection. The red bulb indicates that one should come to a complete halt, and the green bulb indicates that one can proceed. In terms of day-to-day living, personal events such as prolonged headaches, stomach aches, muscle pain, toothaches, etc., are used as indicators of some form of health distress and that one should visit one's health care specialist. Numbers, statistical information, events, etc., can all be used as indicators. Unemployment rates, gross domestic product (GDP), and financial indices such as those used for tracking the ups and downs of various equities are all indicators of the health of the economy. In a sense, indicators have been used by humans for as long as there has been human life on the planet.

From the geoenvironmental perspective, *the indicators* discussed in this chapter are considered to be markers or benchmarks specified or prescribed by individuals or organizations interested in tracking (1) the progress of events, (2) operations and performance of systems, (3) the outcome of actions on something specific, and (4) the status of a particular set of events, actions, process, activity, and situation — all of which are with specific reference to the terrestrial environment. The indicators can also be used (1) for performance assessment of various systems, processes, actions, etc., and (2) to set goals or targets for operation and performance of systems, processes, activities, etc., operations of systems, regulatory bodies, professional organizations, and consultants. There are many varieties and types of indicators that can be used in determining or tracking the sustainability of the terrestrial environment. For convenience, we can group all of these into two major groups: (1) system status indicators and (2) material performance indicators. System status indicators refer to the status of a system and any time period. They are performance indicators since the status at any one time period is the result of the performance of the system under consideration. In that same sense, material performance indicators are designed to provide information on the performance of materials at any specified or required time period — quantitatively or qualitatively. Since material properties and characteristics may be functions of system processes, it follows that material performance indicators are also material properties' indicators. When the term *indicators* is used, it is understood that this refers to both system status and material performance indicators. The greatest benefit of status indicators comes when they are used in relation to some predetermined or defined criteria, target, or objective.

Not all indicators are sustainability indicators. A good example is the measurement of acid precipitation onto the ground where the sources of SO_x and NO_x are known. The acidity measurements themselves are material indicators; i.e., they are acidity indicators. This begs the question of "So what?" Four courses of action come to mind: (1) do absolutely nothing, (2) implement a ground surface neutralization program using perhaps a surface spray technique, (3) monitor target components such as soil and water quality and environmental mobility of pollutants in the affected area, and (4) determine the source of the gaseous discharge and rectify the situation through implementation of better process technology and more effective scrubber systems. Figure 9.1 illustrates this example from a terrestrial environment perspective.

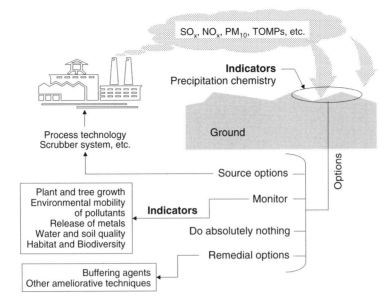

FIGURE 9.1
Illustration of the role of indicators in a typical precipitation situation in an atmosphere with noxious gaseous and other airborne pollutants — in the context of impacts on the land surface. The indicators shown do not cover the entire impact problem. Note that changes to process technology and scrubber system do not fall within the purview of this book — and neither do plant–tree growth, habitat, and biodiversity. PM_{10} = particulate matter less than 10 μm in size; TOMPs = toxic organic micropollutants, such as obtained PAHs, PCBs, dioxins, and furans from incomplete combustion of fuels.

Figure 9.1 shows that *indicators* (1) inform us on where we are in respect to specific indices and prescribed parameters, (2) provide specific information to be used to compare with (or contrast with) baseline values or other target values, (3) provide monitoring and tracking information for use in determining the transient state of items being monitored, and (4) tell us if we are on target or off-track. The SO_x, NO_x, PM_{10}, and toxicorganic micropollutants (TOMPs) shown in the top right-hand corner of the diagram are *material indicators*. They provide information on the noxious gases and other airborne pollutants that are responsible for the chemistry of precipitation onto the land surface. This information gives one the opportunity to exercise any one of four different options, as shown in the diagram. Depending on the chemistry of the precipitation recorded, one of the options is to improve the scrubber system in the industrial plant and shown in the top left-hand corner of the diagram — assuming that the source for the noxious precipitation is traceable to that particular industry.

The other options available, in view of the information obtained from the material indicators, are (1) do nothing, (2) monitor, and (3) undertake remedial actions. In the case of the monitor option, the associated indicators are *status indicators*. These indicators essentially track the status of the various items, such as water and soil quality at various times, to determine whether negative changes occur. More importantly, these system status indicators are important markers along the road to sustainability, and where appropriate, they may be called sustainability indicators. *Sustainability indicators* (SIs) are those indicators that point the way toward achievement of the goals and objectives of sustainability. For these indicators to have value and meaning, it is necessary to provide or specify the sustainability goals or objectives. Because true sustainability may never be attained, as for example in the case of diminishing nonrenewable natural resources, the sustainability indicators provide us with a measure of nonrenewable natural resource depletion (depletion rate). By targeting

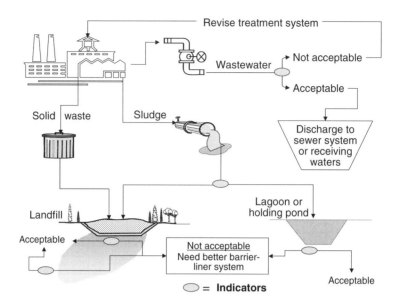

FIGURE 9.2
The role of indicators in assessing status and sustainability of operations in a plant discharging liquid and solid wastes. The indicators in the diagram are shown as shaded ovals. Note that the discharge of acceptable wastewater to the sewer system or receiving waters can also be directed back to the plant as recycled water — if provisions are made to accept recycling as a step toward sustainability of operations.

a depletion rate that would allow for a longer time before exhaustion of the resource, SIs can alert us to whether we are on target or whether drastic remedial or corrective steps need to be taken. However, as will be evident in further discussions on indicators, SIs are really a collection of many intermediate indicators situated as markers along the road to sustainability — with the final set of indicators located at the terminal point of the road. The various intermediate indicators should set the requirements and conditions of performance at stages along the way, as shown, for example, in Figure 9.2. The combination of these indicators can be considered requirements to satisfy sustainability objectives.

9.2.2   Contaminants and Geoenvironment Indicators

As another example of impact from industrial production, the discharge of liquid waste or sludge and solid waste is shown in Figure 9.2. Not all the indicators shown in the diagram (depicted by the shaded ovals) are sustainability indicators (SIs). The indicators shown in Figure 9.2 have several functions. In the top right-hand corner, the information from the wastewater discharge from the industrial plant (wastewater chemistry indicators) tells us whether the wastewater discharge meets the regulatory requirements permitting discharge into the sewer system or receiving waters. The wastewater chemistry indicators, which are material indicators, could include, for example, suspended solids, alkalinity, metals, fats, oils, grease, organic chemicals, etc. Failure of the wastewater chemistry to meet discharge standards will require a reengineering of the treatment system.

Information obtained from analyses of the sludge being discharged (from the discharge end of the sludge pipe, i.e., the sludge discharge indicators) can be used to determine whether the toxicity and sludge characteristics meet lagoon (holding pond) capabilities or whether the sludge should be contained in a secure landfill. This assumes that lagoons and holding ponds do not generally have the same type of secure impervious barrier–liner systems that are required for landfills. In this instance, the sludge discharge indicators are not sustainability indicators, but are status indicators that provide one with the information required to make the necessary judgment for disposal of the sludge.

The contaminant indicators located below the lagoon and the landfill serve several purposes. In the lagoon case, assuming that the barrier–liner system for the lagoon is not as secure as the landfill system, the contaminant indicators will tell us whether the contaminants escaping from the lagoon are in the range of acceptable concentrations. One needs to rely on the attenuation characteristics and properties of the subsoil strata to further ameliorate the concentration of contaminants as the contaminant plume travels further into the soil substratum. If the contaminant indicators show unacceptably high concentrations of contaminants immediately under the lagoon, regulations would require corrective action to be taken.

The same situation applies to the contaminant indicators immediately adjacent to or under the barrier–liner system of the landfill. Corrective action is needed if the contaminant indicators exceed specified concentration trigger levels. Acceptable concentration levels reported by the contaminant indicators immediately under the barrier–liner system do not mean that the other contaminant indicators located farther away from the landfill would report favorably. Much depends on the nature of the contaminants and the transport processes. Contaminant transport modeling, using the information from the first set of indicators under the barrier–liner system, can be useful. Predicted transport values can be used to compare with the second set of contaminant indicators. The contaminant indicators are seen to be markers that show progress of the contaminant front (i.e., trackers of transport), and if design specifications and expectations of attenuation are correct, the indicators should accord well with predictions — provided that the transport models accurately predict performance. The transport of contaminants has been briefly discussed in Chapter 2. A further detailed discussion of these transport and fate processes will be found in Section 9.3.

9.2.3 Prescribing Indicators

Where and how are indicators prescribed? To a large degree, material performance and system status indicators are specified or determined on the basis of how or what one needs to know and undertake to meet the specific targets identified in *sustainability status indicators*. There are two starting points for delineation of indicators:

1. Starting with the objective itself: In this instance, as an example, we could begin with a particular item in a natural capital component, e.g., maintaining the quality of the soil in a tract of land or a particular site. This is important because of the life support role of the specific tract of land. In this instance, one begins by (a) defining or establishing what indicators are needed as sustainability status indicators, e.g., specifics of soil quality, as shown, for example, in Figure 9.3; (b) determining the sources and nature of interactions with the tract of land and their impacts; and (c) establishing the required actions to ameliorate, mitigate, avoid, and protect the desired soil quality. Material and macrostatus indicators are established to determine and track the results of the corrective and protective measures. Figure 9.3 shows the use of status indicators in tracking anticipated or predicted outcomes from analyses or modeling of the processes initiated by the corrective actions. Failure to meet tracking results from the status indicators requires one to decide to (a) ignore indicators or (b) modify, add, or correct the actions previously prescribed to manage the impacts.

2. Starting with the external sources of interactions with the terrestrial environment itself: With this starting point, one follows a reverse sequence. Figure 9.4 shows

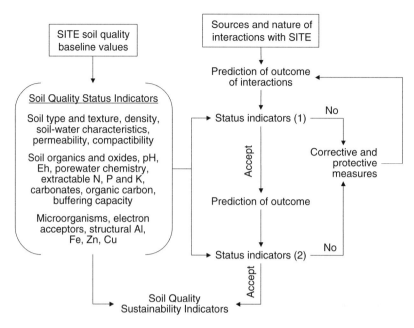

FIGURE 9.3

Example of specification and utilization of system status and material performance (properties) indicators using sustainability of soil quality as a goal. Note that the physical, chemical, and biological indicators listed in the *soil quality status indicators* are meant to portray the kinds of indicators that might be used. These need to be, and can be, expanded as required.

a typical protocol to determine or evaluate sustainability capabilities for the actions or impacts resulting from a specific project, activity, or industry. The status indicators 1 and 2 are indicators that may refer to different time periods, intervals, circumstances, or locations. Prediction of the outcome of preventative or amelio-rative actions (corrective actions) is generally obtained via modeling of the pro-cesses involved in the impact and corrective interactions — as also in the case of the actions shown in Figure 9.3.

The importance of system status and material performance indicators is evident in the *soil quality indicators* shown in the left-hand side of Figure 9.3. The list of physical, chemical and biological indicators shown in the diagram are not meant to be comprehensive. They are prescribed according to the specifics of the natural capital component under consid-eration. The status indicators (1) and (2) are prescribed in accordance with tracking or monitoring requirements. These are both spatial and temporal in nature and can include more than the numbers shown in the diagrams. Even though true sustainability may not be attained, it is necessary for the objectives and goals for sustainability to be properly articulated. These will serve to establish what, where, how many, and how often status indicators will be used or required.

9.3   Assessment of Interaction Impacts

Figure 9.5 shows many of the pollutants and contaminants found as discharge waste from activities associated with urbanization and industrial manufacturing and production.

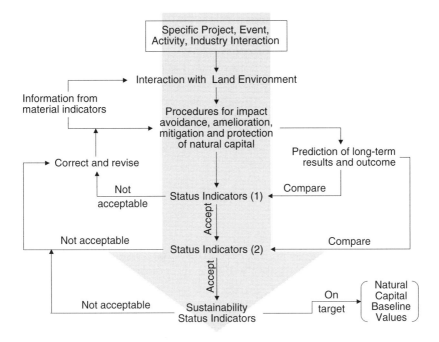

FIGURE 9.4

Prescription of material performance (properties) and system status indicators for determination of whether procedures for impact avoidance, amelioration, mitigation, and protection of natural capital have met, or will meet, sustainability requirements for natural capital — using the baseline values as the comparative base. The *status indicators* (1) and (2) are system status and material performance indicators, with (2) referring to either a time much later than (1) or a circumstance much different from (1).

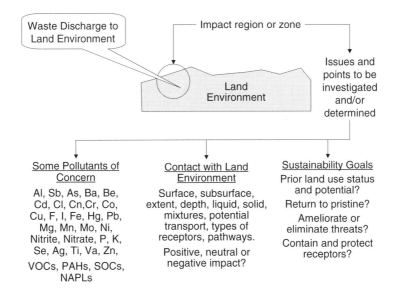

FIGURE 9.5

Simple schematic showing the first sets of determination for assessment of impact of waste discharge to the land environment.

Knowledge of the terrestrial environment impact interactions with specific projects and industrial manufacturing and production processes is required in structuring defensive and mitigation actions. In the case of contaminants in the terrestrial environment, this means that knowledge of the transport and fate of contaminants in the subsoil must be obtained. The aim of the outcome of defensive and mitigation actions is to provide the best opportunity to subscribe to the principles reflected in the requirements for sustainability of the natural capital component under consideration. Contaminant fate and transport modeling is a commonly used tool to predict the outcome of these actions. The basic elements of fate and transport processes and modeling have been discussed in Chapter 2. We will develop these in greater detail in this section.

9.3.1   Sustainability Concerns

The previous chapters have shown that the major sets of interactions between the various external agents (activities, projects, industries, etc.) that have the potential to severely impact the terrestrial environment are the liquid and solid discharges from these agents. The necessary pieces of information and knowledge required to generate preventive and mitigation solutions (Figure 9.3 and Figure 9.4) relate to the local geological and hydrogeological settings, the types and nature of waste discharges, the various environmental and biotic receptors, and, most importantly, the status and sustainability indicators. A useful rule of thumb in prescribing sustainability status indicators (the last set of indicators shown in Figure 9.3 and Figure 9.4) is to set objectives and targets that do not yield negative impact results; i.e., interaction and preventive mitigation impacts should not diminish the natural capital component under consideration. While activities associated with upstream and downstream industries may inevitably generate impacts that cannot be fully ameliorated, minimized, avoided, or totally remediated, it is nevertheless necessary to set status indicators that target sustainability of the natural capital of the terrestrial environment.

9.3.2   Surface Discharge: Hydrologic Drainage, Spills, and Dumping

Surface discharge or liquid/solid contaminant loading of land surface occurs under circumstances that include (1) hydrologic drainage occurring at a mine site at intersections of mine openings with the water table, (2) percolation from waste rock and mill tailings piles and surface runoffs, (3) inadvertent spills or deliberate dumping, (4) leaking pipelines, (5) deposition of airborne noxious pollutants from rainfall and snow loads, (6) application of agricultural chemicals (including pesticides) and irrigation, and (7) designed land farming treatment of contaminated materials. Figure 9.6 shows a schematic of some of the situations leading to contaminant and pollutant loading of the ground surface. The bottom left diagram in the figure concerns the problems discussed in Chapter 6. The land farming scheme shown in the bottom right of the diagram assumes that a prepared impermeable base is in place before placement of the material to land farm. Techniques for land farming will be discussed in a later section when we discuss mitigation and treatment alternatives.

The principal issues for contaminant loading of the land surface relate to the transport of pollutants from the contaminant source. By and large, spills and deliberate dumping of waste materials involve small surficial areas (i.e., small areas on the land surface). These will serve as point sources for transport of pollutants into the ground. Surface runoffs from these regions tend to be small. On the other hand, surface runoffs for other land applications of contaminants resulting from deposition of airborne noxious substances

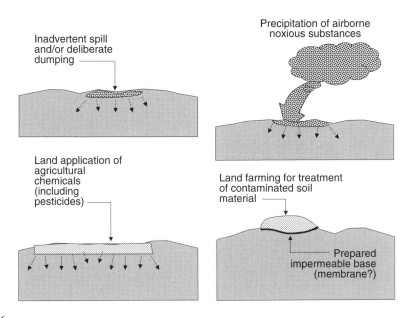

FIGURE 9.6
Scenarios showing contaminant loading on land surface of the geoenvironment. For small spills and dumps, point source pollution is assumed. For precipitation and land application of control agents, etc., nonpoint source pollution is generally assumed to occur.

and use of agricultural chemicals (including pesticides) can be serious issues, because of the extent of the land surface affected. The surface runoffs of pollutants arise if rainfall occurs or continues before the noxious substances and agricultural chemicals have a chance to effectively infiltrate or leach into the subsurface. Since the receiving sites for surface runoffs are topographically lower than the source location, they will be lowlands, wetlands, and receiving waters. Pollution of these regions (from surface runoffs) is commonly identified as *nonpoint source pollution*.

Contamination of the subsurface material (subsurface geologic material) and the underlying aquifers from pollutant sources constituting surface discharges occurs because of the transport of the pollutants in the subsurface material. Laboratory studies on the transport of pollutants in soils have been conducted for a large variety of soil types and pollutants. Field studies have also been performed in support of remediation projects and as due diligence work. Procedures and analytical computer models have been developed to provide one with the capability to determine or predict the movement, distribution, and concentration of pollutants in the subsurface soil. A brief summary of these can be found in Sections 2.5.3 and 2.5.4 in Chapter 2. As also emphasized previously, detailed treatment of these subjects can be found in textbooks dedicated to the study of pollutant fate and transport in soils. A more detailed discussion of contaminant transport processes and predictions will be found in the latter portion of this section.

The essential elements required for assessment or evaluation of the impacts from surface discharge phenomena such as spills, dumping, etc., are shown in Figure 9.7. In addition to determination of the size of affected area and quantity of material spilt, discharged, or dumped, one is required to determine:

1. Nature and composition of the discharged material
2. Hydrogeological setting in the region
3. Subsoil profile, material composition, and properties

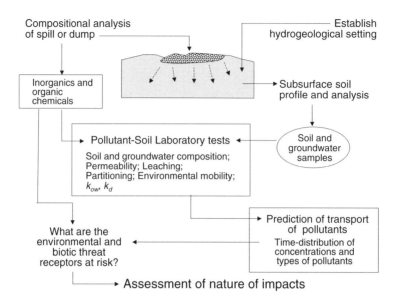

FIGURE 9.7

Schematic diagram showing procedures, factors, tests, and analyses required to begin the process for determination of consequence of surface discharge.

4. Transport processes involving the types of pollutants found in the compositional analyses — from laboratory tests

5. Transport and fate of pollutants (emanating from the surface discharge) in the subsoil — using predictive models and test information on partitioning of pollutants and transmissivity characteristics of the pollutants

6. The environmental and biotic receptors and how will these be affected

7. Source–path–receptor relationships

9.3.3   Subsurface Discharges

Subsurface discharges of pollutants arise when leaks or breakages in underground liquid or gaseous storage facilities and underground liquid or gaseous pipelines occur. Subsurface discharges of pollutants also occur when leachates escape from waste landfills (Figure 9.8). The left-hand portion of Figure 9.8 shows the leachate plume escaping at the bottom of the constructed waste landfill. When properly designed, with adequate barrier and liner systems underlying and surrounding the sides of the landfill, and leachate collection intercepts placed in the bottom barrier system, leakages are not expected. Furthermore, with proper impermeable caps placed on the top of the landfills, water is not permitted to enter into the system. This subject will be considered further in the next chapter. A comprehensive discussion of waste landfills can be found in textbooks dealing with the subject of landfills.

Leachates emanating from landfills result from (1) landfills not properly constructed, (2) breaks in the liner and barrier system, (3) deterioration of the barrier system, (4) incompatible interactions between barrier material and leachate chemistry, resulting in failure of the barrier to perform effectively, and (5) historic (old) landfills built without any real site preparation, and therefore not constructed with attention to requirements for proper and secure containment with barrier and liner systems. Typical constituents in municipal solid waste (MSW) leachates consist of various metals and salts, such as Cd,

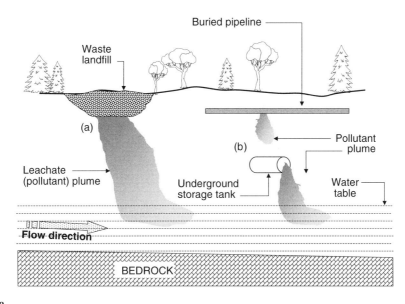

FIGURE 9.8
Schematic showing (a) leachate plume with pollutants and (b) pollutant plumes from a leaking underground storage tank and leaking buried pipeline.

Fe, Pb, Zn, Mn, Na, Ca, Mg, K, and other inorganics, including Cl^-, P, $CaCO_3$, and SO_4^{2-}. Biological and other parameters characterizing MSW leachates include biochemical oxygen demand (BOD), chemical oxygen demand (COD), total suspended solids (TSSs), total dissolved solids (TDSs), pH, total nitrogen (N), and total organic carbon (TOC). Typical organics found in MSW leachates include phenols, volatile acids, organic nitrogen, tannins, lignins, oil and grease, and chlorinated hydrocarbons. One of the many problems encountered when various kinds of organic and inorganic chemicals and elements are combined in a mixture is the reactions that can occur between some of the constituents. For example, reactions between halogenated organics and metals will generate heat in a leachate stream. If this occurs in a relatively dry environment, fire will result. The same occurs when saturated aliphatic hydrocarbons and phenols interact with oxidizing mineral acids.

The constituents of leachate streams in hazardous waste landfills are intimately associated with the composition of the hazardous waste contained in the landfill. The pollutants of concern in the leachate streams are both inorganic and organic in nature. Heavy metals constitute the primary inorganic pollutants, whereas the organic pollutants can cover a whole range of organic chemicals — from volatile organic chemicals (VOCs) to synthetic organic chemicals (SOCs). For both the MSW and hazardous waste landfills, the principal sets of problems, outside of the control and remediation-type problems, concern determination of the transport and fate of the pollutants. Knowledge or prediction of where, when, and extent of leachate penetration into the subsurface surroundings will provide one with (1) the requirements for a proper and effective subsurface monitoring program using monitoring wells and other sensing devices, and (2) the possible scenarios for various management options and remedial actions. The procedures and requirements shown in Figure 9.7 in respect to surface discharge problems also apply to subsurface discharges. A summary of the various transport processes involving sorption and partitioning of inorganic and organic pollutants has been given in Chapter 2, together with a brief mention of the widely used transport relationship. A detailed treatment of these subjects can be found in Yong (2001).

9.4 Pollutant Transport and Fate

Potential sources of pollution of subsurface water or groundwater (pore water and aquifers), other than inadvertent spills and deliberate dumping of hazardous materials, include landfills, underground storage tanks, waste piles and waste sites, underground injection wells, unplugged oil and gas wells, various kinds of surface impoundments and settling ponds, lands treated with pesticides, insecticides, and fertilizers, and pipelines transporting carbon resources. Figure 9.8 shows many of these potential sources. The types of pollutants, their concentrations and proportions, and the transport of these in the subsoil and their fate are critical to the structuring of protective measures necessary for the protection of public health and the health of the terrestrial environment. The need for predictive tools is obvious.

Questions posed by regulators, investigators, containment facility designers, site remediation technologists, and all others working with containment and remediation of site contamination are summarized as follows:

- Source of pollutants? Type and nature of pollutant plume?
- Size and distribution of pollutants in pollutant plume? Dominant toxic elements and pollutants in the pollutant plume?
- Where are the receptors? Paths to receptors? Source–pathways–receptors linkage?
- Rate of transport of pollutants? Can we predict their rate and extent of advance?
- Fate or persistence of the pollutants in the plume? Can we predict?
- Will pollutant plumes threaten or contaminate water resources? Will pollutants pose threats to environment and health of biotic receptors?
- Measures for risk management? Risk tolerance?

9.4.1 Analytical and Predictive Tools

Analytical and predictive tools dealing with the fate and transport of pollutants must account for the following:

- Concentrations of the various target pollutant species
- Hydraulic conductivity of the subsurface material (soil)
- Diffusive capabilities of the target pollutants
- Hydrogeologic setting
- Partitioning potential of the target pollutants
- Solubility of the target pollutants
- Speciation, complexations, and products formed
- Abiotic and biotic reactions and transformations

The factors and elements to be considered fall conveniently into two groups: (1) transport and (2) reactions. Two types of analytical computer models have been developed: (1) models dealing with fate and transport of pollutants, and (2) models that take into account geochemical reactions and their products. By and large, the models dealing with fate and transport of pollutants are nonreactive models. This means to say that other than using

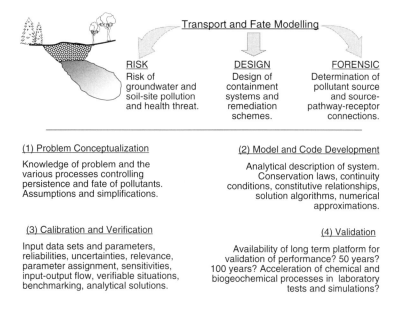

FIGURE 9.9
Principal objectives, issues, and requirements for pollutant transport and fate modeling.

the partitioning coefficients to account for sorption of pollutants from the pore water, no attempt is made to account for the chemical reactions in the soil–water system. In particular, speciation and complexations are not included in the structuring of the basic functions. Attempts have been made (and are being made) to develop reactive fate and transport models. In the second type of models, we have geochemical models that pay attention to geochemical speciation equilibria between the various phases (solids, liquid, and gaseous) in the subsurface setting. This includes the dissolved and adsorbed elements in the various phases.

Assessment and prediction of the transport and fate of pollutants commonly rely on analytical or numerical (computer) models designed to take into account the various processes, site contamination situations, and properties of the pollutants and subsurface materials. These models are useful for regulatory agencies in risk management and performance assessment of target sites and situations, for operators of landfills, and for those involved in site remediation (Figure 9.9). The quality of models, i.e., how accurately their predictions accord with real performance, depends on how well they represent the real problem situation and processes involved. This requires not only proper and accurate problem conceptualization, but also the capability to render these into the appropriate mathematical relationships.

Problem conceptualization includes:

1. Problem recognition: This encompasses more than an adequate description of the site and interacting elements. To structure the output requirements for the model and the manner in which the results need to be expressed, knowledge of the end purpose (use) of the results is required. The three different areas of application of models shown at the top of the illustration in Figure 9.9 (*risk, design, forensic*) will not have the same output requirements inasmuch as the decision-making process and the decisions required are different between them. Taking the problem of the pollutant plume emanating from the landfill, shown in the top left-hand corner of the figure, the differences between the outputs obtained from analysis (or

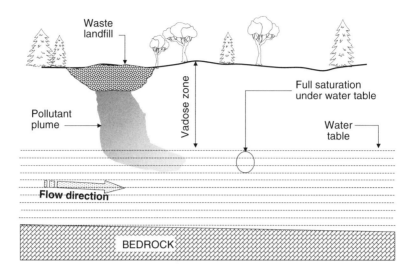

FIGURE 9.10
Schematic showing leachate plume as a pollutant plume spreading downwards toward the water table and spreading in the direction of flow of the groundwater. The vadose zone is unsaturated, and the bottom of the vadose zone contains the capillary fringe.

prediction) of plume advance in respect to risk management, design of barrier systems, or determination of source–pathway–receptor are obvious.

2. Elements involved in the total system within the problem domain: For the pollutant plume emanating from the landfill, Figure 9.10 shows that the plume passes through the vadose zone before meeting the saturated zone under the water table. The elements involved in the problem include not only the interacting elements typified by the soil fractions, pore water and pollutants, and microorganisms, but also the problem setting, i.e., the vadose and saturated zones.

3. Mechanisms and processes of interactions between participating elements.

4. Initial and boundary conditions.

5. Output requirements.

9.4.2 Basic Elements of Interactions between Dissolved Solutes and Soil Fractions

Interactions occurring between the pollutants in the pore water (dissolved solutes) and reactive soil particle surfaces are responsible for the transfer of these solutes from the pore water to the soil particle surfaces (partitioning). Molecular interactions governing sorption of pollutants are essentially electrostatic in nature. They are Coulombic interactions between nuclei and electrons. Of particular importance are the interatomic bonds, such as the ionic, covalent, hydrogen, and van der Waals. Ionic forces are Coulombic forces. These are forces between positively and negatively charged atoms, and the bonds formed are called ionic or electrovalent bonds. The simplest example of ionic bonding is between a sodium atom and a chlorine atom — resulting in the formation of NaCl. The strength of the attractive forces, and hence the strength of the ionic bonds, decreases as the square of the distance separating the atoms.

Another example of ionic bonding is the bond established between the oxygen from a water molecule and the oxygen on a clay particle's surface. This is due to the hydrogen atom, which can attract two electronegative atoms, and the ionic bond formed is called the *hydrogen bond*. In comparison with other bonds between neutral molecules, the hydrogen bond is a strong bond. Hydrogen bonding between two oxygen atoms, which are electronegative, is important in bonding layers of clay minerals together, in holding water at the clay surface, and in bonding organic molecules to clay particle surfaces.

Van der Waals forces of attraction can be categorized into three components: (1) Keesom forces developed as a result of dipole orientation, (2) Debye forces developed due to induction, and (3) London dispersion forces. Adsorption of organic anions onto clay particle surfaces can be in the form of (1) anion associated directly with cation or (2) anion associated with cation via a water bridge — referred to as a *cation bridge*. The process consists of replacement of a water molecule from the hydration shell of the exchangeable cation by an oxygen or an anionic group, e.g., carboxylate or phenate of the organic polymer. Hydrogen bonding to the oxygens of siloxane (mica-type) surfaces of clay particles is generally weak. Adsorption of the organic anion is readily reversible by exchange with chloride or nitrate ions.

Cation exchange in soils refers to the exchange of positively charged ions associated with clay particle surfaces. The process is stoichiometric and electroneutrality needed at the clay particle surfaces must be satisfied. Cations will be attracted to the reactive soil particle surfaces in accordance with the relationship

$$\frac{M_s}{N_s} = \frac{M_o}{N_o} = 1$$

where M and N represent the cation species and the subscripts s and o represent the surface and bulk solution. *Exchangeable cations* are cations that can be readily replaced by other cations of equal valence, or by two of one half the valence of the original one. Thus, for example, if a clay containing sodium as an exchangeable cation is washed with a solution of calcium chloride, each calcium ion will replace two sodium ions, and the sodium can be washed out in the solution.

The quantity of exchangeable cations held by the soil is called the *cation exchange capacity* (CEC) of the soil and is expressed as milliequivalents per 100 g of soil (meq/100 g soil). One milliequivalent is equal to 6.023×10^{20} cation exchange sites in the soil. The CEC is a measure of the amount of negative sites associated with the soil fractions. The predominant exchangeable cations in soils are calcium and magnesium, with smaller amounts of potassium and sodium. The valence of cations plays a significant role in the exchange process. Higher-valence cations will show greater replacing power. The higher the charge, the higher is its attraction to exchange sites. The converse also holds true; i.e., higher-valence cations at the surfaces of clay particles will be more difficult to replace. The replacing power or the strength of attraction of cations to the soil particle surfaces is given by the lytropic series. An example of some typical cations and the replacing power is given as follows:

$$Th^{4+} > Fe^{3+} > Al^{3+} > Cu^{2+} > Ba^{2+} > Ca^{2+} = Mg^{2+} > Cs^{+} > K^{+} = NH_4^{+} > Li^{+} > Na^{+}$$

Exchange–equilibrium equations can be used to determine the proportion of each exchangeable cation to the total cation exchange capacity (CEC) as the outside ion concentration varies. The simplest of these is the Gapon relationship:

$$\frac{M_e^{+m}}{N_e^{+n}} = K \frac{\left[M_o^{+m}\right]^{\frac{1}{m}}}{\left[N_o^{+n}\right]^{\frac{1}{n}}} \tag{9.1}$$

where m and n refer to the valence of the cations and the subscripts e and o refer to the exchangeable and bulk solution ions. The constant K is dependent on the effects of specific cation adsorption and the nature of the clay surface. K decreases in value as the surface density of charges increases.

The adsorption of ions due to the mechanism of electrostatic bonding is called *physical adsorption* or *nonspecific adsorption*. The ions involved in this type of process are identified as *indifferent ions*. The other mechanism of ion adsorption is a chemical reaction that involves covalent bonds and activation energy in the process of adsorption. This type of adsorption process is generally identified as *chemisorption* or *specific adsorption* and occurs at specific sites. As stated in Section 2.5.1 in Chapter 2, *specific cation adsorption* refers to the situation where the ions penetrate the coordination shell of the structural atom and are bonded by covalent bonds via O and OH groups to the structural cations. It is useful to note that when the energy barrier is overcome by the activation energy (in the chemisorption process), desorption of the ions will not be easily accomplished since desorption energy requirements may be prohibitively large. This has considerable significance in the evaluation of contaminant–soil interaction, especially with respect to the environmental mobility of sorbed pollutants.

Chemisorption and adsorption on soil particle surfaces involving siloxane cavities are generally confined to the surface layer and the monolayer next to the electrified interface. To obtain a better picture of the adsorption processes at the surface monolayer and beyond, we need to look more closely into the various energies of interaction developed. In short, the net energy of interaction due to adsorption of a solute ion or molecule onto the surfaces of the soil fractions is the result of both short-range chemical forces, such as covalent bonding, and long-range forces, such as electrostatic forces. Furthermore, sorption of inorganic contaminant cations is related to their valences, crystallinities, and hydrated radii.

9.4.3 Elements of Abiotic Reactions between Organic Chemicals and Soil Fractions

Abiotic adsorption reactions or processes involving organic chemicals and soil fractions are governed by (1) the surface properties of the soil fractions, (2) the chemistry of the pore water, and (3) the chemistry and physical chemistry of the pollutants. Mechanisms pertaining to ion exchange involving organic ions are essentially similar to those between inorganic pollutants and soil fractions. The adsorption of the organic cations is related to the molecular weight of the organic cations. Large organic cations are adsorbed more strongly than inorganic cations by clays because they are longer and have higher molecular weights. Organic chemical compounds develop mechanisms of interactions at the molecular level that include:

- London–van der Waals forces: These have been described previously and include three types — Keesom, Debye, and London dispersion forces.
- Hydrophobic reactions and bonding: Organic chemical molecules bond onto hydrophobic soil particle surfaces because this requires the least restructuring of the original water structure in the pore spaces of the soil. For soil organic matter (SOM), the fulvic acids are by and large hydrophilic. These kinds of organic matter

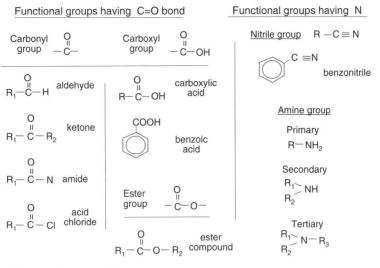

FIGURE 9.11
Some common functional groups for organic chemical pollutants.

have the least influence on the structuring of water. On the other hand, the SOM humins are highly hydrophobic. These have considerable influence on the restructuring of the water structure.

- Hydrogen bonding and charge transfer: Hydrogen bonding is a special case of charge transfer complex formation. These are complexes formed between electron–donor and electron–acceptor. The bonding between the aromatic groups in soil organic matter and organic chemicals is an example of charge transfer.

- Ligand and ion exchange: The bonding process requires that the organic chemical possesses a higher chelating capacity than the replaced ligand.

- Chemisorption: For soils and pollutants, the process of chemical adsorption involves chemical bonding between the pollutant molecule or ion in the pore water and the reactive soil particle surfaces. The process is sometimes called *specific adsorption* and the bonds are covalent bonds.

The information from Section 2.5.2 in Chapter 2 shows that the functional groups for organic chemical compounds (organic chemical pollutants) are either acidic or basic. The characteristics and properties of these groups in organic molecules, such as shape, size, configuration, polarity, polarizability, and water solubility, are important in the adsorption of the organic chemicals by the soil fractions. The various functional groups associated with organic chemical pollutants have been described in Section 2.5.2 in Chapter 2. A summary of some of these groups is shown in Figure 9.11. These include the hydroxyl group (alcohols and phenols), the carboxyl group (COOH), carbonyl (CO), and the amines (primary, secondary, and tertiary). The carbonyl group technically includes the carboxyl group and is considered to be the most important functional group in organic chemistry. Most of the organic chemical pollutants found in the ground contain the carbonyl group. These chemicals are associated with production of pharmaceuticals, synthetic chemicals, and synthetic materials.

Not shown in Figure 9.10 are the hydroxy (OH) compounds. These are the compounds that contain the hydroxyl functional group. The two main groups are (1) aliphatic and (2) aromatic. The aliphatic compounds are the alcohols and the aromatics are the phenols. Alcohols are hydroxyl alkyl compounds (R– OH), with a carbon atom bonded to the hydroxyl group. The more familiar ones are CH_3OH (methanol) and C_2H_5OH (ethanol). Adsorption of the hydroxyl groups of alcohol can be obtained through hydrogen bonding and cation–dipole interactions. Most primary aliphatic alcohols form single-layer complexes on the negatively charged surfaces of the soil fractions, with their alkyl chain lying parallel to the surfaces of the soil fractions. Phenols, on the other hand, are compounds that possess a hydroxyl group attached directly to an aromatic ring. As with the carbonyl functional group, the various hydroxy compounds are widely used and can be found in the manufacture of industrial products and pharmaceutical agents. Both alcohols and phenols can function as weak acids and weak bases.

Functional groups exert considerable influence on the characteristics of organic compounds and contribute significantly to the processes that control accumulation, persistence, and fate of the organic chemical compounds in soil. Organic chemicals with C=O bond functional groups and nitrogen-bonding functional groups (see Figure 9.11) are fixed or variable-charged organic chemical compounds. They can acquire a positive or negative charge through dissociation of H^+ from or onto the functional groups, depending on the dissociation constant of each functional group and the pH of the soil–water system. An increase in the pH of the soil–water system will cause these functional groups (i.e., groups having a C=O bond) to dissociate. The outcome of the release of H^+ is a development of negative charges for the organic chemical compounds. Charge reversal (i.e., from positive to negative charges) could lead to the release of organic chemical pollutants held originally by cation bonding to the negatively charged reactive surfaces of the soil particles. The phenomenon is a particular case of environmental mobility of previously sorbed pollutants.

9.4.4   Reactions in Pore Water

Since pollutants consist of both inorganic and organic chemicals, it is more convenient to use the Brønsted–Lowry concepts of acids and bases to describe the various reactions and interactions occurring in a soil–water–pollutant system. In the Brønsted–Lowry concept, an *acid* is a substance that has a tendency to lose a proton (H^+), and conversely, a *base* is a substance that has a tendency to accept a proton. With this acid–base scheme, an *acid* is a *proton donor*. It is a *protogenic* substance. Similarly, a *base* is a *proton acceptor*; i.e., it is a *protophillic* substance. Water is both a *protophillic* and a *protogenic* solvent; i.e., it is *amphiprotic* in nature. It can act as either an acid or a base. It can undergo self-ionization, resulting in the production of the conjugate base OH^- and conjugate acid H_3O^+. The self-ionization of water is called *autoprotolysis. Neutralization* is the reverse of autoprotolysis. Substances that have the capability to both donate and accept protons such as water and alcohols are called *amphiprotic* substances.

Chemical reactions in the pore water include (1) acid–base reactions and hydrolysis, (2) oxidation–reduction (redox) reactions, and (3) speciation and complexations. Acid–base reactions and equilibrium in the pore water have important consequences on the partitioning and transport of pollutants in the soil. Acid–base reactions are *protolytic* reactions resulting from a process called *protolysis*, i.e., proton transfer between a proton donor (acid) and a proton acceptor (base).

To assess the bonding and partitioning relationships between heavy metals and soil solids, it is useful to use the Lewis (1923) concept of acids and bases. This concept defines an acid as a substance that is capable of accepting a pair of electrons for bonding, and a base as a substance that is capable of donating a pair of electrons. This means that *Lewis acids* are electron acceptors and *Lewis bases* are electron donors. All metal ions M^{nx} are

Lewis acids. The Lewis acid–base concept permits us to treat metal–ligand bonding as acid–base reactions. Hydrated metal cations can act as acids or proton donors, with separate *pk* values for each. The dissociation constant *k* is a measure of the dissociation of a compound. This constant *k* is generally expressed in terms of the negative logarithm (to the base 10) of the dissociation constant, i.e., $pk = -\log(k)$. The smaller the *pk* value, the higher is the degree of ionic dissociation, and the more soluble is the substance. A comparison of the various *pk* values between compounds will tell us which compound would be more or less soluble in comparison with a target compound.

Oxidation–reduction reactions involve the transfer of electrons between the reactants, and the activity of the electron *e* in the chemical system plays a significant role. There is a link between redox reactions and acid–base reactions since the transfer of electrons in a redox reaction is accompanied by proton transfer. Redox reactions involving inorganic solutes result in a decrease or increase in the oxidation state of an atom. Organic chemical pollutants, on the other hand, show the effects of redox reactions through the gain or loss of electrons in the chemical. Biotic redox reactions are of greater significance than abiotic redox reactions. These reactions are significant factors in the processes that result in the transformation, persistence, and fate of organic chemical compounds in soils.

The stability of inorganic solutes in the pore water is a function of such factors as pH, the presence of ligands, temperature, concentration of the inorganic solutes, and the Eh or *pE* of the pore water. *Eh* is the redox potential and *pE* is a mathematical term that represents the negative logarithm of the electron activity e^-. The redox potential Eh is a measure of electron activity in the pore water and is described by the following relationship:

$$Eh = E^0 + \left(\frac{RT}{nF}\right)\ln\frac{a_{i,ox}}{a_{i,red}} \tag{9.2}$$

where E^0 is the standard reference potential, *n* is the number of electrons, *R* is the gas constant, *T* is the absolute temperature, *F* is the Faraday constant, a_i is the activity of the *i*th species, and the subscripts *ox* and *red* refer to the oxidized and reduced *i*th species. At a temperature of 25°C, the relationship between *Eh* and *pE* will be obtained as $Eh = 0.0591$ *pE*. Figure 9.12 shows the pE-pH diagram for Fe and water with a maximum soluble Fe concentration of $10^{-5}\,M$. As can be seen, the valence state of the reactants is a function of the pH-pE status.

Speciation and complexations are central to the processes that control the fate of heavy metals in soils. Speciation refers to the formation of complexes between heavy metals and ligands in the aqueous phase (pore water). In a soil–water system, speciation provides competition between the ligands and the reactive soil solids for sorption of heavy metals. Various dissolved solutes in the pore water participate in the aqueous and surface complexations that are characteristic of the interactions between the solutes and the reactive soil particles. These interactions have a direct impact on the predictions of contaminant transport, especially modeling procedures that rely on the use of simple partition coefficients. For example, Cl$^-$ ions, sulfates, and organics can form complexes with heavy metals. The end result of this is seen as a lesser amount sorbed onto the soil particles, i.e., a lower adsorption isotherm performance, and a larger amount of the target heavy metal transported in the pore water. In other words, the environmental mobility of heavy metals is enhanced with speciation and complexation. Studies on Cd adsorption by kaolinite soil particles indicate that the Cd that were not adsorbed by the soil were in the form of $CdCl_2^0$, $CdCl_3^-$, and $CdCl_4^{2-}$ (Yong and Sheremata, 1991). The amount of Cd not adsorbed by the kaolinite soil particles in the presence of Cl$^-$ was due to (1) a decrease in activity

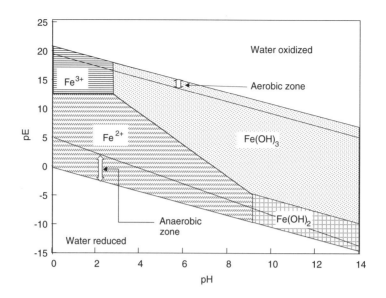

FIGURE 9.12

pE-pH chart for Fe and water with maximum soluble Fe concentrations of 10^{-5} M. The zone sandwiched between the aerobic and anaerobic zones is the transition zone. (From Yong, R.N., *Geoenvironmental Engineering: Contaminated Soils, Pollutant Fate, and Mitigation*, CRC Press, Boca Raton, FL, 2001.)

due to the presence of NaCl, (2) competition from Na^+ for adsorption sites, and (3) complexation of Cd^{2+} as negative and neutral chloride complexes. With increasing pH values, there is a tendency for Cd to be removed from the pore water as hydroxides, and if the pH is above the precipitation pH of Cd, precipitation of the Cd onto clay particle surfaces is likely. At pH values higher than the isoelectric point (iep), Cd is more likely to remain in solution in the presence of Cl^- than in the presence of ClO_4^-. This is the result of competition of Cl^- with OH^- for formation of complexes with the Cd^{2+} that were not sorbed by the kaolinite soil.

9.5   Surface Complexation and Partitioning

As opposed to speciation and complexations in the aqueous phase (pore water), surface complexations refer to the complexes formed by the inorganic pollutants and the reactive sites on the soil particle surfaces. Surface complexations include several mechanisms of solute–particle surface interaction, described broadly as sorption mechanisms. These have been discussed previously as nonspecific adsorption, specific adsorption, and chemisorption. The processes involved have been described as Coulombic molecular interactions, with bonds formed that include ionic, covalent, and van der Waals.

The result of surface complexations is partitioning. We describe *partitioning* of pollutants as the transfer of pollutants in the pore water to the soil solids as a result of sorption mechanisms between the two. This is also called mass transfer (of pollutants). Section 2.5.3 in Chapter 2 has considered the partitioning of pollutants in a very general manner. Partitioning, as a phenomenon, includes the transfer of both inorganic and organic chemical pollutants. This is an important phenomenon since this is the outcome of one of the

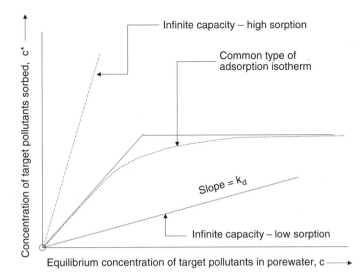

FIGURE 9.13
Illustration of partitioning of target pollutants between pollutants in pore water and pollutants sorbed by soil particles. Partitioning is determined in terms of concentrations of pollutants sorbed by particles and equilibrium concentration of target pollutants remaining in the pore water. Note that unless limits are placed on the maximum sorption capacity of the soils, specification of constant k_d values means infinite sorption capacity of the soil–water system.

fundamental processes that determine the persistence and fate of pollutants. In the case of organic chemicals, the processes involved include London–van der Waals forces, hydrophobic reactions, hydrogen bonding and charge transfer, ligand and ion exchange, and chemisorption. In this section, we will be examining partitioning of both inorganic pollutants and organic chemicals in greater detail. While there may be some similarity in mass transfer mechanisms in the partitioning of inorganic solutes and organic chemicals, it is generally more convenient to consider these separately.

9.5.1 Partitioning of Inorganic Pollutants

A popular measure of partitioning of inorganic and organic chemical pollutants is the partition coefficient k_p. Section 2.5.3 in Chapter 2 has given a very brief description of the general types of partition coefficients. In brief, partition coefficients describe the relationship between the amount of pollutants transferred onto soil particles (sorbed by the soil particles) and the equilibrium concentration of the same pollutants remaining in the pore water (Figure 9.13). The common types of relationships obtained generally through batch equilibrium tests for determination of adsorption isotherms are shown in Figure 2.14 in Chapter 2. The popular relationships, such as Langmuir and Freundlich, are shown in the diagram. To distinguish between partition coefficients obtained using different laboratory techniques, the term *distribution coefficient k_d* is often used to denote partitioning of pollutants obtained with batch equilibrium adsorption isotherm procedures. By and large, the distribution coefficient is the partition coefficient most commonly used to describe partitioning of heavy metals and other inorganic solutes.

What is the significance of partitioning? It is necessary to bear in mind that partitioning is the result of mass transfer of pollutants from the pore water. How one measures the results of mass transfer (i.e., partitioning) can be a contentious issue and can severely affect one's prediction of the transport and fate of pollutants under consideration. A quantitative determination of partitioning, such as the distribution coefficient k_d, is needed in many mathematical relationships structured to evaluate the fate of pollutants transporting in a soil system. For example, Equation 2.3 in Chapter 2 uses the distribution coefficient k_d as a key parameter in the relationship used to predict the transport of contaminants (pollutants) in a saturated soil.

There are at least two broad issues regarding the determination and use of the distribution coefficient k_d: (1) types of tests used to provide information for determination of k_d and (2) range of applicability of k_d in transport and fate predictions. We will discuss the former in this section and leave the latter discussion for a later section, where the problem of prediction of transport and fate of pollutants is addressed. Laboratory tests used to provide information on the mass transfer of pollutants from the pore water onto soil solids are the most expedient means to provide one with information on the partitioning of pollutants. By and large, these tests provide only the end result of the mass transfer, and not direct information on the basic mechanisms responsible for partitioning.

The distribution coefficient k_d is determined from information gained using batch equilibrium tests on soil solutions. Ratios of 10 parts or 20 parts solution to 1 part soil are generally used, and the candidate or target pollutant is part of the aqueous phase of the soil solution. In many laboratory test procedures, the candidate soil is used for the soil in solution, and the candidate or target pollutant is generally a laboratory-prepared pollutant, e.g., $PbNO_3$ for assessment of sorption of Pb as a pollutant heavy metal. Since the soil particles are in a highly dispersed state in the soil solution, one would expect that all the surfaces of all the particles are available for interaction with the target pollutant in the aqueous phase of the soil solution. By using multiple batches of soil solution where the concentration of the target pollutant is varied, and by determining the concentration of pollutants sorbed onto the soil solids and remaining in the aqueous phase, one will obtain characteristic adsorption isotherm curves such as those shown in Figure 9.13 and Figure 2.14 in Chapter 2. The slope of the adsorption isotherm defines k_d.

Consider Figure 9.14 in Chapter 2; it compares the loose structure or dispersed state of soil particles in a soil suspension (bottom illustration) with an aggregation of a multitude

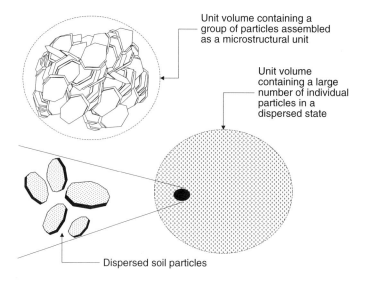

Unit volume containing a group of particles assembled as a microstructural unit

Unit volume containing a large number of individual particles in a dispersed state

Dispersed soil particles

FIGURE 9.14

Influence of soil structure on availability of exposed reactive soil particle surfaces for sorption of pollutants. The dispersed particles are typical of soil suspensions used for batch equilibrium tests for determination of adsorption isotherms. The microstructural unit is typical of the soil structure in the compact soil state in the soil subsurface.

of particles constituting a microstructural unit that is representative of a soil structure in a natural soil subsurface environment (top illustration). The schematic shows that while the two identical unit volumes do not necessarily contain the same number of particles, three points of substance can be identified: (1) a relatively small proportion of the reactive particle surfaces in the microstructural unit shown at the top of the diagram are available for interaction with pollutants in the pore water; (2) it is virtually impossible to determine or define a relationship between the total available surface areas for the dispersed particles in the soil suspension and the microstructural unit — for the same unit volume; and (3) it is also almost impossible to determine or predict the nature and distribution of microstructural units in a natural soil subsurface environment.

Distribution coefficients k_d obtained from adsorption isotherms using the batch equilibrium with soil solutions and prepared target pollutants are very useful in that they define the upper limit of partitioning of the target pollutant. Problems or pitfalls arising from the application of k_d values reported in the literature for use in models to predict actual fate and transport of pollutants in the natural subsoil can be traced to:

- Inappropriate use of the coefficient, i.e., using the reported k_d value to represent partitioning effects in a natural compact subsoil. This can arise from a lack of appreciation or knowledge of the particulars of the batch equilibrium tests. If the tests were conducted to provide the upper limit of partitioning, as has been described, the model would overpredict sorption and therefore underpredict transport.

- Major differences in the composition of the leachate and pollutants in the leachate plume being modeled. Unless the batch equilibrium tests were conducted with actual leachates from the site under consideration, it is inappropriate to use results from single-species pollutant tests to represent multispecies behavior. In any event, it needs to be remembered that the k_d values would be upper-limit values. Competition for sorption sites, preferential sorption, and speciation–complexation are some of the major factors that would directly affect the nature of the adsorption isotherm obtained.

For assessment of partitioning using soils in their natural compact state, it is necessary to conduct column leaching or cell diffusion tests. In these kinds of tests, the natural soil is used in the test cell or column, and either laboratory-prepared candidate pollutants or natural leachates are used. The partition coefficient deduced from the test results is not the distribution coefficient identified with the adsorption isotherms obtained from batch equilibrium tests. Instead, the partition coefficients obtained from column leaching or cell diffusion tests need to be properly differentiated from the traditional k_d. Yong (2001) has suggested that these partition coefficients be called *sorption coefficients* — to reflect the sorption performance of the soils in their natural state in the column or cell. The disadvantages in conducting column leaching and cell diffusion tests are (1) the greater amount of effort required to conduct the tests, (2) the much greater length of time taken to obtain an entire suite of results, and (3) the inability to obtain exact replicate soil structures in the companion columns or cells. The results indicate that the characteristic curves obtained from column leaching tests, for example, are much lower than corresponding adsorption isotherms. Figure 9.15 gives an example.

9.5.2 Organic Chemical Pollutants

The partitioning of organic chemical pollutants is a function of several kinds of interacting mechanisms between the organic chemicals and the soil solids in the natural soil–water

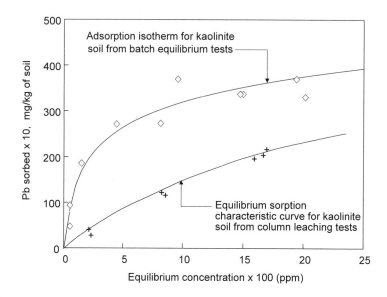

FIGURE 9.15
Comparison of Pb sorption curves obtained from batch equilibrium and column leaching tests for kaolinite soil.
(From Yong, R.N., *Geoenvironmental Engineering: Contaminated Soils, Pollutant Fate, and Mitigation*, CRC Press,
Boca Raton, FL, 2001.)

system that constitutes the subsoil. A key factor in the development of the kinds of
interaction mechanisms is the type or class of organic chemicals. The degree of water
miscibility of the organic chemical appears to be a key element. A good example of this
is the difference between nonaqueous phase liquids (NAPLs) and water-miscible alcohols.
The family of NAPLs includes those that are denser and lighter than water. The dense
NAPLs (DNAPLs) include the organohalides and oxygen-containing organic compounds,
and the light NAPLs (LNAPLs) include gasoline, heating oil, kerosene, and aviation fuel.
Most NAPLs are partially miscible in water. Consideration of the transport of NAPLs in
the saturated zone requires attention to two classes of substances: (1) miscible or dissolved
substances and (2) immiscible substances.

The basic processes involved in the transport and fate of NAPLs are demonstrated in
Figure 9.16. The chemical properties that affect NAPL transport and fate include (1)
volatility, (2) relative polarity, (3) affinity for soil organic matter or organic contaminants,
and (4) density and viscosity. The higher the vapor pressure of the substance, the more
likely it is to evaporate. Movement in the vapor phase is generally by advection. At
equilibrium between NAPLs and the vapor phase, the equilibrium partial pressure of a
component is directly related to the mole fraction and the pure constituent vapor pressure,
as described by Raoult's law. Designating P_i as the partial pressure of the constituent, x_i
as the mole fraction of the constituent, and P_i^0 as the vapor pressure of the pure constituent,
Raoult's law states that when equilibrium conditions are obtained, and when the mole
fraction of a constituent is greater than 0.9, $P_i = x_i P_i^0$.

As shown in Figure 9.16, an organic chemical compound in the soil may be partitioned
between the soil water, soil air, and soil constituents. The rate of volatilization of an organic
molecule from an adsorption site on the solid phase in the soil (or in solution in the soil
water) to the vapor phase in soil air, and then to the atmosphere, is dependent on many
physical and chemical properties of both the chemical and the soil, and on the process
involved in moving from one phase to another. The three main distribution or transport
processes involved are:

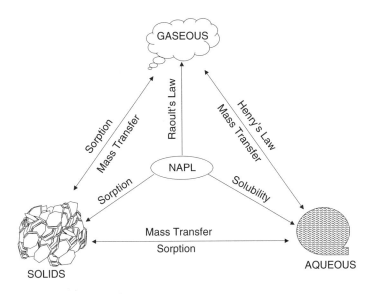

FIGURE 9.16
Processes involved in partitioning and fate of NAPLs.

1. Compound in soil ↔ compound in solution
2. Compound in solution ↔ compound in vapor phase in soil air
3. Compound in vapor phase in soil ↔ air compound in atmosphere

Partitioning of a chemical among the three phases can be estimated from either vapor-phase or solution-phase desorption isotherms. The process by which a compound evaporates in the vapor phase to the atmosphere from another environmental compartment is defined as volatilization. This process is responsible for the loss of chemicals from the soil to the air, and is one of the factors involved in the persistence of an organic chemical. Determination of volatilization of a chemical from the soil to the air is most often achieved using theoretical descriptions of the physical process of volatilization based on Raoult's law and Henry's law. The rate at which a chemical volatilizes from soil is affected by soil and chemical properties and environmental conditions. Some of the properties of a chemical involved in volatilization are its vapor pressure, solubility in water, basic structural type, and the number, nature, and position of its basic functional groups. Other factors affecting the volatilization rate include adsorption, vapor density, and water content of the soil in the subsurface.

Adsorption has a direct impact on the chemical activity by reducing it to values below that of the pure compound. In turn, this affects the vapor density and the volatilization rate since vapor density is directly related to the volatilization rate. Vapor density is the concentration of a chemical in the air, the maximum concentration being a saturated vapor. The role of water content is seen in terms of competition for adsorption sites on the soil. Displacement of nonpolar and weakly polar compounds by water molecules can occur because of preferential sorption (of water). Hydrates, i.e., hydration layer on the soil particle surfaces, will increase the vapor density of weakly polar compounds. If dehydration occurs, the compound sorbs onto the dry soil particles. This means that the chances for volatilization of the organic chemical compound are better when hydrates are present.

When a vapor is in equilibrium with its solution in some other solvent, the equilibrium partial pressure of a constituent is directly related to the mole fraction of the constituent in the aqueous phase. Once again, designating P_i as the partial pressure of the constituent, X_i as the mole fraction of the constituent in the aqueous phase, and H_i as Henry's constant for the constituent, Henry's law states that $P_i = H_iX_i$. By and large, as long as the activity coefficients remain relatively constant, the concentrations of any single molecular species in two phases in equilibrium with each other will show a constant ratio to each other. This assumes ideal behavior in water and the absence of significant solute–solute inter-actions, and also the absence of strong specific solute–solvent interactions.

Partitioning of organic chemicals is most often described by the partition coefficient k_{ow}. This is the octanol–water partition coefficient and has been widely adopted in studies of the environmental fate of organic chemicals. The octanol–water partition coefficient is some-times known as the *equilibrium partition coefficient*, i.e., coefficient pertaining to the ratio of the concentration of a specific organic pollutant in other solvents to that in water. Results of countless studies have shown that this coefficient is well correlated to water solubilities of most organic chemicals. Since *n*-octanol is part lipophilic and part hydrophilic (i.e., it is amphiphilic), it has the capability to accommodate organic chemicals with the various kinds of functional groups. The dissolution of *n*-octanol in water is roughly eight octanol mole-cules to 100,000 water molecules in an aqueous phase. This represents a ratio of about 1 to 12,000 (Schwarzenbach et al., 1993). Since water-saturated *n*-octanol has a molar volume of 0.121 l/mol, as compared with 0.16 l/mol for pure *n*-octanol, the close similarity permits one to ignore the effect of the water volume on the molar volume of the organic phase in experiments conducted to determine the octanol–water equilibrium partition coefficient. The octanol–water partition coefficient, k_{ow}, has been found to be sufficiently correlated not only to water solubility, but also to soil sorption coefficients. In the experimental measure-ments reported, the octanol is considered to be the surrogate for soil organic matter.

Organic chemicals with k_{ow} values less than 10 are considered to be relatively hydrophilic — with high water solubilities and small soil adsorption coefficients. Organic chemicals with k_{ow} values greater than 10^4 are considered to be very hydrophobic and are not very water soluble. Chiou et al. (1982) have provided a relationship between k_{ow} and water solubility S as follows:

$$\log k_{ow} = 4.5 - 0.75 \log S \text{ (ppm)}$$

Aqueous concentrations of hydrophobic organics such as polyaromatic hydrocarbons (PAHs) in natural soil–water systems are highly dependent on the adsorption/desorption equilibrium, with sorbents present in the systems. Studies of compounds, which included normal PAHs, nitrogen, and sulfur heterocyclic PAHs, and some substituted aromatic compounds suggest that the sorption of hydrophobic molecules (benzidine excepted) is governed by the organic content of the substrate. The dominant mechanism of organic adsorption is the hydrophobic bond established between a chemical and natural organic matter in the soil. The extent of sorption can be reasonably estimated if the organic carbon content of the soil is known (Karickhoff, 1984) by using the expression $k_p = k_{oc}f_{oc}$, where f_{oc} is the organic carbon content of the soil organic matter, k_{oc} is the organic content coefficient, and k_p is the linear Freundlich isotherm obtained for the target organic chemical. This approach works reasonably well in the case of high organic contents (e.g., $f_{oc} > 0.001$). Relationships reported in the literature relating k_{ow} to k_{oc} show that these can be grouped into certain types of organic chemicals.

For PAHs, the relationship given by Karickhoff et al. (1979) is

$$\log k_{oc} = \log k_{ow} - 0.21$$

For pesticides, Rao and Davidson (1980) report that

$$\log k_{oc} = 1.029 \log k_{ow} - 0.18$$

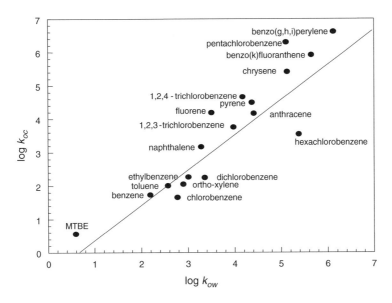

FIGURE 9.17
Relationship of log k_{oc} with log k_{ow} for several organic compounds. (Adapted from Yong, R.N. and Mulligan, C.N., *Natural Attenuation of Contaminants in Soils*, Lewis Publishers, Boca Raton, FL, 2004.)

For chlorinated and methylated benzenes, the relationship given by Schwarzenbach and Westall (1981) is

$$\log k_{oc} = 0.72 \log k_{ow} + 0.49$$

The graphical relationship shown in Figure 9.7 uses some representative values reported in the various handbooks (e.g., Verscheuren, 1983; Montgomery and Welkom, 1991) for log k_{ow} and log k_{oc}. The values used for log k_{ow} are essentially mid-range results reported in the handbooks and in many studies. Not all log k_{oc} values are obtained as measured values. Many of these have been obtained through application of the various log k_{oc} – log k_{ow} relationships reported in the literature (e.g., Kenaga and Goring, 1980; Karickhoff et al., 1979). The linear relationship shown by the solid line in Figure 9.17 is given as

$$\log k_{oc} = 1.06 \log k_{ow} - 0.68$$

This graphical relationship is useful in the sense of partitioning of the organic chemical compounds shown in the diagram. Yong and Mulligan (2004) have discussed some of the pertinent correlations, stating, for example, in regard to the k_{oc} values shown in Figure 9.17 for dichlorobenzene, that they indicate that it partitions well to sediments, and particularly to the organic fractions (SOM). Because of its resistance to anaerobic degradation, it is very persistent.

9.6   Persistence and Fate

One of the many necessary actions needed along the path toward terrestrial environment sustainability is to return contaminated land to its uncontaminated state. This requires one to clean up contaminated sites. The problem of concern in remediation of contaminated sites is the persistence of pollutants in the site. In many instances, natural attenuation can be relied upon as a remediation tool — a sort of natural self-remediation process that is intrinsic to the site properties (see Chapter 10). There exists, however, several kinds of pollutants that are not easily "self-remediated" with natural attenuation processes. These pollutants fall under the general class of persistent pollutants. The term *persistence* has been defined as *the continued presence of a pollutant in the substrate*. The persistence of inorganic and organic pollutants differs in respect to meaning and application. The persistence of heavy metals refers to their continued presence in the subsurface soil regime in any of their individual oxidation states and in any of the complexes formed. Organic chemical pollutants, on the other hand, can undergo considerable transformations because of microenvironmental factors. We define *transformation* to mean the conversion of the original organic chemical pollutant into one or more resultant products by processes that can be abiotic, biotic, or a combination of these. The intermediate products obtained from transformation of organic chemical compounds by biotic processes along the pathway toward complete mineralization are generally classified as degradation products. Transformed products resulting from abiotic processes in general do not classify as being intermediate products along the path to mineralization. They are, however, not easily distinguished because some of the abiotic transformed products themselves may become more susceptible to biotic transformations. When this occurs, the process is known as a combination transformation process.

As discussed briefly in Section 2.3.2 in Chapter 2, a characteristic term used to describe organic chemicals that persist in their original form or in altered forms is *persistent organic chemical pollutants*, or persistent organic pollutants (POPs). These include dioxins, furans, pesticides and insecticides, polycyclic aromatic hydrocarbons (PAHs), and halogenated hydrocarbons. The persistence of organic chemical pollutants in soils is a function of at least three factors: (1) the physicochemical properties of the organic chemical pollutant itself, (2) the physicochemical properties of the soil, and (3) the microorganisms in the soil. Resultant abiotic reactions and transformations are sensitive to factors 1 and 2, and all factors are important participants in the dynamic processes associated with the activities of the microorganisms in the biologically mediated chemical reactions and transformation processes.

9.6.1 Biotransformation and Degradation of Organic Chemicals and Heavy Metals

The various types of organisms and microorganisms responsible for the biotransformation (this includes degradation) of organic chemical compounds can be classified under the Whittaker (1969) five-kingdom classification scheme shown in Figure 9.18. The reader should consult the regular textbooks on microbiology for detailed descriptions of these. The descriptions given by Yong and Mulligan (2004) are summarized as follows:

- Protozoa include pseudopods, flagellates, amoebas, ciliates, and parasitic protozoa. Their sizes can vary from 1 to 2000 mm. They are aerobic, single-celled chemoheterotrophs, and are eukaryotes with no cell walls. They are divided into four main groups: (1) the Mastigophora, which are flagellate protozoans; (2) the Sarcodina, which are amoeboid; (3) the Ciliophone, which are ciliated; and (4) the Sporozoa, which are parasites of vertebrates and invertebrates.
- Fungi are aerobic, multicellular eukaryotes and chemoheterotrophs that require organic compounds for energy and carbon. They reproduce by formation of asexual spores. In comparison to bacteria, they (1) do not require as much nitrogen,

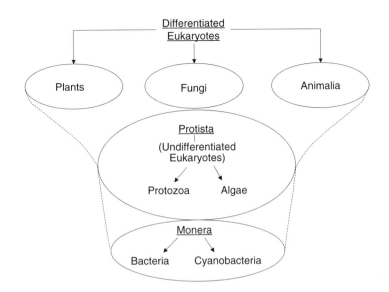

FIGURE 9.18
Organisms and microorganisms participating in natural bioremediation processes grouped according to the Whittaker five-kingdom hierarchial system. (From Yong, R.N. and Mulligan, C.N., *Natural Attenuation of Contaminants in Soils*, Lewis Publishers, Boca Raton, FL, 2004.)

(2) are more sensitive to changes in moisture levels, (3) are larger, (4) grow more slowly, and (5) can grow in a more acidic pH range (less than pH 5). Fungi mainly live in the soil or on dead plants and are sometimes found in freshwater.

- Algae are single-celled and multicellular microorganisms that are green, greenish tan to golden brown, yellow to golden brown (marine), or red (marine). They grow in the soil and on trees or in fresh or salt water. Those that grow with fungi are called lichens. Seaweeds and kelps are examples of algae. Since they are photosynthetic, they can produce oxygen, new cells from carbon dioxide or bicarbonate (HCO_3^-), and dissolved nutrients, including nitrogen and phosphorus. They use light of wavelengths between 300 and 700 nm. Red tides are indicative of excessive growth of dinoflagellates in the sea. The green color in a body of lakes and rivers is eutrophication due to the accumulation of nutrients such as fertilizers in the water.

- Although viruses are smaller than bacteria and require a living cell to reproduce, their relationship to other organisms is not clear. In order for them to replicate, they have to invade various kinds of cells. They consist of one strand of DNA and one strand of ribonucleic acid (RNA) within a protein coat. A virus can only attack a specific host. For example, those that attack bacteria are called bacteriophages.

- The most significant animals in the soil are millimeter-size worms. Nematodes are cylindrical in shape and are able to move within bacterial flocs. Flatworms such as tapeworms, eel worms, roundworms, and threadworms, which are nematodes, can cause diseases such as roundworm, hookworm, and filarisis.

- Bacteria are prokaryotes that reproduce by binary fission by dividing into two cells, in about 20 minutes. The time it takes for one cell to double, however, depends on the temperature and species. For example, the optimal doubling time for *Bacillus subtilis* (37°C) is 24 minutes, and for *Nitrobacter agilis* (27°C), 20 hours. Classification is by shape, such as the rod-shaped bacillus, the spherical-shaped coccus, and the spiral-shaped spirillium. Rods usually have diameters of one half to one micron and lengths of three to five microns. The diameter of spherical cells varies from 0.2 to 2 microns. Spiral-shaped cells range from 0.3 to 5 microns in diameter and 6 to 15 microns in length. The cells grow in clusters, chains, or in single form and may or may not be motile. The substrate of the bacteria must be soluble. In most cases, classification is according to the genus and species (e.g., *Pseudomonas aeruginosa* and *Bacillus subtilis*). Some of the most common species are *Pseudomonas, Arthrobacter, Bacillus, Acinetobacter, Micrococcus, Vibrio, Achromobacter, Brevibacterium, Flavobacterium,* and *Corynebacterium*. Within each species, we will have various strains. Each of these can behave differently. Some strains can survive in certain conditions that others cannot. The ones that are better adapted will survive. For survival, strains, called mutants, originate due to problems in the genetic copying mechanisms. Some species are dependent on other species, though, for survival. Degradation of chemicals to an intermediate stage by one species of bacteria may be required for the growth of another species that utilizes the intermediate.

9.6.1.1 Alkanes, Alkenes, and Cycloalkanes

Alkanes, alkenes, and cycloalkanes, among others (PAHs, asphaltenes, etc.), are components of petroleum hydrocarbons (PHCs). Low-molecular-weight alkanes are the most easily degraded by microorganisms. As the chain length increases from C_{20} to C_{40}, hydrophobicity increases and both solubility and biodegradation rates decrease. Alkenes with a double bond on the first carbon may be more easily degradable than those alkenes with the double bond at other positions (Pitter and Chudoba, 1990). Cycloalkanes are not as degradable as alkanes due to their cyclic structure, and their biodegradability decreases as the number of rings increases.

9.6.1.2 Polycyclic, Polynuclear Aromatic Hydrocarbons

As with cycloalkanes, the compounds become more difficult to degrade as the number of rings of PAHs increases. This is due to decreasing volatility and solubility and increased sorption properties of these compounds. They are degraded one ring at a time in a manner similar to that of single-ring aromatics.

9.6.1.3 Benzene, Toluene, Ethylbenzene, and Xylene

Benzene, toluene, ethyl benzene, and xylene (BTEX) are volatile, water-soluble components of gasoline. Aromatic compounds with benzene structures are more difficult to degrade than cycloalkanes. Aerobic degradation of all components of BTEX occurs rapidly when oxygen is present. Aromatic compounds can also be degraded under anaerobic conditions to phenols or organic acids to fatty acids to methane and carbon dioxide (Grbic-Galic, 1990). Degradation is less assured and is slower than under aerobic conditions.

9.6.1.4 Methyl Tert-Butyl Ether

Methyl tert-butyl ether (MTBE), which is an additive to gasoline, is highly resistant to biodegradation. It is reactive with microbial membranes.

9.6.1.5 Halogenated Aliphatic and Aromatic Compounds

Halogenated aliphatic compounds are pesticides such as ethylene dibromide (DBR) or $CHCl_3$, $CHCl_2Br$, and industrial solvents, including methylene chloride and trichloroethylene. Halogenated aromatic compounds are also pesticides, and they include such pesticides as DDT, 2,4-D and 2,4,5-T, plasticizers, pentachlorophenol, and polychlorinated biphenyls. The presence of halogen makes aerobic degradation of the halogenated aliphatic compounds difficult to achieve, due to the lower energy and the higher oxidation state of the compound. Anaerobic biodegradation is easier to achieve. This is particularly true when the number of halogens in the compound increases — making aerobic degradation more difficult. In the case of the halogenated aromatic compounds, the mechanisms of conversions include hydrolysis (replacement of halogen with hydroxyl group), reductive dehalogenation (replacement of halogen with hydrogen), and oxidation (introduction of oxygen into the ring, causing removal of halogen).

9.6.1.6   *Heavy Metals*

Microbial cells can accumulate heavy metals through ion exchange, precipitation, and complexation on and within the cell surface containing hydroxyl, carboxyl, and phosphate groups. Processes involving bacterial oxidation–reduction will alter the mobility of the heavy metal pollutants in soil. An example of this is the reduction of Cr(IV) in the form of chromate (CrO_4^{2-}) and dichromate ($Cr_2O_7^{2-}$) to Cr(III). Conversion can also be indirect by microbial production of Fe(II), sulfide, and other components that reduce chromium. Oxidation of selenium in the four naturally occurring major species of selenium — selenite (SeO_3^{2-}, IV), selenate (SeO_4^{2-}, VI), elemental selenium (Se, 0), and selenide (–II) — can occur under aerobic conditions. Transformation of selenate can occur anaerobically to selenide or elemental selenium, and methylation of selenium detoxifies selenium for the bacteria by removing the selenium from the bacteria.

9.7   Prediction of Transport and Fate of Pollutants

A key factor in the decision-making process in structuring sustainability objectives for the terrestrial environment is the ability to have knowledge of the movement and spread (transport) of pollutants in the ground. Apart from procurement of field information required to delineate the parameters of the pollution problem at hand, this requires development of techniques for prediction of the transport and fate of the pollutants in the ground. The main elements in mass transport and mass transfer of pollutants in a soil–water system in the ground are shown in Figure 9.19. Prediction of the transport and fate of pollutants in the subsoil requires consideration and incorporation of these elements in the analytical and mathematical analyses. Mass transport refers to transport of the dissolved solutes by advective, diffusive, and dispersion forces.

Mass transfer of pollutants in the soil refers to chemical mass transfer processes. These have been discussed in the previous sections. They include sorption, dissolution and precipitation, acid–base reactions and hydrolysis, oxidation–reduction (redox) reactions, speciation–complexation, and biologically mediated transfer. For aspects of design, containment of high-level nuclear waste, where prediction of the transport of radionuclides is of particular concern, radioactive decay is one of the important chemical transfer processes that needs to be considered. Finally, biologically mediated transfer of pollutants completes the three main categories (physical, chemical, and biological) of pollutant transfer from pore water to soil solids. Transformation and degradation of the organic chemical pollutants require as much attention as sorption and transport of the organic chemicals.

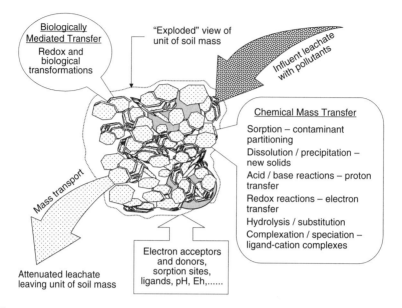

FIGURE 9.19

Elements of basic mass transport and mass transfer in attenuation of pollutants in leachate transport through a soil element.

Many of the chemical mass transfer processes are kinetic processes, and reactions are not instantaneous. However, in a majority of the analytical considerations, most of the reactions are considered instantaneous and chemical equilibrium is immediately obtained. This leads to modeling of transport of pollutants as a nonreactive process — an outcome that is not always appreciated by users of developed models. Consideration of transport and fate as a reactive process requires incorporation of the many transfer mechanisms that are time dependent, and also the transformations and degradations of the chemical. This is not easily accomplished. In the recent attempts to incorporate the reactive processes, incorporation of geochemical models into the standard transport models has been attempted — with differing degrees of complexity and success.

9.7.1   Mass Transport

The three mechanisms for mass transport include advection, diffusion, and dispersion. Advection refers to the flux generated by the hydraulic gradient and is given in terms of the advective velocity, v. Transport of dissolved solutes in the pore water solely by advective means will progress in concert with the advective velocity. In situations where the solutes possess kinetic energy and demonstrate Brownian activity, diffusion of these solutes will occur. Using the pore water as the carrier, the solutes will combine their diffusive capability with advective velocity to produce the combined mass transport. This permits the dissolved solutes to move ahead of the advective front; i.e., the general tendency is for the diffusion front to precede the advection front. In situations where tortuosity and pore size differences combine with pore restrictions to create local mixing in the movement of the dissolved solutes, dispersion results. The degree of dispersion and the resultant effect on mass transport is not readily quantified.

The diffusive movement of a particular solute (contaminant or pollutant) is characterized by its diffusion coefficient. This diffusion coefficient, D_c, is most often considered as equivalent or equal to the effective molecular diffusion coefficient. In dilute solutions of a single ionic species, the diffusion coefficient of that single species is termed the *infinite solution diffusion coefficient*, D_o. The infinite solution diffusion coefficients are dependent on such factors as ionic radius, absolute mobility of the ion, temperature, viscosity of the fluid medium, valence of the ion, equivalent limiting conductivity of the ion, etc. A useful listing of these coefficients for a range of solutes and for various sets of conditions can be found in many basic handbooks and other references (e.g., Li and Gregory, 1974; Lerman, 1979). From a theoretical point of view, the studies of molecular diffusion by Nernst (1888) and Einstein (1905) show the level of complex interdependencies that combine to produce the resultant coefficient obtained. From studies dealing with the movement of suspended particles controlled by the osmotic forces in the solution, the three expressions most often cited are

$$\text{Nernst–Einstein} \qquad D_o = \frac{uRT}{N} = uk'T \qquad (9.3)$$

$$\text{Einstein–Stokes} \qquad D_o = \frac{RT}{6\pi N \eta r} = 7.166 \times 10^{-21} \frac{T}{\eta r} \qquad (9.4)$$

$$\text{Nernst} \qquad D_o = \frac{RT\lambda^o}{F^2 |z|} = 8.928 \times 10^{-10} \frac{T\lambda^o}{|z|} \qquad (9.5)$$

where D_o = diffusion coefficient in an infinite solution, u = absolute mobility of the solute under consideration, R = universal gas constant, T = absolute temperature, N = Avogadro's number, k' = Boltzmann's constant, λ^o = conductivity of the target ion or solute, r = radius of the hydrated ion or solute, λ = absolute viscosity of the fluid, z = valence of the ion, and F = Faraday's constant. A large listing of experimental values for λ^o for major ions can be found in Robinson and Stokes (1959).

We define a dimensionless Peclet number as $P_e = v_L d/D_o$, where v_L is the longitudinal flow velocity (advective flow). From the information reported by Perkins and Johnston (1963), it is seen that for Peclet numbers less than 1 ($P_e < 1$), diffusive transport of the contaminant solutes in a contaminant plume travels faster than the advective flow of water. For $P_e > 10$, advective flow constitutes the dominant flow mechanism for the movement of solutes. In between the values of 1 and 10, there is a gradual change from diffusion-dominant to advection-dominant transport (Figure 9.20). The longitudinal diffusion coefficient, D_L, consists of both the molecular diffusion coefficient, D_m, and the hydrodynamic (mechanical) dispersion coefficient, D_h. This is written as $D_L = D_m + D_h = D_o\tau + \alpha v$, where D_m = molecular diffusion = $D_o\tau$, $D_h = \alpha v$, α = dispersivity parameter, and τ = tortuosity factor.

The tortuosity factor is introduced to modify the infinite solution diffusion coefficient to acknowledge that we do not have an infinite solution, and that diffusion of a single-solute species in a soil–water system is subject to constricting pore volumes and nonlinear paths. Figure 9.20 shows that in the diffusion-dominant transport region, we can safely neglect the v_L term since v_L is vanishingly small. Under those circumstances, in the diffusion-dominant transport region we will have $D_L = D_o\tau$. In the advection-dominant transport region, if we consider that diffusion transport is negligible, then we will have $D_L = v_L$. In the transition region, the relationship for D_L will be given as $D_L = D_o\tau + \alpha v_L$.

The significance of a correct choice or specification of a diffusion coefficient cannot be overstated. Figure 9.21 is a schematic illustration showing the variation of D (or D_L)

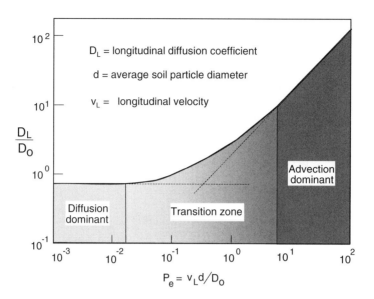

FIGURE 9.20
Diffusion- and advection-dominant flow regions for solutes in relation to the Peclet number. (Adapted from Perkins, T.K. and Johnston, O.C., *J. Soc. Pet. Eng.*, 17, 70–83, 1963.)

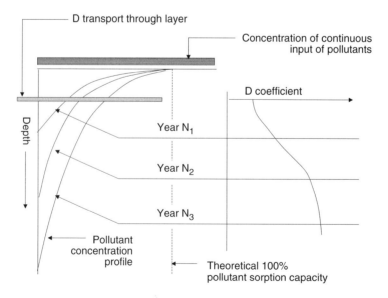

FIGURE 9.21
Pollutant concentration profiles in subsoil in relation to time elapsed, given as years N_n. The diffusion coefficient profile (right-hand curve) is obtained from calculations reflecting changes in pollutant concentration with time. Note that the concentration of the continuous input of pollutants is higher than the theoretical 100% sorption capacity of the soil.

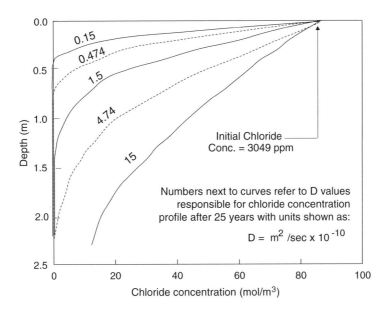

FIGURE 9.22
Effect of different values of coefficients of diffusion D on chloride concentration profile after 25 years of continuous input of chloride at 3049 ppm.

coefficients calculated using Equation 2.3 in Chapter 2 and using the concentration profiles shown at the left-hand side of the diagram. The Ogata and Banks (1961) solution of the transport equation, similar to the one shown in Equation 2.3, for an initial chloride concentration of 3049 ppm as the input source (Figure 9.22), shows the different chloride concentration profiles obtained in relation to variations in the D value used in the calculations. The differences are significant.

9.7.2 Transport Prediction

For pollutants that can be partitioned in the transport process in the subsoil system, it is not uncommon to use the transport relationship given as Equation 2.3 in Chapter 2. For some situations, such as equilibrium-partitioning processes, this is an adequate method for predicting the transport and fate of pollutants that can be partitioned. The assumption is generally made that the rate of reactions is independent of the concentration of contaminants, i.e., a zero-rate reaction process. However, for many other situations relating to pollutants that are partitioned during transport, such as nonequilibrium partitioning, this relationship needs to be knowledgeably applied and perhaps modified to meet the conditions of partitioning. The relationship given as Equation 2.3 in Chapter 2 can be written in its original expanded form as

$$\frac{\partial c}{\partial t} = D_L \frac{\partial^2 c}{\partial x^2} - v \frac{\partial c}{\partial x} - \frac{\rho}{n\rho_w} \frac{\partial c^*}{\partial t} \tag{9.6}$$

where c = concentration of pollutants or contaminants of concern, t = time, D_L = diffusion coefficient, v = advective velocity, x = spatial coordinate, ρ = bulk density of soil media, ρ_w = density of water, n = porosity of soil media, and c^* = concentration of pollutants or contaminants adsorbed by soil fractions (see ordinate in graph shown in Figure 2.14 in

Chapter 2). For ease in discussion, since all pollutants are also contaminants, we will use the general term *contaminants* to indicate both pollutants and contaminants. We recall that the adsorption isotherms portrayed in Figure 2.14 in Chapter 2 are derived from batch equilibrium tests with soil solutions, and we further recall the discussion in the previous Section 9.5.1 and Figure 9.14 that the distribution coefficient k_d obtained from the adsorption isotherms refers directly to a maximum reactive surface reaction process. Equation 2.3 in Chapter 2 is obtained when c^* is assumed to be equal to $k_d c$, i.e., if a linear adsorption isotherm is assumed. Substituting for c^* in Equation 9.6 gives us

$$\frac{\partial c}{\partial t} = D_L \frac{\partial^2 x}{\partial x^2} - v\frac{\partial c}{\partial x} - \frac{\rho}{n\rho_w}\frac{\partial\left(k_d c\right)}{\partial x} \tag{9.7}$$

Collecting terms and defining R as the *retardation* $= \left[1 + \dfrac{\rho}{n\rho_w}k_d\right]$, the equation previously seen as Equation 2.3 is obtained:

$$R\frac{\partial c}{\partial t} = D_L \frac{\partial^2 c}{\partial x^2} - v\frac{\partial c}{\partial x} \tag{9.8}$$

If the partition isotherms obtained from actual laboratory tests do not show linearity, the obvious solution is to use the proper function that describes c^*, e.g., the Freundlich or Langmuir (see Figure 2.14 in Chapter 2) or some other equivalent relationship. In this instance, the opportunity to properly reflect the partitioning process through a compact soil mass should be taken. Instead of using the nonlinear adsorption isotherm obtained from batch equilibrium tests on soil solutions, the sorption relationship obtained from column leaching tests through compact soil should be used. This has been discussed in Section 9.5.1 and illustrated in terms of adsorption curve differences in Figure 9.15.

Equations of the type shown as Equation 2.3 or Equation 9.5 have been described as nonreactive transport relationships. This description has been applied to such equations to reflect the observation that biotic and abiotic chemical reactions in the aqueous phase have not been factored into the structuring of the relationship. The chemical reactions discussed in Section 9.4 cannot be ignored because they not only compete with the soil solids for partitioning of the contaminants, but they may also transform the contaminants — especially the organic chemicals. The reactions and transformations will change not only the character of the contaminants, but also the distribution of contaminants in the zone of interest. The problem is magnified in field situations because of multispecies contaminants and a mixture of both inorganic and organic chemicals. Unlike laboratory leaching column and batch equilibrium tests, real field situations provide one with a complex mix of contaminants transporting through an equally complex subsoil system.

9.7.2.1   *Chemical Reactions and Transport Predictions*

To meet the objectives of sustainability of the terrestrial environment, proper prediction of transport and fate of pollutants requires knowledge of how the abiotic and biotic reactions affect the long-term health of the terrain system, especially the subsoil system. From the myriad of possibilities in handling the complex problem of chemical reactions and reaction rates, and transformations, there exist at least four simple procedures that provide some accounting, to a greater or lesser degree, of the various processes controlling transport. These include (1) adding a reaction term r_c in the commonly used advection–diffusion equation given as Equation 2.3 or Equation 9.6, (2) accounting for the contaminant adsorption–desorption process, (3) using first- or second-order or higher-order reaction rates, and (4) combining transport models with geochemical speciation models. None of these appear to handle biotransformations and their resultant effect on the transport and fate processes.

Addition of a reaction term r_c to Equation 9.8 is perhaps the most common method used to accommodate a kinetic approach to fate and transport modeling. The resultant formulation is a linearly additive term to Equation 9.8 as follows:

$$R\frac{\partial c}{\partial t} = D_L \frac{\partial^2 c}{\partial x^2} - v \frac{\partial c}{\partial x} + r_c \qquad (9.9)$$

The last term in Equation 9.9 can be expressed in the form of a general rate law as follows:

$$r_c = -k\vartheta |A|^a |B|^b \qquad (9.10)$$

where r_c in this case is the rate of increase in concentration of a contaminant of species A, k is the rate coefficient, ϑ represents the volume of fluid under consideration, A and B are the reactant species, and a and b are the reaction orders.

The use of an adsorption–desorption approach to fate and transport modeling recognizes that in field situations, desorption (or displacement) occurs as part of the ion exchange process. Determination of the transport and fate of pollutants using the partitioning approach typified by the advection–diffusion relationship with adsorption–desorption consideration requires that one can write a relationship for c^* in Equation 9.6. Curve-fitting procedures are commonly used to deduce information obtained from batch equilibrium or flow-through (leaching column) tests. The Freundlich and Langmuir curves, for example (see Figure 2.14 in Chapter 2), are specific cases of such procedures.

Prediction and modeling for biotransformation and biodegradation, and their effects on fate and transport, require a different approach. Yong and Mulligan (2004) have provided an accounting of some of the more popular analytical computer models used in application of natural attenuation schemes. These in effect are fate and transport models, since biotransformation and biodegradation are the primary attenuation processes — a principal feature in fate and transport of organic chemicals. For example, the analytical model BIOSCREEN (Newell et al., 1996) developed for the Air Forces Center for Environmental Excellence by Groundwater Services, Inc. (Houston, TX) assumes a declining source concentration with transport and biodegradation processes for the soluble hydrocarbons that include advection, dispersion, adsorption, and aerobic and anaerobic degradation. Most of the available analytical computer models developed to handle biotransformation and biodegradation in fate and transport modeling have the essential items contained in BIOSCREEN. The principal distinguishing factors among the available computational packages, such as BIOPLUME III (Rifai et al., 1997), MODFLOW and RT3D (Sun et al., 1996), BIOREDOX (Carey et al., 1998), and BIOCHLOR (Aziz et al., 2000), include (1) structure of the outputs, (2) manner in which degradation is handled, such as order of degradation and degradation rates, (3) types of organic chemicals, (4) inclusion of heavy metals and some other inorganics, (5) availability and types of electron acceptors, and (6) adsorption–desorption.

Solution of the transport relationships shown as Equations 9.5 and 9.6, and other similar relationships, can be achieved using analytical or numerical techniques. For well-defined geometries, initial and boundary conditions, and processes, analytical techniques provide exact solutions that can give further insight into the processes involved in the problem under consideration. Numerical techniques such as finite difference, finite element, and boundary element are useful and are perhaps the techniques favored by many because of their capability to handle more complex geometries and variations in material properties and boundary conditions.

9.7.3　Geochemical Speciation and Transport Predictions

Abiotic reactions and transformations, together with the biotic counterparts, form the suite of processes that are involved in the transport and fate of contaminants in the subsoil. The reactions between the chemical species in the pore water and also with the reactive soil particle surfaces discussed in the previous sections and chapters constitute the basic platform. Because individual chemical species have the ability to participate in several types of reactions, the equations to describe the various equilibrium reactions can become complicated, particularly since one needs to be assured that all the reactions are captured.

Geochemical modeling provides a useful means for handling the many kinds of calculations required to solve the various equilibrium reactions. Specific requirements are a robust thermodynamic database and simultaneous solution of the thermodynamic and mass balance equations. Appelo and Postma (1993) provide a comprehensive treatment of the various processes and reactions, together with a user guide for the geochemical model PHREEQE, developed by Parkhurst et al. (1980). As with many of the popular models, the model is an aqueous model based upon ion pairing, and includes elements and both aqueous species and mineral phases (fractions).

Other available models include the commonly used MINTEQ (Felmy et al., 1984) and the more recent MINTEQA2, which includes PROFEFA2 (Allison et al., 1991), a preprocessing package for developing input files; GEOCHEM (Sposito and Mattigod, 1980); HYDROGEOCHEM (Yeh and Tripathi, 1990); and WATEQF (Plummer et al., 1976). By and large, most of the geochemical codes assume instantaneous equilibrium; i.e., kinetic reactions are not included in the calculations. In part, this is because reactions such as oxidation–reduction, precipitation–dissolution, substitution–hydrolysis, and, to some extent, speciation–complexation can be relatively slow. To overcome this, some of the models have been able to provide analyses that point toward possible trends and final equilibria. The code EQ6 (Delaney et al., 1986), though, does provide for consideration of dissolution–precipitation reactions. Transformations, however, are essentially not handled by most of the codes.

9.8　Concluding Remarks

1. The concern in this chapter is for terrestrial environment sustainability as it pertains to the presence of various waste discharges originating from industrial and urban activities. We have focused our attention on developing concepts that consider the natural capital of terrestrial environment, and have specified the objectives for sustainability as clearly as we can: to ensure that each natural capital component maintains its full and uncompromised functioning capability without loss of growth potential.

2. For the objectives to be fulfilled, actions, reactions, and management techniques require specification of *indicators* that mark the path toward sustainability of the

natural capital. This recognizes that absolute sustainability is not generally attainable.

3. The first half of this chapter has examined the nature of indicators and has distinguished between system status and material performance indicators. This distinction is necessary since various situations demand a proper accounting of the relationship between the two. Figure 9.4 provides the protocols that assist in this type of accounting.

4. Pollutants and contaminants and the manner in which they are handled in respect to the terrestrial environment have a great impact on whether we can hope to reach sustainability of the terrestrial environment or its natural capital. The impact of these (pollutants and contaminants) and the implementation of indicators as a technique for assessment need proper consideration. Figure 9.5 provides the sustainability goals in respect to waste discharge onto the terrestrial environment, and the subsequent figures provide examples. To some extent, we can learn how to prescribe the necessary indicators to achieve the objective. Obviously, real field situations are both site specific and industry specific.

5. In the assessment of waste impacts, it is clear that this cannot be achieved without an understanding of both the interactions with the subsoil system and the goals of sustainability. Central to the various issues is the problem of gaining a proper knowledge of the health status of the subsoil system. This requires one to be able to predict the transport and fate of pollutants in the subsoil system. The problem of prediction is not a simple problem that can be handled with one set of tools. Analytical computer modeling is perhaps the most common technique used to provide information that allows one to predict system behavior.

6. For pollutants that can partition between the aqueous phase and the soil solids in the subsoil system, we have well-developed advection–diffusion transport models that can address the problem. The pitfalls in implementation of such models include the availability of appropriate and realistic input parametric information (especially partition and distribution coefficients), and chemical reactions that affect the status of the pollutants in the system.

7. The use of geochemical speciation modeling allows one to determine these reactions. However, since kinetic reactions are not readily handled in the present available geochemical models, and since most of these models are not coupled to the regular transport models, much work remains at hand to obtain a reactive prediction model that can tell us about the fate and transport of pollutants in the subsoil system. Present research into coupling between geochemical models and advection–dispersion models has identified the complex and highly demanding computational requirements for a coupled model. Nevertheless, a realistic reactive coupled model is needed if we are to reach the stage where knowledge of the fate and persistence of pollutants in the subsoil system is to be obtained.

References

Allison, J.D., Brown, D.S., and Novo-Gradac, K.J., (1991), *MINTEQA2/PRODEFA2, A Geochemical Assessment Model for Environmental Systems*, U.S. EPA, Washington, DC.

Appelo, C.A.J. and Postma, D., (1993), *Geochemistry, Groundwater and Pollution*, Balkema, Rotterdam, 536 pp.

Aziz, C.E., Newell, C.J., Gonzales, J.R., Haasm, P.E., Clement, T.P., and Sun, Y., (2000), *BIOCHLOR: Natural Attenuation Decision Support System, User's Manual, Version 1.1*, EPA/600/R-00/008, U.S. EPA Office of Research and Development, Washington, DC.

Carey, G.R., van Geel, P.J., Murphy, J.R., McBean, E.A., and Rover, F.A., (1998), Full-scale field application of a coupled biodegradation-redox model BIOREDOX, in *Natural Attenuation of Chlorinated Solvents*, Wickramanayake, G.B. and Hinchee, R.H. (Eds.), Batelle Press, Columbus, OH, pp. 213–218.

Chiou, G.T., Schmedding, D.W., and Manes, M., (1982), Partition of organic compounds on octanol-water system, *Environ. Sci. Technol.*, 16:4–10.

Delaney, J.M., Puigdomenech, I., and Wolery, T.J., (1986), *Precipitation Kinetics Option of the EQ6 Geochemical Reaction Path Code*, Lawrence Livermore National Laboratory Report, UCRL-56342, Livermore, CA, 44 pp.

Einstein, A., (1905), Uber die von der Molekularkinetischen theorie der Warme Geoforderte Bewegung von in Rubenden Flussigkeiten Suspendierten Teilchen, *Annalen Phys.*, 4:549–660.

Felmy, A.R., Girvin, D.C., and Jeene, E.A., (1984), *MINTEQ: A Computer Program for Calculating Aqueous Geochemical Equilibria*, PB84-157148, EPA-600/3-84-032, U.S. EPA, Washington, DC, February.

Grbic-Galic, D., (1990), Methanogenic transformation of aromatic hydrocarbons and phenols in groundwater aquifers, *J. Geomicrobiol.*, 8:167–200.

Karickhoff, S.W., (1984), Organic pollutants sorption in aquatic system, *J. Hydraul. Eng.*, 110:707–735.

Karickhoff, S.W., Brown, D.S., and Scott, T.A., (1979), Sorption of hydrophobic pollutants on natural sediments, *Water Resour.*, 13:241–248.

Kenaga, E.E. and Goring, C.A.I., (1980), Relationship between Water Solubility, Soil Sorption, Octanol-Water Partitioning and Concentration of Chemicals in Biota, ASTM-STP 707, ASTM, West Conshokocken, PA, pp. 78–115.

Lerman, A., (1979), *Geochemical Processes: Water and Sediment Environments*, John Wiley & Sons, New York, 481 pp.

Lewis, G.N., (1923), *Valences and the Structure of Atoms and Molecules*, The Chemical Catalogue Co., New York.

Li, Y.H. and Gregory, S., (1974), Diffusion of ions in sea water and in deep-sea sediments, *Geochem. Cosmochim. Acta*, 38:603–714.

Montgomery, J.H. and Welkom, L.M., (1991), *Groundwater Chemicals Desk Reference*, Lewis Publishers, Ann Arbor, MI, 640 pp.

Nernst, W., (1888), Zur Kinetik der in Losung befinlichen Korper, *Z. Phys. Chemie*, 2:613–637.

Newell, C.J., McLeod, R.K., and Gonzales, J., (1996), *BIOSCREEN Natural Attenuation Decision Support Systems*, EPA/6000/R-96/087, U.S. EPA, Washington, DC, August.

Ogata, A. and Banks, R.B., (1961), *A Solution of the Differential Equation of Longitudinal Dispersion in Porous Media*, U.S. Geological Survey Paper 411-A, USGS, Washington, DC.

Parkhurst, D.L., Thorstenson, D.C., and Plummer, L.N., (1980), *PHREEQE: A Computer Program for Geochemical Calculations*, U.S. Geological Survey Water Resources Investigation, 80-96, USGS, Washington, DC, 210 pp.

Perkins, T.K. and Johnston, O.C., (1963), A review of diffusion and dispersion in porous media, *J. Soc. Pet. Eng.*, 17:70–83.

Pitter, P. and Chudoba, J., (1990), *Biodegradability of Organic Substances in the Aquatic Environment*, CRC Press, Boca Raton, FL, 167 pp.

Plummer, L.N., Jones, B.F., and Truesdell, A.H., (1976), *WATEQF: A FORTRAN IV Version of WATEQ, a Computer Code for Calculating Chemical Equilibria of Natural Waters*, U.S. Geological Survey Water Resources Investigation, 76-13, USGS, Reston, VA, 61 pp.

Rao, P.S.C. and Davidson, J.M., (1980), Estimation of pesticide retention and transformation parameters required in nonpoint source pollution models, in *Environmental Impact of Nonpoint Source Pollution*, Overcash, M.R. and Davidson, J.M. (Eds.), Ann Arbor Sciences, Ann Arbor, MI, pp. 23–27.

Rifai, H.S., Newell, C.J., Gonzales, J.R., Dendrou, S., Kennedy, L., and Wilson, J., (1997), *BIOPLUME III Natural Attenuation Decision Support System, Version 1.0, User's Manual*, prepared for the U.S. Air Force Center for Environmental Excellence, Brooks Air Force Base, San Antonio, TX.

Robinson, R.A. and Stokes, R.H., (1959), *Electrolyte Solutions*, Butterworths, London, 571 pp.

Schwarzenbach, R.P., Gschwend, P.M., and Imboden, D.M., (1993), *Environmental Organic Chemistry*, John Wiley & Sons, New York, 681 pp.

Schwarzenbach, R.P. and Westall, J., (1981), Transport of non-polar organic compounds from surface water to groundwater: laboratory sorption studies, *Environ. Sci. Technol.*, 15:1360–1367.

Sposito, G. and Mattigod, S.V., (1980), *GEOCHEM: A Computer Program for the Calculation of Chemical Equilibria in Soil Solutions and Other Natural Water Systems*, Department of Soils and Environment Report, University of California, Riverside, 92 pp.

Sun, Y., Petersen, J.N., Clement, T.P., and Hooker, B.S., (1996), A Monitoring Computer Model for Simulating Natural Attenuation of Chlorinated Organics in Saturated Groundwater Aquifers, RPA/540/R-96/509, paper presented at the Proceedings of the Symposium on Natural Attenuation of Chlorinated Organics in Groundwater, Dallas, TX.

Verscheuren, K., (1983), *Handbook of Environmental Data on Organic Chemicals*, 2nd ed., Van Norstrand Reinhold, New York, 1310 pp.

Whittaker, R.H., (1969), New concepts of kingdoms or organisms: evolutionary relations are better represented by new classifications than by the traditional two kingdoms, *Science*, 163:150–160.

Yeh, G.T. and Tripathi, V.S., (1990), *HYDROGEOCHEM, a Coupled Model of HYDROlogic Transport and GEOCHEMical Equilibria in Reactive Multicomponent Systems*, Oak Ridge National Laboratory, Oak Ridge, TN.

Yong, R.N., (2001), *Geoenvironmental Engineering: Contaminated Soils, Pollutant Fate, and Mitigation*, CRC Press, Boca Raton, FL, 307 pp.

Yong, R.N. and Mulligan, C.N., (2004), *Natural Attenuation of Contaminants in Soils*, Lewis Publishers, Boca Raton, FL, 319 pp.

Yong, R.N. and Sheremata, T.W., (1991), Effect of chloride ions on adsorption of cadmium from a landfill leachate, *Can. Geotech. J.*, 28:378–387.

10

Pollutant Impact Mitigation and Management

10.1 Introduction

Pollutants are chemical stressors that can severely affect the quality of water and ground-water resources and soil quality. These pollutants include nonpoint source pollutants such as herbicides, pesticides, fungicides, etc., spread over large land surface areas and point source pollutants from effluents, waste treatment plants, and liquid discharges as wastes and spills from industrial plants (e.g., heavy metals and organic chemicals). The previous chapters dealing with urbanization and industries have shown that liquid and solid waste discharges, together with rejects, debris, and inadvertent spills in the plants, all combine to create significant threats to the health of biotic receptors and also the environment. To demonstrate the magnitude of the problem, we can give the example of sites contaminated with hazardous wastes and other material discards. Russell et al. (1991) suggests that over the next three decades, the expenditure required to clean up all the sites in the United States contaminated with hazardous materials may lie in the order of between $370 billion and $1.7 trillion.

Methods and procedures for mitigating some of the major impacts from pollutant stressors, together with treatment and remediation options, will be discussed in the later sections of this chapter. The discussion in this chapter recognizes that the impact from the presence of pollutants in the ground needs to be mitigated and managed — as a beginning step toward protection of the resources and the natural capital in the geoenvironment and as a necessary step toward achievement of a sustainable geoenvironment. The emphasis will be on using the properties and characteristics of the natural soil–water system as the primary agent for such purposes. The motivation for this is not because of the high expenditures incurred with the use of various technological remediation schemes and processes, but because it allows one to address pollution sources that encompass the range from point source to nonpoint source. Managing the impact from nonpoint source pollution with technological solutions can be prohibitive because of the extent of the source (if such is known), and the extent of pollution resulting from such a source. A good example of this is the transport of pollutants in the ground and on the ground surface in conjunction with pesticide and herbicide use. This is schematically illustrated in Figure 10.20 in the concluding section of this chapter (Section 10.8), where the point is made that more attention needs to be paid to the impacts from both atmospheric-based and land-based nonpoint sources of pollution on the health of both soil and water resources. If the transport of pollutants is to receiving waters such as streams and rivers, how does one use technological aids and engineered systems to manage and control the advance of the pollutants? Erecting barriers that run a certain length of the stream can be prohibitively costly. A good practical solution is to invoke the properties of the natural soil system as a partner in mitigation management. Amending and enhancing the properties to make it more effective as a control tool would also be a good tactic since this allows the subsoil to remain in place as a mitigation management tool. This tactic is now being used in a limited way in passive remediation treatment of contaminated sites.

10.2 Soils for Pollution Mitigation and Control

The latter part of Chapter 2 dealt with the nature and basic properties of soils as they relate to the transport and fate of pollutants in the soil. In this section, we will be dealing with the aspects of soils as they relate to resource materials for mitigation of impacts from containment and transport of pollutants and other hazardous substances. The importance of soil as a resource material for management of contaminants and waste products is due to its physical, mechanical, chemical, and biological properties. These properties constitute the basic tools for the many different strategies and measures available for passive and aggressive management of the land and water resources in the geoenvironment. The short discussion of these tools in Section 3.4.1 in Chapter 3 referred to the total actions of the various soil properties in management of waste streams as the *natural attenuation* process of soils. In this section, we will examine the basic properties and attributes of soils in respect to "why and how" they can function as tools for mitigation and management of waste streams in soils.

The properties of soil directly involved as a pollutant mitigation and control tool are:

1. Those that refer directly to the soil solids themselves. These are primarily the physical and mechanical properties of the soil, and also the surface properties of the soil solids responsible for sorption of pollutants. These include the density, macro- and microstructure, porosity and continuity of void spaces, exposed surface areas in the void channels, cation exchange capacity (CEC), and functional groups associated with the soil solids.

2. Those that depend on the interactions between the pollutants and the soil solids and the chemical constituents in the pore water. In this respect, the properties of the complete soil–water system become more important. These would be the chemistry of the pore water, the presence and types of inorganic and organic ligands in the pore water, pH and Eh or pE, exchangeable ions, and CEC. All these properties, together with the biological properties of the soil–water system, will define the initial state of the soil–water system, and hence the capability of the soil to react with incoming waste leachate streams and pollutants.

10.2.1 Physical and Mechanical Properties

In Section 2.4.3 in Chapter 2, we pointed out that the two primary types of interactions between the soil particles and liquid waste streams and pollutants in transport in the soil subsurface are physical and chemical in nature. The major physical interactions shown in Figure 2.10 in Chapter 2 apply to transport in the *in situ* state. For soil conditions both in the *in situ* state and in the prepared state (i.e., engineered soil barriers, for example), we need to have good mechanical performance characteristics from the soil. The physical and mechanical properties useful for mitigation and control of liquid and solid waste substances and pollutants are those that impede or prevent the flow or passage of liquid and solid substances. We need to distinguish between (1) the natural *in situ* soil condition, where the soil is in the landscape as a surface and subsurface soil, and (2) the situation where human intervention and manipulation of the soil is possible, i.e., placement of a prepared soil in the ground as an active mitigation or treatment tool. In the case of natural soils in the landscape and subsurface, transport of liquid waste streams, leachates, and pollutants is controlled by the *in situ* physical properties of the subsurface soil. Without human intervention, it becomes a case of *getting what the natural situation dictates*. Changes in the physical and mechanical properties of the subsoil that will likely occur because of chemical and physicochemical interactions with pollutants and contaminants will be discussed in the next subsection.

In the case where prepared soil is used as a sole treatment tool or as one of the tools in a designed mitigation treatment process, control on the soil physical and mechanical properties can be exercised. At this stage, the design physical and mechanical properties of the soil to be used are important factors in the mitigation and prevention of pollutant transport. The soil properties of significance include physical properties such as density, permeability, and porosity, and mechanical properties such as compactibility, compressibility, consolidation, and strength. All of these properties depend on the texture, grain (particle) morphology, particle size distribution, and composition. These all combine to control the packing of the particles and density of the compacted material. Figure 10.1 shows the relationship between all of these and the physical and mechanical properties obtained in relation to the compactibility of the soil material. The mechanical properties are of importance in liner, buffer, and barrier systems, and the physical properties feature prominently in fluid and gas transport through the soil. The common assumption that transport of contaminants and pollutants is halted when fluid transport is stopped is wrong since transport mechanisms for contaminants and pollutants are via diffusive processes. As long as there is water in the soil barrier system, diffusive transport will occur. The water in the soil, even if it is immobile, serves as the carrier for the contaminants and pollutants.

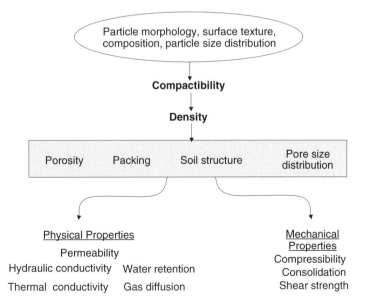

FIGURE 10.1
Control of physical and mechanical properties by soil composition and particle properties and characteristics.

10.2.1.1   Soil Microstructure Controls on Hydraulic Transmission

Figure 2.10 in Chapter 2 illustrates the main points of physical interactions between a liquid waste stream and the soil particles through which it permeates. The physical properties of the soil immediately involved in defining the nature of the fluid permeation are known as the transmission properties of the soil. These are essentially linked to the permeability of the soil to aqueous and gaseous phases, as shown in the bottom left compartment in Figure 10.1. Considering only fluid flow, the factors that affect hydraulic conductivity can be conveniently divided into three distinct categories: (1) external environmental factors, such as hydraulic head and temperature; (2) fluid-phase factors; and (3) soil structural factors. The fluid-phase factors of significance in the rate of fluid movement in the soil include the density, viscosity, and chemistry of the solutes contained in the fluid phase. Soil structural features are very influential in controlling flow rate and partitioning of pollutants. These include the microstructure and the micropores in the soil, the pore size distribution, and the continuity of pores. All of these are functions of the density of the soil.

For any given density of soil, there is an almost infinite number of arrangements of soil particles in a typical unit volume of soil. The sketches shown in Figure 2.9 and Figure 2.10 in Chapter 2 indicate that individual particles acting as single units are rarely found — except for granular soils. Figure 10.2 shows a scanning electron microscopy (SEM) picture of a typical clay soil unit composed of aggregate groups of clay particles, and depending on the sizes of these groups, they are generally called domains, clusters, peds, or micro-structural units. Because of the variety in sizes, we will use the general term *microstructural units* in our discussion of the aggregate groups that make up the microstructure of the soils. The importance of soil structure in defining flow through a soil is evident from Figure 10.2. Since flow occurs through void spaces that are connected, it is evident that the nature of the void spaces and how these spaces are connected will be influential in determining the flow rate of the liquid waste and pollutants. Greater densities of soil will show smaller void spaces. Note that the micropores in the microstructural units will not show the same characteristics of flow as found in the macropores, i.e., the pores between peds. Because of the infinite variations in sizes and types of microstructural units and their distribution, it would not be surprising to find that soils with similar compositions can have different densities and correspondingly different hydraulic conductivities. If soil is to be used as a tool to control flow and distribution of pollutants, i.e., to mitigate pollution, it is important to determine what key factors are involved in controlling pollutant partitioning and distribution in the soil.

Soil permeability to liquids is determined as the hydraulic conductivity of a soil. This holds true for saturated soils. However, for unsaturated soils, movement of liquids is generally identified as diffusive flow, even though this may not be exactly correct. The permeability of a soil is generally expressed in terms of a permeability coefficient, k. The common technique is to perform laboratory permeability tests, as shown for example in Figure 10.3, which depicts a constant head permeability test. Procedures for conducting permeability tests using constant head and falling head techniques, and also with flexible wall permeameters, have been written as standards, e.g., ASTM D5084-03 (2003). While the double-Mariotte tube system shown in Figure 10.3 is not the prescribed or specified system for administering the constant hydraulic head for permeameter tests, it is nevertheless a proper and useful system to use. It permits flexibility in adjusting the hydraulic head required for constant head permeation. The Darcy coefficient of permeability, k, is obtained from the relationship $Q = kiA = k(h/L)A$. The hydraulic gradient, i, is the ratio of the hydraulic head h and L, the spatial distance, and A is the cross-sectional area of the test sample.

Since the Darcy model for determination of the permeability coefficient k from experiments such as those shown in Figure 10.3 does not consider the properties of the permeant

FIGURE 10.2
Scanning electron microscopy (SEM) picture showing typical aggregate grouping of particles forming the structure of a clay soil. The black band in the bottom middle portion of the picture represents a scale of 10 μm. Note the variety and sizes of voids, ranging from microvoids in the aggregate group (cluster, ped) to the macrovoids between aggregate groups of particles.

and the microstructure of the soil, Yong and Mulligan (2004) have proposed a relationship that uses a modification of the combined form of the Poiseuille and Kozeny–Carman (PKC). This takes into account the influence of the properties of pore channels defined by the structure of a soil, and the fact that the wetted soil particles' surface area is controlled by the microstructure of the soil. The relationship obtained is shown as

$$v = k * i = \frac{C_s n^3 \gamma}{\eta T^2 S_w^2} \frac{\Delta \psi}{\Delta l} \tag{10.1}$$

where k^* = PKC permeability coefficient (which considers permeant and soil microstructure properties) $= \dfrac{C_s n^3 \gamma}{\eta T^2 S_w^2}$. C_s = shape factor, with values ranging from 0.33 for a strip cross-

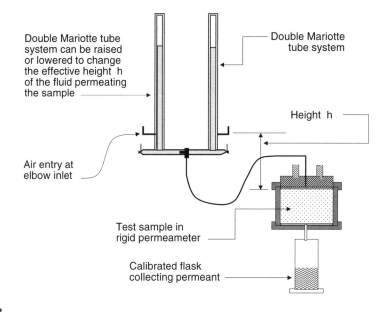

FIGURE 10.3
Water entry experiment with constant hydraulic head, *h*, for permeating fluid in a rigid permeameter.

sectional face to 0.56 for a square face. Yong and Warkentin (1975) have suggested that a value of 0.4 for C_s may be used as a standard value, with a possible error of less than 25% in the calculations for an applicable value of k^*. i = hydraulic gradient = ratio of the potential or hydraulic head difference between the entry and exit points of the permeant, and the direct path length Δl of the soil mass being tested. T = tortuosity = ratio of effective flow path Δl_e to thickness of test sample Δl, and which is quite often taken to be $\approx \sqrt{2}$. δ and η = density and viscosity of the permeating fluid, respectively. n = porosity of the unit soil mass. S_w = wetted surface area per unit volume of soil particles.

Equation 10.1 uses the soil property parameters C_s, T, and S_w in structuring the relationship that describes permeability of a soil. These soil property parameters are dependent on soil composition and soil structure. Assuming that the physical properties of a leachate permeant are not too far distant from those of water at about 20°C, and further assuming a tortuosity T value of $\sqrt{2}$ and $C_s = 0.4$, the graphical relationships shown in Figure 10.4 and Figure 10.5 can be obtained. These graphs show the relationship between the PKC permeability coefficient, k^*, and the amount of surface area wetted in fluid flow through the soil. A comparison of the calculated wetted surface areas, S_w, for the soils shown in Figure 10.5 shows that the wetted surface areas vary from about 3 to 7% of the specific surface area of the soils. This indicates that microstructural units such as those shown in Figure 10.2 encompass a large number of soil particles — to the extent that the effective surface areas presented to a contaminant leachate stream represents only a small fraction of surface areas present. For soils that are prepared for use as barriers and liners, the sizes and distribution of microstructures are significant factors in determining the effectiveness of the barriers and liners.

10.2.1.2   Microstructure, Wetted Surfaces, and Transport Properties

Hydraulic conductivity through a soil engineered barrier is greatly facilitated when interconnected voids and their connecting channels are large. Large voids in a compact soil generally mean large grain sizes or large microstructural units. In addition to the advantage of larger flow paths, large interconnected voids generally mean that the surfaces

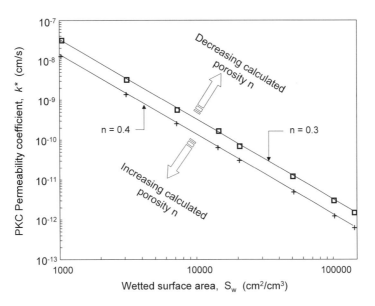

FIGURE 10.4
Variation of PKC permeability coefficient, k^*, with wetted surface area, S_w, and calculated porosity, n.

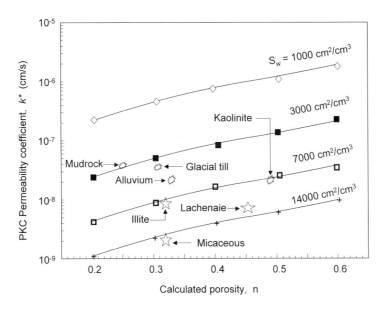

FIGURE 10.5
Variation of PKC permeability coefficient, k^*, with calculated porosity — in relation to the wetted surface area, S_w.

presented as the surrounding surfaces in the void spaces are lesser than if the void volumes were smaller. The surrounding flow path contact surfaces are important factors because they (1) offer drag or boundary resistance to flow and, more importantly (2) provide the surface areas and corresponding functional groups for chemical reactions that include sorption, ion exchange, and complexations. We will discuss this latter aspect in greater detail in the next section.

Figure 10.6 shows two scanning electron micrographs of the same clay soil. The clay soil was obtained as a core sample from a recently cut slope in northern Québec, Canada, where the winters can be cold and harsh. The picture on the left shows the structure of the soil

Width of black
band is 10 μm

Same clay soil sample after 32 cycles
of freeze-thaw. Note development of
large microstructural units.

Natural clay soil before
freeze-thaw cycles

FIGURE 10.6
SEM picture showing formation of large microstructural units in a natural clay soil after 32 cycles of freezing and thawing.

obtained from a core sample to consist of small microstructural units apparently uniformly distributed in the cross-section. The right-hand picture shows the same soil in the thawed state after 32 cycles of freezing and thawing (Yong et al., 1984). The dramatic increase in sizes of the microstructural units is obvious, testifying to the marked decrease in surface areas presented to a permeating fluid, and also to the significant increase in void spaces. Transmission or transport through the soil material shown on the right-hand side will be considerably facilitated by the freeze–thaw effect. The lesson to be learned from the pictures in Figure 10.6 is that environmental effects can alter the initial conditions to such an extent that design of mitigation and treatment procedures must anticipate such events.

10.2.2   Chemical Properties

The chemical properties of significance include those that promote ion exchange, sorption and precipitation of solutes in the fluid phase (including pore water) in the soil, and complexation. These are properties that are more appropriately defined as soil–water system properties. These have been discussed briefly in Section 2.5.1 in Chapter 2 in respect to partitioning processes involving heavy metals, and in Sections 9.4 and 9.5 in Chapter 9. To fully utilize soil as a resource material for management of waste leachate streams and pollutants, a broader discussion of the important chemical properties and interactions between pollutants and soil particles or fractions is needed.

10.2.2.1   Sorption

As discussed in the previous chapter, sorption processes involving molecular interactions are (1) Coulombic in nature, (2) interactions between nuclei and electrons, and (3) essentially electrostatic in nature. The major types of interatomic bonds are ionic, covalent, hydrogen, and van der Waals. Ionic forces hold together the atoms in a crystal. The various types of bonds formed from various types of forces of attraction include (1) ionic bonds, i.e., electron transfer between the atoms, which are subsequently held together by the opposite charge attraction of the ions formed; (2) covalent bonds, developed as a result of electron sharing between two or more atomic nuclei; and (3) Coulombic bonds, developed from ion–ion interaction.

For interactions between instantaneous dipoles, we have the three types of *van der Waals* forces: Keesom, Debye, and London dispersion forces. Bondings developed by van der Waals forces are by and large the most common type of bonding between organic chemicals and mineral soil fractions. Electrical bonds can be formed between negatively charged organic acids and positively charged clay mineral edges. Sorption of organic anions can occur if polyvalent exchangeable cations are present. The polyvalent bridges formed will be due to (1) an anion associated directly with a cation or (2) an anion associated with a cation in the form of a cation bridge (water bridge).

10.2.2.2 Cation Exchange

Cation exchange involves those cations associated with the negative charge sites on the soil solids, largely through electrostatic forces. Ion exchange reactions occur with the various soil fractions, i.e., clay and nonclay minerals. This process, which has been discussed in detail in Section 9.4.2 in Chapter 9, is set in motion because of the need to satisfy electroneutrality and is stoichiometric. Calculations or determinations of the proportion of each type of exchangeable cation to the total cation exchange capacity of the soil can be made using exchange equilibrium equations, such as the Gapon relationship shown in Equation 9.1.

From the electrostatic point of view, physical adsorption (or sorption) of pollutants in the pore water (or from incoming leachate) by soil fractions is due to the attraction of positively charged pollutants such as the heavy metals to the negatively charged surfaces of the soil fractions. As we have pointed out in Section 9.4.2, this type of adsorption is called nonspecific adsorption. By definition, we can refer to *nonspecific adsorption* as the case where ions are held by the soil particles primarily by electrostatic forces. This distinguishes it from *specific adsorption*, which is another way of identifying *chemisorption*, a process that involves covalent bonding between the pollutant and the soil particle (generally mineral) surface. Examples of nonspecific adsorption are the adsorption of alkali and alkaline earth cations by the clay minerals. By and large, cations with smaller hydrated size or large crystalline size would be preferentially adsorbed.

10.2.2.3 Solubility and Precipitation

The pollutants affected by solubility and precipitation processes are mostly heavy metals. The pH of the soil–water system plays a significant role in the fate of heavy metal pollutants because of the influence of pH on the solubility of the heavy metal complexes. According to Nyffeler et al. (1984), the pH at which maximum adsorption of metals occurs varies according to the first hydrolysis constant of the metal (cationic) ions. When the ionic activities of heavy metal solutes in the pore water of a soil exceed their respective solubility products, precipitation of heavy metals as hydroxides and carbonates can occur. The two stages in precipitation are nucleation and particle growth. This will generally be under slightly alkaline conditions. The precipitate will either form a new separate substance in the pore water or be attached to the soil solids. Gibbs phase rule restricts the number of solid phases that can be formed.

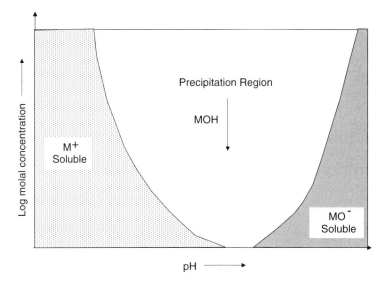

FIGURE 10.7
Solubility–precipitation chart for a metal hydroxide complex.

Factors involved in formation of precipitates include soil–water system pH, type and concentration of heavy metals, presence of inorganic and organic ligands, and the individual precipitation pH of heavy metal pollutants. In the solubility–precipitation diagram shown in Figure 10.7 for a metal hydroxide complex, the left-shaded area marked as *soluble* identifies the zone where the metals are in soluble form with positively charged complexes formed with inorganic ligands. The right-shaded soluble area contains the metals in soluble form with negatively charged compounds. The precipitation region in between the two shaded areas contains various metal hydroxide species. Figure 10.8 shows heavy metal precipitation information using data reported by MacDonald (1994). Transition from soluble forms to precipitate forms occurs over a range of pH values for the three heavy metals, and that onset of precipitation can be as early as a pH of about 3.2 in the case of the single heavy metal species (Pb). The process of precipitation is a continuous process that begins with onset at some early pH and finally concludes at some pH value — generally around pH 7 for most metals. The influence of other metal species in the precipitation process is felt not only in terms of when onset pH occurs, but also in the rate of precipitation in relation to pH change. Figure 10.8 shows that the onset of precipitation of Zn as a single species is about pH 6.4, and that reduces to about pH 4.4 when other metals are present. Given that the experiments were conducted with equal amounts of each of the three heavy metals, it is expected that the concentrations of the other metals would also have an effect on modification of the onset pH. We must note that the precipitation boundaries are not distinct separation lines, and that transition between the two regions or zones occurs in the vicinity of the boundaries throughout the entire pH range.

The role of pH in the soil–water system is important because of the various complexes formed in relation to pH. For example, when a heavy metal contaminant solution such as a $PbCl_2$ salt enters a soil–water system at pH values below the precipitation pH of Pb, a portion of the metals will be adsorbed by the soil particles. The ions remaining in solution would either be hydrated or form complexes such as Pb^{2+}, $PbOH^+$, and $PbCl^+$. These would be contained in the left-shaded area of Figure 10.7. When the pH is raised to the pH levels shown in the right-shaded area of Figure 10.7, one would form complexes such as PbO_2H^- and PbO_2^{2-} that would reside in the right-shaded area.

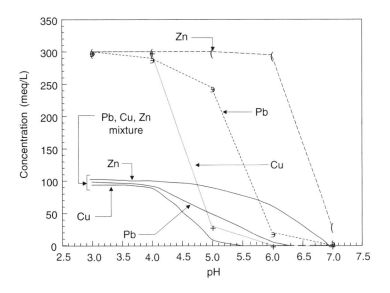

FIGURE 10.8
Precipitation of heavy metals Pb, Cu, and Zn in aqueous solution. Bottom curves are precipitation of individual metal from a mixture of Pb, Cu, and Zn in equal proportions of metal nitrate solution (100 meq each). Top curves are for single solutions of individual metals at 300 meq/l concentration. (Data from MacDonald, E., Aspects of Competitive Adsorption and Precipitation of Heavy Metals by a Clay Soil, M.Eng. thesis, McGill University, Montreal, Canada, 1994.)

10.2.2.4 Speciation and Complexation

The processes associated with speciation and complexation apply primarily to inorganic pollutants in the liquid waste streams, particularly the heavy metals. The term *speciation* refers to the formation of complexes between heavy metals and ligands in the aqueous phase. Ligands are defined as anions that can form coordinating compounds with metal ions. Inorganic and organic ligands include, for example, CO_3^{2-}, SO_4^{2-}, Cl^-, NO_3^-, OH^-, SiO_3^{2-}, PO_4^{3-}, and humic and fulvic acids. For the complexes formed between heavy metals and humic and fulvic acids, these would generally be chelated compounds. From the preceding, we note that the chemistry of the pore water in the soil–water system is an important factor in mitigation and control of pollutant transport in the soil–water system — through competition between the ligands and the soil solids for sorption of the heavy metals.

10.2.3 Biological Properties

Biological properties of soils (soil–water systems) are very important factors in the passive and aggressive treatment and management of organic chemical pollutants in the subsurface soil regime. These properties are determined by the large variety of microorganisms that reside in the soil–water system. These microorganisms consist of viruses, bacteria, protozoa, fungi, and algae. Microorganisms are the key to biological treatment and management of contaminants and pollutants.

10.2.3.1 Protozoa

Protozoa are aerobic, single-celled chemoheterotrophs. They are classified as eukaryotes with no cell walls and with sizes that vary from 1 to 2000 mm. They include pseudopods, flagellates, amoebas, ciliates, and parasitic protozoa. The four primary groups of protozoa are (1) Mastigophora, flagellate; (2) Sarcodina, amoeboid; (3) Ciliophone, ciliated; and (4) Sporozoa, parasites of vertebrates and invertebrates. Protozoa are found in water and soil, feed on bacteria, and need water in order to move. Although they do not generally biodegrade contaminants, they are useful in reducing bacterial numbers near injection wells that become clogged due to excessive growth. Soil protozoa are heterotrophic, and although their general food source is bacteria, they are known to feed on soluble and even insoluble organic material. They mineralize nutrients and release excess nitrogen as NH_4^+, which is beneficial to plants and others on the food web.

10.2.3.2   Fungi

Fungi include slime molds (filamentous fungus), yeasts, and mushrooms, and are aerobic, multicellular eukaryotes and chemoheterotrophs that require organic compounds for energy and carbon. They are larger than bacteria and do not require as much nitrogen. They grow more slowly and in a more acidic pH range than bacteria, and are more sensitive to changes in moisture levels. Yeasts are unicellular organisms that are larger than bacteria and are shaped like eggs, spheres, or ellipsoids.

10.2.3.3   Algae

Algae are single-celled and multicellular microorganisms and are considered to be the abundant photosynthetic microorganisms in soils. According to Martin and Focht (1977), (1) the availability of inorganic nutrients such as C, N, P, K, Fe, Mg, and Ca is said to be responsible for soil algae's abundance, and (2) their principal functions in soil are nitrogen fixation, colonization of new rock and barren surfaces, supplying of organic matter and nitrogen for humus formation, weathering of rocks and minerals, and binding of soil particles through surface bonding.

10.2.3.4   Viruses

Viruses are the smallest type of microbe and can be 10,000 times smaller than bacteria. They require a living cell to reproduce. It is said that their primary function is to reproduce, and they do it well by taking over a host cell. The have a direct influence on bacterial abundance, and through lysis (cell destruction) and transduction, i.e., transfer of viral DNA from one cell to another through viruses that attack bacteria (bacteriophages), they can alter bacterial genetic diversity. Beyond their direct attack on the various microbial cells and their influence on community composition, their other functions in soil are not very well known or established.

10.2.3.5   Bacteria

Bacteria are single-celled microorganisms that vary in size and shape from very small spheres to rods that can vary from 1 μm to a few microns in length and width. There are literally many thousands of different bacterial species coexisting in the soil. With favorable conditions of temperature and nutrient availability, it is reported that the bacterial population in soil can be in the order of 10^8 to 10^{10} per gram of soil. They are both autotrophic and heterotrophic. Most bacteria used for bioremediation treatment of organic chemicals are chemoorganotrophs and heterotrophs. Those requiring organic substrates for energy are called chemoorganotrophs, and those using organics as a carbon source are called heterotrophs. Those that use inorganic compounds as an energy source are named chemolithotrophs. Nitrifying bacteria (*Nitrosomonas* and *Nitrobacter*) that use carbon dioxide as a carbon source, instead of organic compounds, are called autotrophs. Nitrifying bacteria produce nitrite from an ammonium ion, which is then followed by conversion to nitrate.

10.3   Natural Attenuation Capability of Soils

By definition, the reduction of toxicity and concentration of contaminants in a contaminant plume during transport in the subsurface soil is called *contaminant attenuation*. We use the general term *contaminants* to include pollutants and all other kinds of hazardous substances in the fluid phase of the soil–water system. If the various processes responsible for contaminant attenuation are naturally occurring, the attenuation process is said to be the result of the *natural attenuation capability* (i.e., *assimilative capacity*) of the subsurface soil. What are these naturally occurring attenuation processes? These are the physical, chemical, and biological properties discussed in the previous section. They all contribute to the assimilative capacity of soil, i.e., the capacity of the soil to attenuate the flux of contaminants through processes that include physical, chemical, and biologically mediated mass transfer and biological transformation.

The American Society for Testing and Materials (ASTM) (1998) defines natural attenuation as the "reduction in mass or concentration of a compound in groundwater over time or distance from the source of constituents of concern due to naturally occurring physical, chemical, and biological processes, such as; biodegradation, dispersion, dilution, adsorption, and volatilization." The U.S. Environmental Protection Agency (U.S. EPA, 1999), on the other hand, considers natural attenuation specifically in the context of a monitored scheme for treatment of polluted sites. Accordingly, it uses the term *monitored natural attenuation* and defines it as

> the reliance on natural attenuation processes (within the context of a carefully controlled and monitored site cleanup approach) to achieve site-specific remediation objectives within a time frame that is reasonable compared to that offered by other more active methods. The "natural attenuation processes" that are at work in such a remediation approach include a variety of physical, chemical, or biological processes that, under favorable conditions, act without human intervention to reduce the mass, toxicity, mobility, volume, or concentration of contaminants in soil or groundwater. These *in-situ* processes include biodegradation; dispersion; dilution; sorption; volatilization; radioactive decay; and chemical or biological stabilization, transformation, or destruction of contaminants.

In the specific context of pollutants, contaminants, and transport of liquid wastes, leachates, etc., in the subsoil, the reduction and detoxification of all of these contaminants from processes associated with natural attenuation is termed *intrinsic remediation*. More specifically, reduction in concentration of contaminants is by processes of partitioning and dilution, and reduction in toxicity of the contaminants is generally achieved by biological transformation (of organic chemicals) and sequestering of the toxic inorganic contaminants. The test results reported by Coles and Yong (2004) in Figure 10.9 show the importance of soil composition on the retention of lead (Pb) and cadmium (Cd). In the particular case shown in the figure, humic matter in the form of fulvic acid, a sulfide mineral called mackinawite (Fe, Ni)$_9$S$_8$, and kaolinite were used as control soil material.

One speaks of the natural attenuation of contaminants as being a set of positive processes that mitigate the impact of contaminants in the ground through a reduction of their

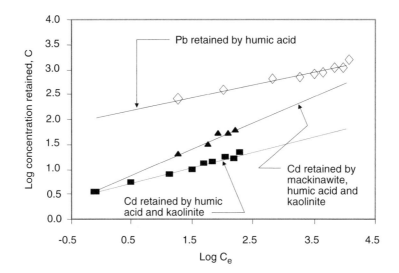

FIGURE 10.9
Freundlich-type adsorption isotherms are showing the retention of Pb humic acid; Cd by kaolinite and humic acid; and mackinawite (Fe, Ni)$_9$S$_8$, humic acid, and kaolinite. Solutions of Pb and Cd at pH 4.3 were obtained as PbCl$_2$ and CdCl$_2$, respectively. (Data from Coles, C.A. and Yong, R.N., in *Geoenvironmental Engineering: Integrated Management of Groundwater and Contaminated Land*, Yong, R.N. and Thomas, H.R., Eds., Thomas Telford, London, 2004, pp. 20–28.)

intensity — as measured in terms of concentration and toxicity of the contaminants. In the past, the use of natural attenuation processes had been considered almost exclusively in connection with remediation of contaminated sites — and more specifically with sites contaminated with organic chemical pollutants, as shown for example in the definition provided by the U.S. EPA. Little distinction was made between intrinsic remediation and intrinsic bioremediation. More recently, with a better appreciation of the assimilative capacity of soils, and especially in view of a growing body of research information on contaminant–soil interactions, more attention is being paid to the use of natural attenuation as a tool for mitigating and managing the transport and fate of contaminants in the ground. A contributing factor has also been the accelerating costs for application of aggressive remediation techniques to treat contaminated sites. The reader is reminded that *contaminants* include all the polluting and health-threatening elements entering into the ground, such as pollutants, toxicants, leachates, liquid wastes, hazardous substances, etc.

10.3.1   Natural Attenuation by Dilution and Retention

We have, up to now, considered natural attenuation of contaminants as being due to the processes associated with physical, chemical, and biological properties of soil. We have considered that these properties contribute directly to the partitioning of contaminants, i.e., the transfer of contaminants in the pore water to the surfaces of the soil solids. Strictly speaking, there is another set of processes that arguably can be considered part of the natural attenuation capacity of soils, except that we would now have to refer to this as the natural attenuation capacity of soil–water systems. While the groundwater and pore water aspects of the soil–water system have heretofore been considered only in respect to their physical and chemical interactions with the soil solids and contaminants, they attenuate contaminants through processes of dilution. Thus, in addition to the processes previously described, reduction in concentrations of contaminants can be accomplished by dilution through mixing of the contaminants with uncontaminated or less contaminated groundwater. In total, natural attenuation of contaminants in soils includes (1) dilution, (2) interactions and reactions between contaminants and soil solids, resulting in partitioning of the contaminants between the soil solids and pore water, and (3) transformations that reduce the toxicity threat posed by the original pollutants. The likelihood of only one mechanism being solely responsible for attenuation of pollutants in transport in the soil is very remote. In all probability, all the various processes or mechanisms will participate to varying degrees in the attenuation of pollutants, with perhaps partitioning being by far the more significant factor in attenuation of contaminants.

10.3.1.1   Dilution and Retention

In the context of contaminant transport in soils, *dilution* refers to the reduction in concentration of contaminants in a unit volume as a result of a reduction of the ratio of number of contaminants, n_c, to the volume, V, of the host fluid. An example of this would be when the original contaminant load is given as 100 ppm, and dilution with groundwater reduces this to 50 ppm. The singular process responsible for the decrease or reduction in concentration is dilution, and transport in the subsoil will likely be consistent with the advective velocity of the groundwater. Except for physical controls on groundwater flow, no other soil properties are involved in the dilution process.

Retention refers to the retention of contaminants by the soil solids through partitioning processes that involve physicochemical and chemical mass transfer. The result of retention is a decrease in the concentration of contaminants in a leachate plume or liquid waste stream as one progresses away from the source. Using the same numerical example as above, one would see a reduction from 100 to 50 ppm in a contaminant transport stream at a point farther downstream from the contaminant source. The difference between this and the previous example is that when reduction of concentration is obtained through retention processes, the contaminants retained will not be readily available for transport. In the case of dilution as a means for reduction in concentration of contaminants, there are no contaminants held by the soil particles. All the contaminants will be delivered downstream in due time. Figure 10.10 illustrates the differences using assumed ideal bell-shaped concentration distribution pulses. In the case of dilution, the diagram shows that eventually, the total contaminant load will be delivered downstream. The areas of the assumed bell-shaped dilution pulses are all constant and equal to the original rectangular distribution shown on the ordinate. In contrast, the assumed bell-shaped retention pulses will show decreasing areas. These retention pulses will diminish to zero as long as the assimilative capacity of the soil is not exceeded. This is a significant point of consideration in the use of the natural attention capacity of soils for mitigation and management of impacts from liquid waste and contaminant discharges.

10.3.2   Biodegradation and Biotransformation

The common perception is that biological activities in the subsoil will degrade organic chemical compounds. Not always understood or perceived is whether this does in fact contribute to attenuation of the concentration and toxicity of the contaminants in the subsoil. Figure 10.11 provides the overall view of various attenuating mechanisms in the soil. The top left-hand corner of the diagram shows the biologically mediated transfer mechanisms participating in the attenuation process. Not well illustrated, or sometimes not fully acknowledged, are the redox reactions (this includes both abiotic and biotic redox reactions) listed under the "Chemical Mass Transfer" box in the bottom right corner of

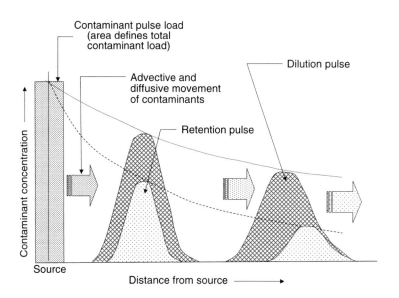

FIGURE 10.10
The difference between retardation and retention of contaminants during transport of a contaminant pulse load. Note that the areas of the assumed bell-shaped dilution pulses are constant, and that they are equal to the original contaminant pulse load. The areas of the assumed bell-shaped retention pulses diminish as one progresses farther away from the source.

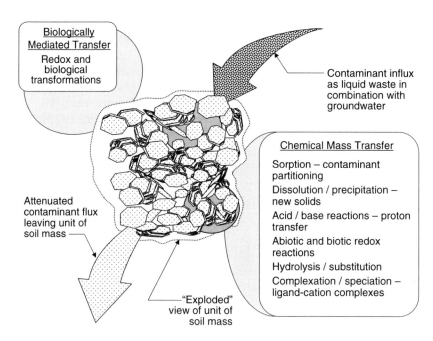

FIGURE 10.11
Processes involved in attenuation of pollutants in leachate transport through a soil element. Dilution of contaminants is not included in the schematic illustration. (Adapted from Yong, R.N. and Mulligan, C.N., *Natural Attenuation of Contaminants in Soils*, Lewis Publishers, Boca Raton, FL, 2004.)

the diagram. The discussions in the latter sections of this chapter will provide case histories dealing with treatments using various procedures and techniques. Among these will be the use of intrinsic bioremediation in management of pollutant transport. Before embarking on these discussions, it is useful to clarify the use of the term *substrate* in the discussions to follow. In the context of biological activities in the subsoil, the term *substrate* is used to mean the food source for microorganisms. This term should not be confused with the term *substrate* used in some soil mechanics and geotechnical engineering literature to mean subsurface soil stratum. The use of *substrate* as a term here is consistent with usage in microbiology, and is meant to indicate that it (substrate) serves as a nutrient source and also a carbon or energy source for microorganisms.

By definition, *biodegradation* refers to the decomposition of organic matter by microorganisms. The end result of the metabolic and enzymatic processes is seen in terms of smaller compounds, and in respect to organic pollutant remediation goals, ultimately as CO_2, CH_4, and H_2O. The organic wastes and chemicals of interest in this book are pollutants such as organic compounds that contain carbon, and organic waste matter. Strictly speaking, biodegradation is a particular form of biotransformation, since *biotransformation* (or biological transformation) means the conversion of a chemical substance into another chemical substance (generally a metabolite) by enzymatic action or other biological processes. The use of the term *biodegradation* implies that the biological transformation process reduces the original organic chemical compound into smaller fragments, with the presumed conclusion that these smaller fragments are less toxic or less threatening to the receptors. To some extent, this is probably valid. However, the well-documented and well-reported example of anaerobic degradation of C_2Cl_4 (perchloroethylene, or PCE) to C_2H_3Cl (vinyl chloride) shows that the latter degraded compound is more of a threat than the original, because C_2H_3Cl is more toxic and volatile than PCE and does not partition well in the soil. In any event, the transformation of original organic chemical compounds in the subsurface soil into smaller units occurs by oxidation and reduction mechanisms (redox reactions) resulting from the metabolic activities of the microorganisms in the soil. In the next few subsections, we will show examples from Yong and Mulligan (2004) of biotransformation and biodegradation of some organic chemical compounds. Greater details of the many kinds of transformations and conversions can be found in specialized texts dealing with bioremediation.

10.3.2.1 *Petroleum Hydrocarbons: Alkanes, Alkenes, and Cycloalkanes*

Petroleum hydrocarbon consists of various compounds, such as alkanes, cycloalkanes, aromatics, polycyclic aromatic hydrocarbons, asphaltenes, and resins. Their biodegradability in the subsoil ranges from very biodegradable to recalcitrant. This is because of the varying degrees of branching, chain lengths, molecular sizes, and substitution with nitrogen, oxygen, or sulfur atoms. Many of the alkanes found in petroleum are branched.

Alkanes (C_2H_{2n+2}) are aliphatic compounds. Low-molecular-weight alkanes are the most easily degraded by microorganisms. However, as the chain length increases from C_{20} to C_{40}, hydrophobicity increases while solubility and biodegradation rates decrease. Conversion of alkanes leads to the formation of an alcohol using a monooxygenase enzyme. This is followed by oxidation to an aldehyde and then to a fatty acid (Pitter and Chudoba, 1990). Further oxidation (β-oxidation) of the fatty acid yields products less volatile than the original contaminants. Anaerobic bacteria such as sulfate-reducing bacteria are capable of degrading fatty acids via this step (Widdel, 1988). Although bacteria that are capable of degrading *n*-alkanes cannot degrade branched ones (Higgins and Gilbert, 1978), *Brevibacterium ethrogenes*, *Corynebacteria* sp., *Mycobacterium fortuitum*, *Mycobacterium smegmatis*, and *Nocardia* sp. have shown to grow on branched alkanes. The first degradation step is the same as that for the unbranched alkanes. However, the β-oxidation is more difficult and less efficient (Pirnik, 1977). In addition, in the presence of *n*-alkanes, the metabolism of the branched alkanes will be repressed, which will cause difficulties during the degradation of mixtures such as petroleum.

Alkenes with a double bond between carbons have not been extensively studied for biodegradation. Those containing the double bond on the first carbon may be more easily degradable than those alkenes with the double bond at other positions (Pitter and Chudoba, 1990). The products of oxidation of 1-alkenes can be either diols or the methyl group.

Because of their cyclic structure, cycloalkanes are not as degradable as alkanes, and they become less degradable as their number of rings increases. Pitter and Chudoba (1990) attribute some of this to their decreasing solubility. Species of *Nocardia* and *Pseudomonas* are able to use cyclohexane as a carbon source. Oxidation of the cycloalkanes with the oxidase enzyme leads to production of a cyclic alcohol and then a ketone (Bartha, 1986).

10.3.2.2 Gasoline Components BTEX and MTBE

Benzene, toluene, ethylbenzene and xylene (BTEX) are volatile, water-soluble, hazardous components of gasoline. Aerobic degradation of all components of BTEX occurs rapidly with available oxygen. Under anaerobic conditions, degradation is less reliable and is slower than under aerobic conditions. Bacterial metabolism proceeds through a series of steps depending on the availability of electron acceptors.

The gasoline additive methyl tert-butyl ether (MTBE) is highly resistant to biodegradation since it is reactive with microbial membranes. Some believe that it is slowly biodegraded (Borden et al., 1997), while others believe that it partially degrades to tert-butyl alcohol, a health hazard (Landmeyer et al., 1998).

10.3.2.3 Polycyclic Aromatic Hydrocarbons

Polycyclic aromatic hydrocarbons (PAHs), $C_{4n+2}H_{2n+4}$, are components of creosote. As with cycloalkanes, they are difficult to degrade, and as the number of rings increases, the compounds become more difficult to degrade — a result of their decreasing volatility and solubility, and increased sorption. They are degraded one ring at a time. As an example, the pathway for biodegradation of anthracene is from anthracene *cis*-1,2-dihydrodiol to salycilate with at least six intermediates, beginning with 1,2-dihydroxy anthracene onward to 1-hydroxy-2-naphthoic acid as the last intermediate before salycilate.

10.3.2.4 Halogenated Aliphatic and Aromatic Compounds

Halogenated aliphatic compounds include (1) pesticides such as ethylene dibromide (DBR), $CHCl_3$, or $CHCl_2Br$ and (2) industrial solvents such as methylene chloride and trichloroethylene. Because of the presence of halogen, the lower-energy and higher-oxidation state makes aerobic degradation more difficult to achieve than anaerobic biodegradation. Methylene chloride, chlorophenol, and chlorobenzoate are the most aerobically biodegradable. Removal of the halogen and replacement by a hydroxide group is often the first step of the degradation process, particularly when the carbon chain length is short. An example of this is methylene chloride, with formaldehyde, 2-chloroethanol, and 1,2-ethanediol as intermediates and carbon dioxide as the final product (Pitter and Chudoba, 1990).

Biodegradation of chlorinated ethenes involves formation of an epoxide and hydrolysis to carbon dioxide and hydrochloric acid. Reductive dehalogenation can occur anaerobically and involves replacement of the halogen with hydrogen or formation of a double bond when two adjacent halogens are removed (dihalo-elimination). This is the particular case for perchloroethylene (PCE). As discussed previously, PCE and trichloroethylene (TCE) can be reduced to form vinylidene and vinyl chloride (VC) that are more toxic and volatile than the original compound. Oxidation of vinyl chloride to carbon dioxide and water occurs under aerobic conditions. Induction of monooxygenase or dioxygenase enzymes can lead to the cometabolism of TCE by methanotrophs (Alvarez-Cohen and McCarty, 1991). However, molecular oxygen and a primary substrate (methane, ethene, phenol, toluene, or other compounds) must be available for natural attenuation by this mechanism.

Halogenated aromatic compounds include pesticides such as DDT, 2,4-D, and 2,4,5-T, plasticizers, pentachlorophenol, and polychlorinated biphenyls. Although polychlorinated biphenyls (PCBs) have been banned since the 1970s, the record shows that they are still found in aqueous and sediment systems. Congeners containing fewer chlorines are degraded more quickly than those with more than four chlorine atoms (Harkness et al., 1993). Soluble forms are much more likely to biodegrade through natural attenuation than those sorbed to solids or entrapped in nonaqueous phase liquids (NAPLs). Mechanisms involved in transformations and conversions of halogenated aromatic compounds include biodegradation, hydrolysis (replacement of halogen with a hydroxyl group), reductive dehalogenation (replacement of halogen with hydrogen), and oxidation (introduction of oxygen into the ring, causing removal of halogen). As the number of halogens rises, reductive dehalogenation will occur. In addition, ring cleavage could occur before oxidation, reduction, or substitution of the halogen.

Bacterial strains of *Pseudomonas* sp., *Acinetobacter calcoaceticus*, and *Alkaligenes eutrophus* have been able to degrade aromatic halogenated compounds by oxidizing them to halocatechols followed by ring cleavage (Reineke and Knackmuss, 1988). Cleavage for chlorobenzene, for example, can occur either at the ortho position to form chloromuconic acid or at the meta position to form chlorohydroxymuconic semialdehyde. Subsequent dehalogenation can be spontaneous (Reineke and Knackmuss, 1988). As reported by Yong and Mulligan (2004), chlorinated benzoates (Suflita et al., 1983), 2,4,5-T pesticides, PCBs (Thayer, 1991), and 1,2,4-trichlorobenzenes (Reineke and Knackmuss, 1988) are known to undergo reductive dehalogenation under anaerobic conditions. *Rhodococcus chlorophenolicus* (Apajalahti and Salkinoja-Salonen, 1987) and *Flavobacterium* sp. (Steiert and Crawford, 1986) can aerobically biodegrade pentachlorophenol, while anaerobic degradation of 3-chlorobenzoate and PCBs has been identified by methanogenic consortia (Nies and Vogel, 1990).

10.3.2.5 Metals

It has long been assumed that biological transformation and degradation applied primarily to organic chemical compounds. More recently, however, research has shown that microbial conversion of metals occurs. The following short account summarizes the discussion from Yong and Mulligan (2004). Microbial conversion includes bioaccumulation, biological oxidation and reduction, and biomethylation (Soesilo and Wilson, 1997). Microbial cells can accumulate heavy metals through ion exchange, precipitation, and complexation on and within the cell surfaces containing hydroxyl, carboxyl, and phosphate groups. Bacterial oxidation and reduction could be used to alter the mobility of the metals. For example, some bacteria can reduce Cr(IV) in the form of chromate (CrO_4^{2-}) and dichromate ($Cr_2O_7^{2-}$) to Cr(III), which is less toxic and mobile due to precipitation above pH 5 (Bader et al., 1996).

Mercury can be found as Hg(II), volatile elemental mercury Hg(0), and in methyl and dimethyl forms. Metabolism occurs through aerobic and anaerobic mechanisms through uptake or conversion of Hg(II) to Hg(0), methyl- and dimethylmercury, or insoluble Hg(II) sulfide precipitates. Although volatilization or reduction during natural attenuation would still render mercury mobile, Hg(II) sulfides are immobile if sufficient levels of sulfate and electron donors are available.

Arsenic can be found as the valence states As(0), As(II), As(III), and As(V). Forms in the environment include As_2S_3, elemental As, arsenate (AsO_4^{3-}), arsenite (AsO_2^-), and other organic forms, such as trimethyl arsine and methylated arsenates. The anionic forms are mobile and highly toxic. Microbial transformation under aerobic conditions produces energy through oxidation of arsenite. Other mechanisms include methylation, oxidation, and reduction under anaerobic or aerobic conditions.

Selenium, which is a micronutrient for animals, humans, plants, and some microorganisms, can be found naturally in four major species, selenite (SeO_3^{2-}, IV), selenate (SeO_4^{2-}, VI), elemental selenium (Se, 0), and selenide (–II) (Frankenberger and Losi, 1995; Ehrlich, 1996). Oxidation of selenium can occur under aerobic conditions, while selenate can be transformed anaerobically to selenide or elemental selenium. Methylation of selenium detoxifies selenium for the bacteria by removing the selenium from the bacteria. Immobilization of selenate and selinite is accomplished via conversion to insoluble selenium. Due to the many forms of selenium, selenium decontamination by microorganisms is not promising.

10.3.2.6 Oxidation–Reduction (Redox) Reactions

It is useful to recall that (1) the chemical reaction process defined as *oxidation* refers to a removal of electrons from the subject of interest, and (2) *reduction* refers to the process where the subject (electron acceptor or *oxidant*) gains electrons from an electron donor (*reductant*). By gaining electrons, a loss in positive valence by the subject of interest results, and the process is called a reduction. Oxidation–reduction (redox) reactions have been briefly discussed in Section 9.4.4 in Chapter 9. Biological transformation of organic chemical compounds results from biologically mediated redox reactions. Bacteria in the soil utilize oxidation–reduction reactions as a means to extract the energy required for growth. They are the catalysts for reactions involving molecular oxygen and organic chemicals (and also soil organic matter) in the ground. Oxidation–reduction reactions involve the transfer of electrons between the reactants. The activity of the electron e^- in the chemical system plays a significant role. Reactions are directed toward establishing a greater stability of the outermost electrons of the reactants, i.e., electrons in the outermost shell of the substances involved. The link between redox reactions and acid–base reactions is evidenced by the proton transfer that accompanies the transfer of electrons in a redox reaction. Manahan (1990) gives the example of the loss of three hydrogen ions that accompanies the loss of an electron by iron (II) at pH 7, resulting in the formation of a highly insoluble ferric hydroxide, as indicated by the following:

$$Fe(H_2O)_6^{2+} \rightarrow Fe(OH)(H_2O)_5^{2+} + H^+ \qquad (10.2)$$

It is not easy to distinguish between abiotic and biotic (biologically mediated) redox reactions. To a large extent, it is not always possible to eliminate or rule out involvement of microbial activity in biotic redox reactions. There does not appear to be a critical need to distinguish between the two in reactions that concern organic chemical pollutants, since it is almost certain that with all the microorganisms in the subsoil, some measure of microbial activity would be involved. In any event, the number of functional groups of organic chemical pollutants that can be oxidized or reduced under abiotic conditions is considerably smaller than those under biotic conditions (Schwarzenbach et al., 1993).

The two classes of electron donors of organic chemical pollutants are (1) electron-rich B-cloud donors, which include alkenes, alkynes, and the aromatics, and (2) lone-pair electron donors, which include the alcohols, ethers, amines, and alkyl iodides. Similarly, in the case of electron acceptors, we have (1) electron-deficient π-electron cloud acceptors, which include the π-acids, and (2) weakly acidic hydrogens, such as s-triazine herbicides and some pesticides.

A measure of the electron activity in the pore water of a soil–water system is the *redox potential Eh*. It provides us with a means for determining the potential for oxidation–reduction reactions in the pollutant–soil–water system under consideration, and is given as

$$Eh = pE\left(\frac{2.3RT}{F}\right) \tag{10.3}$$

where E is the electrode potential, R is the gas constant, T is the absolute temperature, and F is the Faraday constant. The electrode potential E is given in terms of the half reaction:

$$2H^+ + 2e^- \Leftrightarrow H_2(g) \tag{10.4}$$

When the activity of $H^+ = 1$ and the pressure H_2 (gas) = 1 atm, we obtain $E = 0$.

10.4 Natural Attenuation and Impact Management

The natural attenuation capacity of soils in the substratum has long been recognized and described by soil scientists as the assimilative capacity of soils. The discussions at the beginning of the previous section and in the earlier chapters of this book show that this is now a tool that can be used as a passive treatment process in the remediation of sites contaminated by organic chemicals. The U.S. EPA has wisely coupled the requirement for continuous on-site monitoring of contaminant presence whenever natural attenuation is to be used as a tool for site remediation, as seen in the definition provided in the first part of Section 10.3. The procedure for application of this attenuation process is called *monitored natural attenuation* (MNA). Guidelines and protocols for application of MNA as a treatment procedure in remediation of contaminated sites have been issued. Since site specificities differ from site to site, the prudent course of action is to adapt the guidelines and protocols for site-specific use. A general protocol, from Yong and Mulligan (2004), for considering MNA as a remediation tool is shown in Figure 10.12. A very critical step in the application of MNA as a site remediation tool is to have proper knowledge of (1) *lines of evidence* indicating natural or intrinsic remediation, (2) contaminants, soil properties, and hydrogeology, and (3) regulatory requirements governing *evidence of success* of the MNA remediation project. Lines of evidence and evidence of success will be discussed later in this section.

The data and information inputs shown on the left-hand side of Figure 10.12 tell us what is required to satisfy site-specific conditions, and whether the *indicators* for natural bioremediation are sufficient to proceed with further examination to satisfy that the use of MNA is a viable treatment option. Negative responses from the first two decision steps will trigger technological or engineered solutions to the remediation problem. As will be seen, laboratory research and transport and fate modeling are needed to inform one about the ability of the site materials and conditions to attenuate the pollutants.

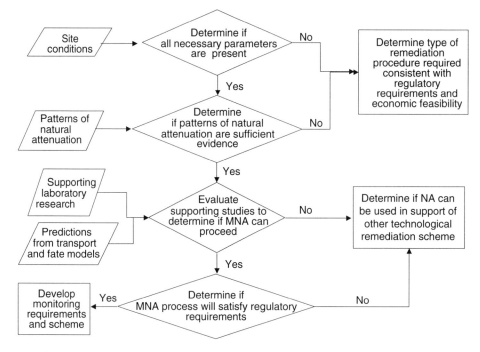

FIGURE 10.12
General protocol for considering moderate natural attenuation (MNA) as a remediation tool. (From Yong, R.N. and Mulligan, C.N., *Natural Attenuation of Contaminants in Soils*, Lewis Publishers, Boca Raton, FL, 2004.)

The use of natural attenuation (NA) as an active tool in the management of contaminant impact and transport, as opposed to the use of MNA as a passive tool, has been hampered because of insufficient knowledge of the many processes that contribute to the natural attenuation process. The designation of MNA as more of a passive tool, as opposed to an active tool, is based on the fact that except for the monitoring requirement, the use of natural attenuation processes as existent *in situ* is essentially a "do nothing" solution. The "do nothing" part refers to the human contribution to the processes resulting in natural attenuation of contaminants. As a clarification, we should point out that the acronym NA is used to denote natural attenuation as a process tool. When we wish to discuss the processes that result in the natural attenuation of contaminants, we will use the complete term *natural attenuation*. To make NA an active tool, we can (1) enhance the processes that contribute to natural attenuation capability and (2) incorporate NA as part of a scheme to mitigate and manage the geoenvironmental impacts from discharge or containment of waste products and pollutants.

10.4.1   Enhancement of Natural Attenuation (NA) Capability

Waste discharge and ground contamination impact mitigation essentially reduce to minimizing damage done to the land environment and its inhabitants. What this means is the reduction or elimination of health threats to the biotic receptors because of ground contamination, and prevention of contamination of receiving waters and groundwater. This is the essence of contaminant impact mitigation and management. The objectives or desired endpoints sought in contaminant impact mitigation fit very well with the capabilities of the processes in soils that contribute to natural attenuation. Accordingly, NA can be used as tool to provide impact mitigation and management since the basic processes involved in NA result in reduction of concentration of contaminants and toxicity. To increase the capability of NA, one can consider enhancement of the natural assimilative capacity of the soil–water system. These can take the form of geochemical and biogeochemical aids, bioaugmentation, and biostimulation. Successful enhancement of NA will produce a subsoil with properties that can be considered *enhanced NA capability*.

10.4.1.1 Soil Buffering Capacity Manipulation

The capability of a soil to accept and retain inorganic and some organic contaminants can in some instances be assessed by determining its chemical buffering potential, particularly if the reactions in the soil–water system result in changes in the pH of the system. The chemical buffering system contributes significantly to the carrying capability of the soil, i.e., the capability of the soil barrier or subsoil to accept and retain contaminants. The main issue is the ability of the soil–water system to maintain a natural pH level (within acceptable limits) in spite of input of acidic or alkaline contaminant leachates. *In situ* soil pH manipulation for the purpose of contaminant impact mitigation requires introduction of buffering agents generally through leaching methods or via injection. In contaminant–soil interaction, the chemical buffering system describes the capability of the system to act as chemical barrier against the transport of contaminants.

The buffering capacity of a soil determines the potential of a soil for effective interaction with leachate contaminants, and is more appropriate for inorganic soils and inorganic contaminant leachates. The principal features that establish the usefulness of buffering capacity assessment center around the *acidity* or *alkalinity* of the initial soil–water system and the solutes in the leachate. Soil conditioning in respect to changes in the natural soil buffering capacity is usually considered in terms of addition of buffering agents, much in the same manner as solution chemistry. In the *in situ* soil conditioning case, however, addition of buffering agents needs to be effectuated through injection wells or leaching, as for example by adding lime to the surface as the leach source to increase the pH of the soil. In a site contaminated with heavy metals, raising the pH of the soil–water system would precipitate the heavy metal pollutants, and thus make them less bioavailable. However, we must recognize that this is not a permanent solution because if the pH of the system is subsequently reduced by environmental forces or external events, the same heavy metals will become mobile again. To avoid subsequent solubilization, the precipitated heavy metals should be removed from the contaminated site.

10.4.1.2 Biostimulation and Bioaugmentation

Section 3.4.1 in Chapter 3 introduced the use of biological aids, *biostimulation* and *bioaugmentation*, as part of the available tools for groundwater management. *Biostimulation* occurs when stimuli such as nutrients and other growth substrates are introduced into the ground to promote increased microbial activity of the microorganisms existent in the site. The intent is to obtain improved capabilities of the microorganisms to better degrade the organic chemical pollutants in the soil. The addition of nitrates, Fe(III) oxides, Mn(IV) oxides, sulfates, and CO_2, for example, will allow for anaerobic degradation to proceed. Biostimulation is perhaps one of the least intrusive of the methods of enhancement of the natural attenuation capacity soils.

Bioaugmentation denotes the process whereby exogenous microorganisms are introduced *in situ* to aid the native or indigenous microbial population in degrading the organic chemicals in the soil. The reason one would use bioaugmentation is presumably because the microorganisms in the soil are not performing up to expectations. This could be because the concentrations of microorganisms are insufficient, or maybe because of inappropriate consortia. The function of the exogenous microorganisms is to augment the indigenous microbial population such that effective degradative capability can be obtained. Frequently, biostimulation is used in conjunction with bioaugmentation. There is the risk that (1) use of microorganisms grown in uncharacterized consortia, which include bacteria, fungi, and viruses, can produce toxic metabolites (Strauss, 1991), and (2) the interaction of chemicals with microorganisms may result in mutations in the microorganisms themselves or microbial adaptations.

10.4.1.3   *Biochemical and Biogeochemical Aids*

Introduction of geochemical aids *in situ* can use the same techniques employed to introduce the various kinds of growth substrate, nutrients, and exogenous microorganisms for biostimulation and bioaugmentation. Manipulations of pH and pE or Eh using geochemical aids can increase the capability of the soil to mitigate the impact of some toxic pollutants. A good case in point is the changes in toxicity for chromium and arsenic because of changes in their oxidation state. Chromium (Cr) as Cr(III) is an essential nutrient that helps the body use sugar, protein, and fat. On the other hand, chromium as Cr(VI) has been determined by the World Health Organization (WHO) to be a human carcinogen. Cr(III) can be oxidized to Cr(VI) by dissolved oxygen, and quite possibly with manganese dioxides. If such a possibility exists in a field situation, management of the potential impact can take the form of *in situ* geochemical or biogeochemical intervention to create a reducing environment in the subsurface. A useful procedure would be to deplete the oxygen in the subsurface to create a reduced condition in the soil. The danger or risk of manipulation of the Eh of the soil–water regime is incomplete knowledge of all the elements in the subsoil that are vulnerable to such manipulation. The case of arsenic in the ground is a good example. It is known that arsenic (As) as As(III) is more toxic than As(V). If ground conditions show that arsenic is present as As(V), creating a reducing environment to prevent oxidation of chromium to the more toxic oxidation state would create the reverse effect on As(V). Reduction of As(V) to As(III) would increase the toxicity of arsenic.

Manipulation of pH can change the nature of the assimilative capacity soils, as stated previously. It addresses the precipitation of heavy metals in solution (pore water) or dissolution of precipitated heavy metals. Changes in pH of the soil–water system will produce changes in the sign of surface electrostatic charges for those soil materials with amphoteric surfaces, i.e., surfaces that show pH dependency charge characterization. Subsoils containing oxides, hydrous oxides, and kaolinites are good candidates for pH-dependent charge manipulation. Changes in surface charge characterization can result in increased bonding of metals or release of heavy metals from disruption of bonds. Both pH and Eh changes will also have considerable influence on acid–base reactions and on abiotic and biotic electron transfer mechanisms. Abiotic transformations of organic chemical pollutants due to acid–base and oxidation–reduction reactions are minor in comparison to biotic transformations.

10.4.2   NA Treatment Zones for Impact Mitigation

Treatment zones are regions in the subsoil that utilize NA and, more specifically, enhanced NA, to attenuate the impact of contaminants during transport in the subsoil. Since it is rare to find source discharges conveniently located directly in a region where NA is very active, the usual procedure is to provide a treatment zone that would capture the pollutant plume during transport. A good knowledge of site hydrogeology is essential for this type of mitigation procedure to function properly. Figure 10.13 shows a simple source–path-

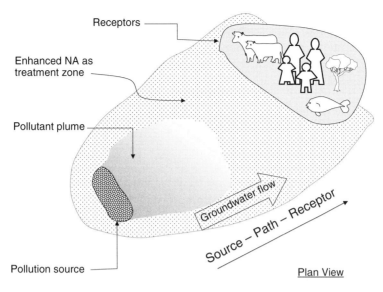

FIGURE 10.13
Simplified plan view of a treatment zone established with enhancement of the natural attenuation capability of the site subsoil — to mitigate impact of pollutants on land environment and to protect receptors. The diagram shows the source–pathway–receptor (SPR) problem, with the direction of groundwater flow as the pathway from source to receptors.

way–receptor (SPR) problem that uses a treatment zone to mitigate the impact of the pollutant plume generated by the pollution source shown at the bottom left of the diagram. To determine if the treatment zone is effective in mitigating the impact of the generated pollutant, the monitoring scheme shown in Figure 10.14 is recommended. By this means, determination of reduction in concentration and toxicity of the pollutants can be obtained. This procedure also allows one to establish evidence of success of the treatment zone. It

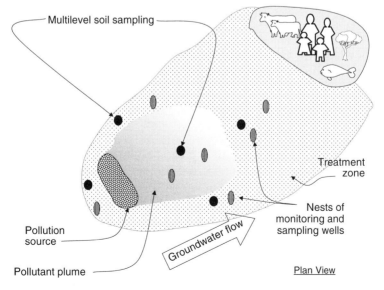

FIGURE 10.14
Plan view of distribution of monitoring wells and soil sampling boreholes for verification monitoring of treatment zone effectiveness and long-term conformance monitoring.

can be argued that one should remove the pollution source as part of the mitigation procedure. Assuming that the source is an industrial facility, the obvious course of action is to (1) implement operational procedures in the facility that will reduce output of polluting items and (2) establish or improve treatment of discharges to capture all the noxious substances discharged into the land environment. The combination of the treatment zone with removal of the pollution source will ensure short-term mitigation and longer-term elimination of the threats posed by the pollution plume — a positive step toward geoenvironmental sustainability.

10.4.2.1 *Permeable Reactive Barriers and NA*

Permeable reactive barriers (PRBs) are engineered material barriers constructed and placed in the ground to intercept pollutant plumes. The material in these barriers generally consists of permeable soil material containing various elements designed to react with the kinds of pollutants entering the barrier. The whole intent of PRBs is to provide a chemical-physical sieve or filter that would capture the pollutants as they pass through the barrier. In a sense, therefore, PRBs are barriers with highly engineered and efficient attenuation properties and characteristics. They are sometimes also known as treatment walls.

The major pollutant capture and immobilization processes needed for the engineered materials in the PRBs to function effectively include (1) sorption, precipitation, substitution, transformation, complexation, and oxidation and reduction for inorganic pollutants, and (2) sorption, biotic and abiotic transformations, and degradation for organic chemical pollutants. As noted in Section 3.4.1 in Chapter 3, the types of reagents, compounds, and microenvironment in the PRBs include a range of oxidants and reductants, chelating agents, catalysts, microorganisms, zero-valent metals, zeolites, reactive clays, ferrous hydroxides, carbonates and sulfates, ferric oxides and oxyhydroxides, activated carbon and alumina, nutrients, phosphates, and soil organic materials. The selection of engineered materials in the PRBs, such as reagents and compounds, and the manipulation of the pH-pE microenvironment in the treatment walls will need to be made on the basis of site-specific knowledge of the nature of the pollutants.

The success of PRBs in mitigating pollutant impacts depends on:

1. Effectiveness of types of engineered material in the PRB: This depends on a proper knowledge of the pollutants, barrier material, and the kinds of processes (interactions and bonding mechanisms) resulting from pollutant–barrier material interactions.

2. Sufficient residence time of the pollutant plume in the PRB: There must be sufficient time in residence in the PRB for pollutant–material interactions to be fully realized. This is a function of both the permeability of the barrier itself and the kinds of reactions needed to occur between pollutants and barrier material. It would be useless if the pollutant passed through the barrier at high rates — rates that would not permit reactions to mature. Conversely, it would be useless if the pollutant would not penetrate the barrier, hence denying any opportunity for reactions to occur.

3. Proper intercept of pollutant plume advance: A thorough knowledge of site hydrogeology is required to allow one to place the barrier for optimum intercept of the pollutant plume. A knowledge of the advective velocity is also required.

For more effective use of PRBs, an enhanced NA treatment zone can be used and placed ahead of the PRB. Such a case has been shown in Figure 3.11 in Chapter 3. The treatment zone in the diagram is shown as an optional tool. Pollutant plumes can be channeled to

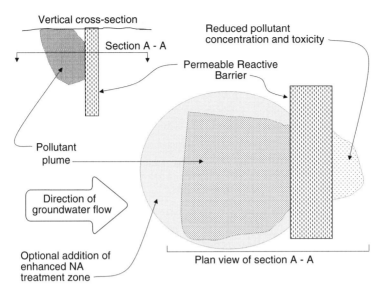

FIGURE 10.15
Cross-section and plan view of permeable reactive barrier (PRB). Plan view shows leachate plume entering the PRB with pollutants and leaving the PRB with reduced pollutant concentration and toxicity. If required, an enhanced NA treatment zone can be situated in front of the PRB — to increase the effectiveness of the pollutant impact mitigation scheme. (Adapted from Yong, R.N. and Mulligan, C.N., *Natural Attenuation of Contaminants in Soils*, Lewis Publishers, Boca Raton, FL, 2004.)

flow through reactive walls by shepherding the plume with a *funnel–gate* technique. In this technique, the plume is essentially guided to the intercepting reactive wall by a funnel constructed of impermeable material, such as sheet-pile walls, and placed in the contaminated ground to channel the plume to the PRB (Figure 10.15). Other variations of the funnel–gate technique exist — obviously in accordance with site geometry and site specificities.

10.5 Lines of Evidence

Lines of evidence is a term that is associated with the use of NA as a tool for mitigation and management of impacts from waste and pollutant discharges to the land environment. *Lines of evidence* (LOE) refers to the requirement to determine whether a soil has capabilities for *in situ* attenuation of contaminants. This requirement originates from procedures associated with the use of monitored natural attenuation (MNA) as a treatment procedure. This is a prudent course of action because there is a need to determine how effective a particular soil will be in attenuating contaminants. The types of information and analyses required for LOE indicators are shown in Figure 10.16. Site and problem specificities will dictate how many pieces of information and what specific kinds of analyses will be needed. The type of information needed to define the site characteristics is shown in the top right-hand corner of the diagram. The physical (geologic and hydrogeologic) setting sets the parameters of the problem to be resolved. Whether the natural attenuation capability of the subsoil is capable of mitigating and managing the pollutant plume anticipated within the site boundaries will be established by the other two categories: patterns of natural attenuation and supporting laboratory tests and analyses.

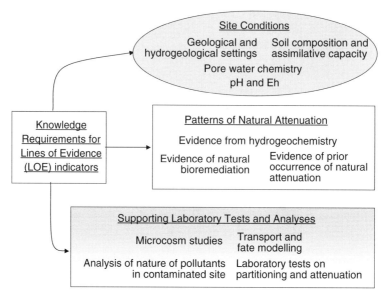

FIGURE 10.16
Required information and analyses for lines of evidence (LOE) indicators.

Knowledge of the patterns of natural attenuation identified in the central box is essential. What is sought in this category is evidence of previous natural (intrinsic) remediation of contaminants. To determine this, it is necessary to recall the various mechanisms and processes that establish retention and transformation of the various kinds of pollutants generally found in the subsoil. In addition, one needs to determine or assess the environmental mobility of the contaminants in the site under consideration. These are necessary pieces of information for prescription of *indicators* for the lines of evidence (LOE).

10.5.1   Organic Chemical Compounds

The previous chapters have shown that organic chemicals such as organic solvents, paints, pesticides, oils, gasoline, creosotes, greases, etc., are responsible for many of the chemicals found in contaminated sites. These chemicals are known generally as xenobiotic compounds. It is not possible to categorize them all in respect to how they would interact in a soil–water system. The more common organic chemicals found in contaminated sites can be grouped into three broad groups:

1. Hydrocarbons: Including the petroleum hydrocarbons (PHCs), various alkanes and alkenes, and aromatic hydrocarbons such as benzene, multicyclic aromatic hydrocarbons (MAHs) (e.g., naphthalene), and polycyclic aromatic hydrocarbons (PAHs) (e.g., benzo-pyrene)

2. Organohalide compounds: Of which the chlorinated hydrocarbons are perhaps the best known. These include trichloroethylene (TCE), carbon tetrachloride, vinyl chloride, hexachlorobutadiene, polychlorinated biphenyls (PCBs), and polybrominated biphenyls (PBBs)

3. Miscellaneous compounds: Including oxygen-containing organic compounds such as phenol and methanol and nitrogen-containing organic compounds such as trinitrotoluene (TNT)

As we have seen in Section 2.3.2 in Chapter 2, the density of these compounds in comparison with that of water has direct control on their transport in the subsoil. We classify nonaqueous phase liquids (NAPLs) into the light NAPLs, identified as LNAPLs, and the dense ones, called the DNAPLS. LNAPLs include gasoline, heating oil, kerosene, and aviation gas. DNAPLs include the organohalide and oxygen-containing organic compounds such as 1,1,1-trichloroethane; chlorinated solvents such as tetrachloroethylene (PCE), trichloroethylene (TCE), and carbon tetrachloride (CT); PCBs; pentachlorophenols (PCPs); and tetrachlorophenols (TCPs).

As shown in Figure 2.7 in Chapter 2, since LNAPLs are lighter than water and the DNAPLs are heavier than water, LNAPLs will likely stay above the water table, and DNAPLs tend to sink through the water table and come to rest at an impermeable bottom (bedrock).

The various results of transformations and biodegradation of organic chemicals have been discussed in various forms in Section 10.3. The significant outcome of NA as a tool for mitigation of impact is the evidence of occurrence of biodegradation and transformation of the target organic chemicals in the NA process. The *indicators* that need to be prescribed in the LOE relate to specific decreases in concentration of the pollutants and transformations (conversions and biodegradation) of organic chemical pollutants. Determination of the nature and composition of the transformed products of the original organic chemical pollutants is required. Knowledge of the products obtained via abiotic and biotic processes is essential. A good example of this is recognizing that abiotic transformation products are generally other kinds of organic chemical compounds, whereas transformation products resulting from biotic processes are mostly seen as stages (intermediate products) toward mineralization of organic chemical compounds. Biologically mediated transformation processes are the only types of processes that can lead to mineralization of the subject organic chemical compound. Complete conversion to CO_2 and H_2O (i.e., mineralization) does not always occur. However, intermediate products can be formed during the mineralization.

10.5.2 Metals

At the very least, prescription of the indicators for the lines of evidence in respect to heavy metals (HMs) requires determination of (1) the nature and concentration of sorbed metal ions, (2) pore water chemistry, including pH and Eh, and (3) the environmental mobility of heavy metals. The environmental mobility of HMs is dependent to a very large extent upon whether they are in the pore water as free ions, complexed ions, or sorbed onto the soil particles. Prescription of *indicators* for LOE should take into account the assimilative capacity of the subsoil and the nature and fate of the HMs in the subsoil. As long as the full assimilative potential of the soil for HMs is not reached, attenuation of the HMs will continue. Metals that are sorbed onto the soil particles are held by different sets of forces — determined to a large extent by the soil fractions and the pH of the soil–water system. The various types of soils and their different soil fractions have different sorption capacities, dependent on the nature and distribution of the HMs and pH of the system.

Precipitation of HMs as hydroxides, sulfides, and carbonates generally classifies as part of the assimilative mechanism of soils because the precipitates form distinct solid material species and are considered part of the attenuation process. Either as attached to soil particles or as void pluggers, precipitates of HMs can contribute significantly to attenuation of HMs in contaminant plumes. Determination of the chemistry, including pH and Eh, for assessment of lines of evidence should not neglect examination of possibilities of precipitation and solubilization of metals as part of the evidence phase. Hydroxide precipitation is favored in alkaline conditions, as for example when Ca(OH)$_2$ is in the groundwater in abundance. With available sulfur and in reducing conditions, sulfide precipitates can be obtained. Sulfide precipitates can also be obtained as a result of microbial activity, except that this will not be a direct route. Sulfate reduction by anaerobic bacteria will produce H$_2$S and HCO$_3^-$, thus producing the conditions for formation of metal sulfides.

10.6 Evidence of Success

Evidence of success (EOS) is a requirement specified by Yong and Mulligan (2004) as testimony to the success of utilization of NA as a tool for remediation of contaminated sites. While monitoring is a necessity in application of MNA, there is a need for one to have knowledge of whether the "signs and signals" registered in the monitoring program testify to a successful MNA treatment program. In essence, EOS takes the role of *indicators* of success or steps toward success by MNA in remediation of contaminated sites.

With the same rationale, EOS can be used in impact mitigation and management programs. Figure 10.17 takes the MNA protocol shown in Figure 10.12 as the basis for determining whether NA can be successfully used as an impact mitigation and management tool. The first two levels of protocol are similar to the MNA steps. At the third step or level, a clear knowledge of the kinds of impact, and mitigation and management requirements needs to be articulated. These are combined with information from laboratory tests that are designed to provide the kinds of information necessary to determine material parameters and interaction processes. Supporting predictions on fate and trans-

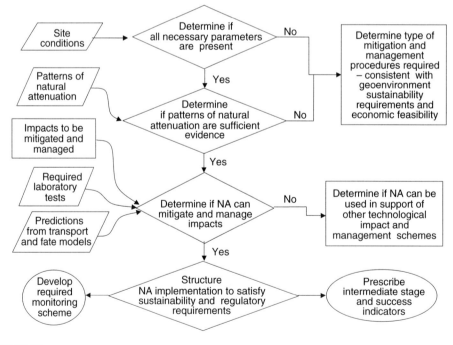

FIGURE 10.17
General protocol to determine feasibility and application of natural attenuation (NA) as a tool for impact mitigation and management.

port are necessary pieces of information. All of these combine to provide one with the tools to determine whether NA can be successfully used to meet the requirements for impact mitigation and management. Negative responses will require that NA be rejected as a tool, or used in conjunction with other technological tools, to provide the necessary impact mitigation and management solution.

A positive response will require structuring implementation procedures and strategies in combination with a competent monitoring scheme to track progress. A very necessary part of the implementation scheme is the specification or prescription of *success indicators* shown in the ellipse at the bottom right-hand corner of Figure 10.17. These indicators also serve as markers for performance assessment of the impact mitigation scheme. We have shown the requirements in terms of intermediate-stage indicators and final success indicators. Since the time required for processes contributing to the natural attenuation to fully complete their functions, it is necessary to prescribe intermediate indicators as tracking indicators and as performance assessment markers can be extensive. Note that we have used *italicized* notation for indicators when we mean them as markers and have left them without italicization when we discuss them as general items.

Monitoring and sampling of pore water and soils are needed in the contaminant attenuation zone. The choice of type of monitoring wells and sampling devices and their spatial distribution or location will depend on the purpose for the wells and devices. At least three separate and distinct monitoring–sampling schemes need to be considered:

1. Initial site characterization studies: Site characterization monitoring and sampling provide information on site subsoil properties and hydrogeology. Subsurface flow delineation provides one with the information necessary to anticipate transport direction and extent of pollutant plume propagation. With a proper knowledge of the requirements of the verification and long-term monitoring–sampling schemes, a judicious distribution of monitoring wells and sampling devices upgradient and downgradient can be made such that the information obtained can be used to service the requirements for all three monitoring–sampling schemes.

2. Verification monitoring: This requires placement of monitoring wells and soil sampling devices within the heart of the pollutant plume and also at positions beyond the plume. Figure 10.14 gives an example of the distribution of the wells and devices. Obviously, assuming that the wells and devices are properly located, the more monitoring and sampling devices there are, the better one is able to properly characterize the nature of the pollutant plume. Monitoring wells and sampling devices placed outside the anticipated pollutant plume will also serve as monitoring wells and sampling devices for long-term conformance assessment.

3. Long-term conformance monitoring: This is essential to verify success of the mitigation scheme and for long-term management of the potential impact.

Analyses of samples retrieved from monitoring wells will inform one about the concentration, composition, and toxicity of the target pollutant. A knowledge of the partition coefficients and solubilities of the various contaminants, together with the monitoring well information, will provide one with the opportunity to check the accuracy of predictions from transport–fate models. For organic chemicals detected in the monitoring–sampling program, laboratory research may be required to determine the long-term fate of the transformed or intermediate products. This is not a necessary requirement if modeling predictions, and especially the *indicators* for the intermediate, show good accord with the sampling values of pollutant concentrations. Tests on recovered soil samples from the sampling program should determine the environmental mobility of the pollutants and also the nature and concentration of pollutants sorbed onto the soil particles (soil solids). Detailed discussions of many of the bonding mechanisms and their reactions to changes in the immediate environment have been developed in Chapters 2 and 9.

10.7 Engineered Mitigation Control Systems

As we have indicated at the outset, the use of technological schemes for mitigation and management of pollutants in the ground is probably best utilized for limited and well-defined source locations of pollutants. Good examples of these are waste landfills, leaking underground storage tanks, spills and discharges, and containment ponds. Many of these are shown in the diagram in Figure 2.6 in Chapter 2. Technological solutions for management and control range from construction of impervious barriers that would intercept the plume to removal of the pollution source and the entire affected region. To a very large extent, the methods chosen or designed to manage and control pollutant advance in the subsoil are necessarily site and situation specific. Also, to a large extent, the nature of the threats posed by the pollutants and the pathways to the various receptors is a consideration that will dictate the type and kind of technological solution sought. Finally, the control management technological solution sought will always be analyzed within the framework of risks–reward and cost-effectiveness.

The record shows that there are some very difficult-to-treat contaminated sites. By and large, these are sites contaminated with organic chemical compounds that are severe threats to human health. For these kinds of contaminated sites, containment with confining structures has been constructed. These allow these sites to be isolated while awaiting effective and economic remediation solutions. In the case of impervious barriers, these are generally constructed from sheet piles lined in the interior with membranes to deny lateral advance of the pollution plume. Difficulties arise in controlling the downward advance of the pollution plume when the plume arrives at the lateral impervious barrier. Suggestions range from driving the sheet piles down into an impervious layer, as shown for example in Figure 10.18, to injection grouting to develop an impervious base at some depth in the ground, to inclined to horizontal placement of sheet piles using techniques similar to those used in the oil industry for inclined drilling.

If an impervious clay layer can be found directly below the contaminated region, the methodology shown in Figure 10.18 is probably the most expeditious means for controlling the escape of fugitive pollutants. By and large, for situations such as the one depicted in the figure, the pollutants resident in the contaminated site would likely be various kinds of organic chemicals. Heavy metals associated with these chemicals will likely be sorbed by the soil solids and will not be very mobile. Hence, the nest of treatment wells sunk into the contaminated region will be geared toward bioremediation of the organic chemical pollutants. Monitoring wells placed outside the confining sheet-pile wall, particularly downstream, will provide continuous information on the efficiency of the containment system. Note that the schematic representation of the nest of wells (treatment and monitoring) is relatively crude. Wells should be placed with varying vertical and horizontal sampling points and locations. If extra precaution is sought, a treatment zone using enhanced NA capability, as described in Section 10.4.1, outside the confining sheet-pile wall can be introduced.

For control of pollution plumes during treatment as part of impact management, several options are available. If the pollutant source is well delineated and defined, and if site

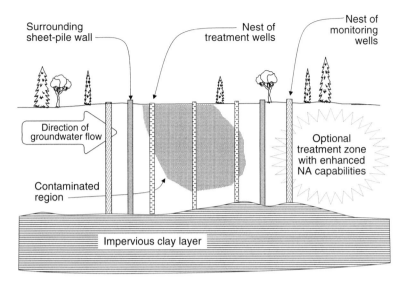

FIGURE 10.18
Containment of contaminants in a contaminated site using a confining sheet-pile wall that surrounds the contaminated region. The sheet-pile wall is sunk into the impervious clay layer to prevent bottom escape of pollutants. Treatment wells are sunk into the contaminated regions, and monitoring wells are placed downstream, with some upstream also. Optional treatment zones using enhanced NA capabilities can be used if needed.

hydrogeology is well understood, the solution shown in Figure 10.19 is one that utilizes the capabilities of enhanced NA in combination with the permeable reactive barrier (PRB) previously described. This procedure is both a mitigation and management tactic for management of pollutant impact. Note again that the monitoring and treatment wells are somewhat simplistic in illustrative portrayal.

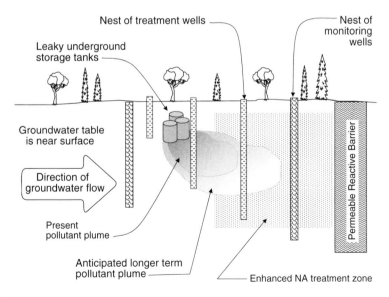

FIGURE 10.19
Use of treatment wells and an enhanced natural attenuation treatment zone in combination with a permeable reactive barrier to mitigate and manage impact from leaky underground storage tanks. Treatments for enhancement can be any or all of the following: geochemical intervention, biostimulation, and bioaugmentation. Treatment occurs in the pollutant plume and downgradient from the plume.

The engineered barrier systems for a municipal and a hazardous waste landfill shown in Figure 1.10 in Chapter 1 are a good demonstration of the extent to which composite barrier systems can be designed and engineered to meet the requirements for management and control of pollutants. The details of the filter, membrane, and leachate collection system are specified by regulatory *command-and-control* requirements or by performance requirements. In the case of the municipal solid waste (MSW) landfill liner system shown in the bottom right-hand corner of Figure 1.10, the soil material comprising the engineered clay barrier underlying the synthetic membrane must possess hydraulic conductivity values that are below the maximum permissible values. The basic idea in the design details of the engineered barrier for the MSW landfill is that if leachates inadvertently leak through the high-density polyethylene (HDPE) membrane and are not captured by the leachate collection system, the pollutants in the leachate plumes will be attenuated by the engineered clay barrier. The engineered clay barrier serves as the second line of defense or containment.

For the hazardous waste (HW) bottom liner system shown in the bottom left-hand corner of Figure 1.10, there are two lines of defense before the soil sub-base. The HDPE acts as the first barrier. Before this, the leachate collection system is designed to collect leachates draining down from the waste pile. If the system functions well, i.e., as designed, there should be very little leachate reaching the HDPE barrier. If, however, leachate does collect at the HDPE barrier, and if this barrier is somehow breached, the underlying synthetic membrane (most likely another HDPE) is designed to prevent the leachate from escaping. Above this synthetic membrane there is a leak detection system that will alert the managers of the landfill that the first HDPE has been breached and that the leachate collection system is most likely malfunctioning. If the second membrane fails, the underlying soil sub-base can be designed as an attenuation barrier using NA principles.

10.7.1   Remediation as Control Management

Technically speaking, remediation of sites and regions contaminated with noxious substances and pollutants belongs to a category separate from pollutant impact mitigation. We have included this here because the *treatment wells* shown in Figure 10.18 and Figure 10.19 are in fact wells or devices that introduce remediation aids. In addition, it can be legitimately argued that remediation of a contaminated site in effect removes the pollutant source — assuming of course that the remediation treatment process is successful.

The priority requirement in remediation treatment of a contaminated site is to eliminate the health and environmental threats posed by the presence of pollutants in the contaminated site. Traditionally, this objective is met with the *dig-and-dump* technique. Replacement with clean fill material will now ensure that all the pollutants have been removed from the affected site or region. If total removal of all pollutants is not an option, minimization of the risk posed by the presence of the pollutants is the next priority. This latter course can take several forms. The basic factors to be considered include:

- <u>Contaminants/pollutants</u>: Type, concentration, and distribution in the ground
- <u>Site</u>: Site specificities, i.e., location, site constraints, substrate soil material, lithography, stratigraphy, geology, hydrogeology, fluid transmission properties, etc.
- <u>Economics and risks</u>: Cost-effectiveness, timing and risk management

The techniques that can be considered fall into five groups: (1) physicochemical, (2) biological, (3) thermal, (4) electrical-acoustic-magnetic, and (5) combination. Physicochemical techniques rely on physical or chemical procedures for removal of the pollutants. These include precipitation, desorption, soil washing, ion exchange, flotation, air-stripping, vapor and vacuum extraction, demulsification, solidification, stabilization, reverse osmosis, etc.

Biological techniques are generally used to treat organic chemicals, but as we have pointed out previously, these can also be used for remediation of heavy metal contaminated sites. The techniques used include bacterial degradation or transformation, biological detoxification, aeration, fermentation, and biorestoration. Thermal procedures include vitrification, closed-loop detoxification, thermal fixation, pyrolysis, supercritical oxidation, etc. Electrical-acoustic-magnetic methods include electrochemical oxidation, electrokinetics, electrocoagulation, ultrasonic, and electroacoustics. Finally, the last group that specifies *combination* implies that any of the four previous groups may be combined in a series-type technical solution to provide the necessary remediation treatment. This is sometimes called a *treatment train*.

Remediation of contaminated sites is a very large challenge that offers innumerable opportunities for technological innovation. The basic means for treatment given in the preceding paragraph have been used in many different technologically clever ways to effect remediation of contaminated sites. The reader is advised to consult specialized manuals and textbooks devoted exclusively to remediation and treatment of contaminated sites. Bioremediation occupies perhaps the greatest attention of most researchers and practitioners attending to remediation. Much research is being conducted and reported in the various specialized journals.

10.8 Concluding Remarks

The previous chapters have shown that the various discards, spills and loss of materials (chemicals, etc.), and discharges of wastes, either in liquid form or as solids, are common to all types of human activities — associated with such entities as (1) households, (2) cities, (3) industries, (4) farms, and (5) mineral and hydrocarbon exploitation. These pose significant threats that are well perceived to the land environment and the receiving waters. Not as well perceived are the threats presented by atmospheric-based nonpoint sources, such as those shown in Figure 10.20. Under rainfall conditions, pesticides, herbicides, and other pest control chemical aids have the potential not only to combine with the rainfall runoff to pollute the receiving waters, but also to infiltrate into the ground and threaten groundwater supplies. In addition to the nonpoint sources that originate on the land surface (land based), there are the nonpoint sources that originate from precipitation through the atmosphere containing noxious gas emissions and airborne pollutants (as particulates) from offending smokestacks and other types of smoke discharges. These can be called atmospheric-based nonpoint pollution or contaminant sources. Included in this list are NO_x (nitrogen oxides), SO_2 (sulfur dioxide), CO (carbon monoxide), Pb and other metals (such as Al, As, Cu, Fe, La, Mg, Mn, Na, Sb, V, and Zn) (Lin et al., 1997) as airborne particulates, volatile organic chemicals (VOCs), polycyclic aromatic hydrocarbons (PAHs) (such as benzene, toluene, and xylene), and particulate matter (PM_{10} and $PM_{2.5}$). $PM_{2.5}$, i.e., particulate matter less than 2.5 μm in size, will in all likelihood remain suspended in the ambient air, while PM_{10} (particulate matter less than 10 μm but greater than 25 μm) will be deposited eventually or with the aid of precipitation. While acid rain is one of the outcomes of precipitation through this type of atmosphere, deposition of the airborne

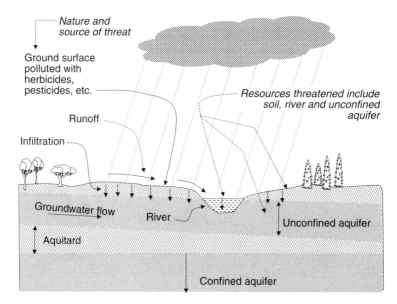

FIGURE 10.20

Demonstration of threats and impacts from atmospheric-based and land-based nonpoint sources. Resources threatened include the river, soil, and unconfined aquifer. We presume that the aquitard is sufficient to protect contamination of the confined aquifer. If this is incorrect, the confined aquifer will eventually be contaminated by the water in the unconfined aquifer.

pollutants has not received the attention that it deserves. Airborne particulate matter is a great concern to public health in the ambient air because of the effect of these particulates on the hearts and lungs of those who are exposed to them.

The impacts to the health and quality of biotic receptors and the land environment presented by atmospheric-based and land-based nonpoint sources of pollution cannot be readily mitigated by limited point-directed technological solutions. We recognize that the broad-based nature of the affected regions, i.e., land surfaces and water bodies, makes containment and management of the spread of pollutants prohibitively difficult and costly. In consequence, the use of the natural in-place soil as a tool for mitigation of the impact of such pollutant sources is a solution that needs to be exploited. To do so, we need to develop a better appreciation of the assimilative properties of soils, and also of the various geochemical and biological aids that will increase or enhance the assimilative capability of the soils. This has been the focus of this chapter. Contaminated land not only detracts considerably from one's ability to provide the necessary food supply, but also compromises the receiving waters and sources of water supply for humans. In summary form, the main issues addressed include:

- Impacts from pollutants in the ground need to be mitigated and managed as a beginning step toward protection of the resources in the environment, and also as a first step toward achievement of a sustainable geoenvironment.

- Using the properties of the natural soil–water system as the primary agent for such purposes allows one to address pollution sources that encompass the range from point source to both land-based and atmospheric-based nonpoint sources.

- Enhancing the natural attenuation properties of the subsoil to make it more effective as a control tool allows the subsoil to remain in place as a mitigation management tool. This is the essence of a semipassive remediation reatment of contaminated sites.

- The physical, mechanical, chemical, and biological properties of soil make it a good resource material for management of contaminants and waste products. These properties are responsible for the assimilative capacity of soils and the natural attenuation capability of the soil.

- Mitigation and management of pollutants in the subsoil should seek to reduce and eliminate the presence of pollutants in the soil. Engineering the natural attenuation capability of soils, through enhancements of the attenuation capability with geochemical, biological, and nutrient aids, will provide greater management options.

- Reduction in the concentration and toxicity of pollutants is the ultimate goal. Achievement of this goal can be obtained using various strategies involving technology and the properties of the soil. The various options discussed are by no means the complete spectrum of capabilities. More innovative schemes are being developed to address the problems of contamination of the land environment.

Until the threats and problems of land environment pollution are successfully managed and essentially rendered harmless, the road to sustainability will not be clear. While the treatment in this book does not address the monumental problem of depletion of nonrenewable resources, it nevertheless argues that the impacts from pollution of the land environment, which is the main focus of the concerns in this book, need to be mitigated and managed as a very necessary step toward the road to sustainability of the geoenvironment.

References

Alvarez-Cohen, L. and McCarty, P.L., (1991), Effects of toxicity, aeration and reductant supply on trichloroethene transformation by a mixed methanotropic culture, *Appl. Environ. Microbiol.*, 57:228–235.

Apajalahti, J.H.A. and Salkinoja-Salonen, M.S., (1987), Complete dechlorination of tetrahydroquinone by cell extracts of pentachloro-induced *Rhodococcus chlorophenolicus*, *J. Bacteriol.*, 169:5125–5130.

ASTM D5084-03, (2003), *Standard Test Methods for Measurement of Hydraulic Conductivity of Saturated Porous Materials Using a Flexible Wall Permeameter*, ASTM International, West Conshohocken, PA.

ASTM E1943-98, (1998), *Standard Guide for Remediation of Ground Water by Natural Attenuation at Petroleum Release Sites*, ASTM International, West Conshohocken, PA.

Bader, J.L., Gonzales, G., Goodell, P.C., Pilliand, S.D., and Ali, A.S., (1996), Bioreduction of Hexavalent Chromium in Batch Cultures Using Indigenous Soil Microorganisms, HSRC/WERC, paper presented at the Joint Conference on the Environment, Albuquerque, NM, April 22–24.

Bartha, R., (1986), Biotechnology of petroleum pollutant biodegradation, *Microbial Ecol.*, 12:155–172.

Borden, R.C., Daniel, R.A., LeBrun, L.E., and Davis, C.W., (1997), Intrinsic biodegradation of MTBE and BTEX in a gasoline-contaminated aquifer, *Water Resour. Res.*, 33:1105–1115.

Coles, C.A. and Yong, R.N., (2004), Use of equilibrium and initial metal concentrations in determining Freundlich isotherms for soils and sediments, in *Geoenvironmental Engineering: Integrated Management of Groundwater and Contaminated Land*, Yong, R.N. and Thomas, H.R. (Eds.), Thomas Telford, London, pp. 20–28.

Ehrlich, H.L., (1996), *Geomicrobiology*, 3rd ed., Marcel Dekker, New York, 719 pp.

Frankenberger, W.T., Jr. and Losi, M.E., (1995), Applications of bioremediation in the cleanup of heavy metals and metalloids, in *Bioremediation Science and Applications*, SSSA Special Publication 43, Skipper, H.D. and Turco, R.F. (Eds.), SSSA, Madison, WI, pp. 173–210.

Harkness, M.R., McDermott, J.B., Abramawicz, D.A., Salvo, J.J., Flanagan, W.P., Stephens, M.L., Mondella, F.J., May, R.J., Lobos, J.H., Carrol, K.M., Brennan, M.J., Bracco, A.A., Fish, K.M., Warmer, G.L., Wilson, P.R., Dietrich, D.K., Lin, D.T., Morgan, C.B., and Gately, W.L., (1993), *In situ* stimulation of aerobic PCB biodegradation in Hudson River sediments, *Science*, 259:503–507.

Higgins, I.J. and Gilbert, P.D., (1978), *The Biodegradation of Hydrocarbons. The Oil Industry and Microbial Ecosystems*, Heyden and Son, Ltd., London, pp. 80–115.

Landmeyer, J.E., Chapelle, F.H., Bradley, P.M., Pankow, J.F., Church, C.D., and Tratnyek, P.G., (1998), Fate of MTBE relative to benzene in a gasoline-contaminated aquifer (1993–1995), *Ground Water Monitoring Rev.*, 18:93–102.

Lin, Z.Q., Schemenauer, R.S., Schuepp, P.H., Barthakur N.N., and Kennedy, G.G., (1997), Airborne metal pollutants in high elevation forests of southern Quebec, Canada, and their likely source regions, *J. Agric. Forest Meteorol.*, 87:41–54.

MacDonald, E., (1994), Aspects of Competitive Adsorption and Precipitation of Heavy Metals by a Clay Soil, M.Eng. thesis, McGill University, Montreal, Canada.

Manahan, S.E., (1990), *Fundamentals of Environmental Chemistry*, 4th ed., Lewis Publishers, Boca Raton, FL, 612 pp.

Martin, J.P. and Focht, D.D., (1977), Biological properties of soils, in *Soils for Management of Organic Wastes and Waste Waters*, Elliot, L.F. and Stevenson, F.J. (Eds.), American Society of Agronomy, Madison, WI, pp. 115–172.

Nies, L. and Vogel, T.M., (1990), Effects of organic substrates on dechlorination of Arochlor 1242 in anaerobic sediments, *Appl. Environ. Microbiol.*, 56:2612–2617.

Nyffeler, U.P., Li, Y.H., and Santschi, P.H., (1984), A kinetic approach to describe trace-element distribution between particles and solution in natural aquatic systems, *Geochim. Cosmochim. Acta*, 48:1513–1522.

Pirnik, M.P., (1977), Microbial oxidation of methyl branched alkanes, *CRC Crit. Rev. Microbiol.*, 5:413–422.

Pitter, P. and Chudoba, J., (1990), *Biodegradability of Organic Substances in the Aquatic Environment*, CRC Press, Boca Raton, FL, 306 pp.

Reineke, W. and Knackmuss, H.J., (1988), Annual review of microbiology, in *Annual Reviews*, Ornston, L.N., Balows, A., and Baumann, P. (Eds.), Annual Reviews, Inc., Palo Alto, CA.

Russell, M., Colglazier, E.W., and English, M.R., (1991), *Hazardous Waste Remediation: The Task Ahead*, Waste Management Research and Education Institute, University of Tennessee, Knoxville.

Schwarzenbach, R.P., Gschwend, P.M., and Imboden, D.M., (1993), *Environmental Organic Chemistry*, Wiley & Sons, New York, 681 pp.

Soesilo, J.A. and Wilson, S.R., (1997), *Site Remediation Planning and Management*, Lewis Publishers, New York, 432 pp.

Steiert, J.G. and Crawford, R.L., (1986), Catabolism of pentachlorophenol by a *Flavobacterium* sp., *Biochem. Biophys. Res. Commun.*, 141:825–830.

Strauss, H., (1991), *Final Report: An Overview of Potential Health Concerns of Bioremediation*, Environmental Health Directorate, Health Canada, Ottawa, 54 pp.

Suflita, J.M., Robinson, J.A., and Tiedje, J.M., (1983), Kinetics of microbial dehalogenation of haloaromatic substrates in methanogenic environments, *Appl. Environ. Microbiol.*, 45:1466–1473.

Thayer, A.M., (1991), Bioremediation: innovative technology for cleaning up hazardous waste, *Chem. Eng. News*, 69:23–44.

U.S. EPA, (1999), *Information on OSWER Directive 9200.4-17p: Use of Monitored Natural Attenuation at Superfund, RCRA Corrective Action, and Underground Storage Tank Sites*, EPA-540-R-99-009, U.S. EPA, Washington, DC.

Widdel, F., (1988), Microbiology and ecology of sulfate- and sulfur-reducing bacteria, in *Biology of Anaerobic Microorganisms*, Zehner, J.B. (Ed.), John Wiley & Sons, New York, pp. 469–585.

Yong, R.N., Boonsinsuk, P., and Tucker, A.E., (1984), A study of frost-heave mechanics of high clay content soils, *Trans. ASME*, 106:502–508.

Yong, R.N. and Mulligan, C.N., (2004), *Natural Attenuation of Contaminants in Soils*, Lewis Publishers, Boca Raton, FL, 319 pp.

Yong, R.N. and Warkentin, B.P., (1975), *Soil Properties and Behaviour*, Elsevier Scientific Publishing Co., Amsterdam, 449 pp.

11

Toward Geoenvironmental Sustainability

11.1 Introduction

Land and water resources are the principal physical capital items that constitute the geoenvironment. Their quality and health are vital issues because they provide the habitat and also the basis for life support systems for plants, animals, and humans. These are natural resources that are basically renewable resources. In the absence of human-created stressors generating negative impacts, the various resource elements have the ability to maintain their natural quality through natural processes or through replenishment. These resource elements have the ability to flourish and grow and are integral parts of the overall environment. We consider a significant part of *sustainability of the geoenvironment* to include sustainability of land and water resources. Sustainability is obtained when all the resource elements of the various land and water ecosystems are renewed, replenished, recharged, and restocked — to a level that will continue to meet the needs of those that depend on these resources. This will only occur when the health and quality of the land and water resources are protected, maintained, and allowed to flourish. Failure to do so will lead to a degradation of the quality of these two major capital resources, and in turn will imperil and diminish the capability of these resources to allow the elements and habitants of the multitude of land and water ecosystems to renew and replenish themselves.

The focus of this book has been directed toward the major impacts or deterrents to sustainability objectives arising from human activities. The primary sustainability concerns relating to industry and urban interactions with the land environment and its receiving waters are impacts from:

- Exploitation and extraction of renewable and nonrenewable resources
- Depletion of nonrenewable resources and misuse and mismanagement of renewable resources
- Noxious discharges from industrial operations and urban activities

A variety of implications and impacts arising from these concerns have been discussed in the previous chapters. As noted in the concluding section of the previous chapter, the point source pollution impacts from urban and industrial activities account for one major component of the large ground contamination problem. The other major component is atmospheric-based and land-based nonpoint source pollution. Figure 11.1 shows some of the elements of these two major components and the natural resources threatened by both kinds of contaminant sources.

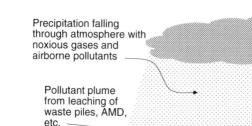
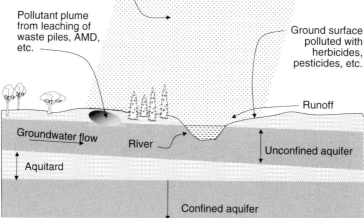

Precipitation falling
through atmosphere with
noxious gases and
airborne pollutants

Pollutant plume
from leaching of
waste piles, AMD,
etc.

Ground surface
polluted with
herbicides,
pesticides, etc.

Runoff

Groundwater flow

River

Unconfined aquifer

Aquitard

Confined aquifer

FIGURE 11.1

Some of the more prominent causes of pollution of recharge water for rivers, other receiving waters, and groundwater (aquifers). Contamination of the confined aquifer depends on whether communication is established with the unconfined aquifer.

11.1.1   Unsustainable Actions and Events

Unsustainable actions refer to those actions and circumstances that generate negative impacts to, in, or on the geoenvironment. Discussions on many of these impacts or activities and events contributing to such impacts have been given in the previous chapters. The following brief discussion in this section provides a set of examples that demonstrate the problem at hand and also why the quality and health of the land and water resources are central issues.

11.1.1.1   Iron and Coal Mining

A recounting of the 1966 major tip failure (slippage) at Aberfan that was responsible for the deaths of more than 100 individuals can be found in McLean and Johnes (2000). The height of the piles together with the slope of the piles and the weather conditions contributed to the disastrous slippage problems. Even without the disaster of tip failure, these tips, heaps, and piles constitute a blight to the land environment, and leaching of these piles will produce conditions that will (1) render the landscape sterile (Haigh, 1978) because of the acidity and deficiencies in nitrogen and phosphate and (2) threaten the groundwater and water resources if and when the leachates reach these resources.

11.1.1.2   Oil and Petroleum

Exploration, drilling, extraction, and production of petroleum can have an impact on the environment at numerous points. Contamination of shores after oil spills can lead to oil that can remain for decades at beaches or marshes. Pressurized hoses with cold or hot water are used for cleanup of spills on beaches, etc., and may create more problems — as for example in the case of the *Exxon Valdez* spill, where the more than 200,000 tonnes of disposable diapers, pads, clothing, and other waste materials used in washing individual stones on the cobble beaches required landfilling or burning (Graham, 1989). In addition, the hot spray used in the cleaning activities is reported to have caused damage to the benthic fauna.

Oil spills and pipeline leaks may also contaminate the soil and groundwater. Drilling mud stored in improperly lined pits may also leak into the soil. Well blowouts from overpressurized zones can lead to spreading of the petroleum on the surface of the land or water if at sea. On March 29, 1980, 468,000 tonnes of crude oil (the largest spill in history) rushed into the Gulf of Mexico after a blowout, thereby affecting Mexican and Texas beaches. Approximately 0.8% of the oil reached the Texas shoreline, with about 5% of the original amount remaining after 1 year (Payne and Phillips, 1985).

According to the Toxic Release Inventory data for the years 1992 to 1995, in 1993, 700,000 tonnes of waste were produced during petroleum refining (U.S. EPA, 1995b). Of this amount, 30% was released to the environment or transferred off-site. Pollutants included volatile organic compounds (VOCs), carbon monoxide, sulfur oxides, nitrogen oxides, particulates, ammonia, hydrogen sulfide, metals, spent acids, and other materials as air emissions, in the water, and sludges. The other 70% was recycled, recovered for energy, or treated. Direct discharges to the environment or disposed of off-site accounted for 4% of the wastes.

11.1.1.3   Medical Wastes

In the United States, medical wastes make up approximately 0.3 to 2% of the municipal solid waste stream (U.S. EPA, 1990). The infectious waste is estimated by the Environmental Protection Agency (EPA) at 2 to 3 million tonnes per year. Pathogen transport and survival is possible if these types of wastes are disposed as municipal solid waste (MSW). However, there is little experimental data available for the determination of pathogens in leachates or in runoff from a landfill. Approximately 15% of hospital waste is infectious. Medical waste incinerators are used by 70% of U.S. hospitals. There is little regulation of these incinerators due to their small size. As a result, in some states, such as Washington, between 48 and 87% of the incinerators have no emission control equipment (Washington State Department of Ecology, 1989). Low-level radioactive waste is generated through various medical procedures involving radiopharmaceuticals, radiology, and nuclear medicines. However, this type of waste from hospitals is less than 5% of the total waste in the United States (U.S. Congress OTA, 1989). Most of these wastes have a very short half-life and can be stored on-site.

11.1.1.4   Pulp and Paper

Between 16,600 and 50,000 l of water per tonne of pulp are required for paper production (U.S. EPA, 2002). Effluent solids, biological oxygen demand, color, and toxicity are present in the wastewater. Approximately 30 to 70% of the contaminants are removed during treatment. The pulping and bleaching stages are the major sources of pollutants. Sludge generation is in the range of 14 to 140 kg per tonne of pulp. In the United States, total generation is about 2.5 million tonnes per year. Chlorinated organic compounds can be released from the sludge. Although these levels have been reduced, the pH of the sludge is typically still higher than 12.5. Energy consumption trends have shifted in recent years. From 1972 to 1999, wood wastes, spent liquor solids, and other internal sources of energy have increased from 41% of the energy used to 58%, thus relying less on fossil fuels.

11.1.1.5   Cement, Stone, and Concrete

Wilson (1993) reviewed the environmental considerations of concrete. In 1991, according to the U.S. Bureau of Mines, more than 1.15 billion tonnes were produced worldwide. Cement is the main ingredient of concrete. A typical mix is 12% Portland cement, 34% sand, 48% crushed stone, and 6% water, all of which are abundant. However, transportation can be a main issue, and water required for washing and reduction of the impact of dust. Fly ash, a waste from coal-fired power plants, is now present at a proportion of about 9% in cement. Energy consumption is considerable in cement production. Addition of fly ash with concrete can make the process more energy efficient. Since coal is often used for energy emissions of carbon dioxide, sulfur and nitrous oxide are high; carbon dioxide emissions are estimated at 1200 kg CO_2 per tonne of cement — 60% from energy use and the other 40% from calcining. Dust emissions are also significant (180 kg/tonne of cement produced), but should be controlled as much as possible by water sprays, hoods, etc. Other pollutants include sulfur dioxide and nitrous oxide from both fuels and raw materials. Water pollution from washwater of high pH is generated. Settling ponds are used to remove the solids. Reduction of the pH below 12 renders the wastewater not hazardous. Water usage has been reduced through recycling in the plant in a closed loop. Concrete has accounted for up to 67% by weight of construction and demolition (C&D) waste. More and more of this waste will be used in road aggregate. Precasting concrete at a central facility can reduce materials used and wastewater generated.

In the stone and concrete products industries, more than 530 million tonnes of wastes were generated in 1993 (U.S. EPA, 1995c). Approximately 96% was recycled, treated, or recovered for energy, while 2.3% was transferred off-site or released to the environment. Off-site disposal, underground injection, and air, land, or water discharge accounted for 2.2% of the waste.

11.1.1.6   *Various Stressors and Impacts*

Waste generation and disposal demonstrate the problems of stressors and impacts. For example, waste production by industry and commerce in England from 1998 to 1999 (DEFRA, 2000) was estimated at 68.8 million tonnes. This does not include the 294 million tonnes produced by demolition, mining, quarrying, agricultural wastes, construction wastes, and sewage sludge. In the Netherlands, 35 million tonnes more manure are produced than can be utilized by arable farming (Tirion, 1999).

Tanacredi (1999) reports that as a result of human development, 115,000 ha of wetlands are removed each year in the United States. Less than 40% of the area remains compared to preindustrial times. A key example of this occurred in Europe. Construction of a dam for energy purposes in Slovakia caused the Szigetkoz marsh (500 m^2) in Hungary to dry up, depriving 5000 species of flora and fauna of a habitat.

More than 360,000 tonnes of waste were generated in 1993 by the fabricated metals industry (U.S. EPA, 1995a). According to the Toxic Release Inventory (TRI), approximately 62% of the waste was either recycled or treated, or the energy was recovered on-site. Another 34% of the waste was either released to the environment or transferred off-site. Direct releases by air, water, land, or underground injection or disposal accounted for 13.2% of the waste.

The impacts from contaminants, together with impacts from utilization of energy and other natural resources, on the geoenvironment are the main geoenvironmental issues addressed as barriers to sustainability goals. In this chapter we will (1) discuss the use or exploitation of nonliving renewable natural resources, (2) look at some typical case histories and examples of sustainability actions, and (3) present the geoenvironmental perspective of the present status of *where we are in the geoenvironmental sustainability framework*, with a view that points toward *where we need to go*.

11.2   Exploitation and State of Renewable Natural Resources

By all accounts, metal and mineral resources together with fossil fuels are nonrenewable natural resources. There is also a case to be made for classifying water from deep-seated aquifers as nonrenewable resources. Their continued extraction will not only result in their depletion, but will ultimately lead to their exhaustion. Fossil fuels are of particular concern since they are not only used for fuel, but also are the main source of raw material for countless numbers of products, ranging from plastics and synthetic fibers to pharmaceutical and other consumer products. This eventuality has been recognized, and efforts have been and are being made by industry and consumers to (1) reduce or find more efficient fuel consumption engines and (2) find renewable substitutes. It is not within the purview of this book to discuss alternate fuels for powering engines and motors. There are numerous textbooks and research literature dealing with advances and innovative ideas in this subject.

In respect to generation of electricity outside of hydroelectric facilities and oil- and coal-generating plants, various alternative and renewable sources of electricity generation are available. These include tidal wave, solar thermal, solar photoelectric, geothermal, winds onshore and offshore, and biomass. It is estimated that all of these combined to produce almost 400 tW-h (400×10^9 kW-h) of electricity in 2002. These important subjects are also not within the purview of this book. The reader is advised to consult the numerous practical articles and books dealing with energy conservation and use of nonrenewable natural resources.

There are two specific classes of renewable natural resources: living and nonliving. Living renewable natural resources include land and aquatic animals, forests, native plants, etc., while nonliving renewable natural resources include water and soil. By definition, renewable natural resources refer to those resources that have the capability to regenerate, replenish, and renew themselves, either naturally or with human intervention, within a reasonable time period. *Sustainability* as an objective requires that full regeneration–replenishment of renewable resources must be obtained. It is recognized that when consumption (use, exploitation, etc.) exceeds the regeneration–replenishment rate, sustainability of the renewable resource is not obtained. This does not mean that the renewable resource will not or cannot renew itself. It simply means that the amount or rate of the resource that is renewed is insufficient to meet the demands placed on it. In recognition of this, we need to distinguish between (1) unsustainable renewable natural resources, i.e., renewable resources that by virtue of circumstances cannot be fully renewed or replenished, and (2) sustainable renewable natural resources, i.e., renewable resources that can be totally regenerated and replenished. When the consumption rate is greater than the rate of regeneration or replenishment, the amount or nature of the particular renewable natural resource will be depleted and may eventually become extinct. Striking examples of this are overfishing and overuse of groundwater (water from aquifers).

11.2.1   Sustainability of Renewable Nonliving Natural Resources

There is a further distinction or differentiation needed in discussing renewable natural resources. One needs to distinguish between *natural* and *developed* resources. Differentiation between renewable *natural* resources and renewable *developed* resources is necessary to distinguish between the renewable natural capital items (water, soil, land and aquatic animals, native plants) and restocking and regeneration of man-made capital, such as fish farms and agricultural output from land farming. As has been noted previously, just because a resource is renewable does not make it sustainable. Two necessary, but not sufficient, conditions for sustainability of renewable natural resources are (1) replenishment and regeneration of the natural capital items in a reasonable time frame, either through natural processes or through sound management practice, and (2) that renewed natural resources are sufficient and will continue to be sufficient to meet the demands placed on them. Impediments to sustainability are due to (1) rate of recharge or regeneration or replenishment being outpaced by overexploitation of the natural resource, and (2) corruption, degradation, or pollution of the natural resource.

The renewable nonliving natural resources of prime importance are water and soil. They are in essence renewable dynamic resources — characterized by recharge and replenishment of these resources. However, when recharge and replenishment cannot overcome the deficits in the nonliving renewable resources, these resources are no longer sustainable. A good example of the preceding is the excessive use, overexploitation, and pollution of water resources. A full treatment of these and other issues relating to sustainable water use will be found in the textbooks dealing with this particular problem. The geoenvironmental concerns for water and soil are in respect to degradation in water and soil quality due to their misuse and also due to pollution of these capital items. Discussions in the previous chapters have shown that water and groundwater pollution, together with soil pollution and loss of soil quality, is the major downfall of sustainability of water and soil resources — other than overuse, abuse, and misuse of these nonliving renewable resources by humans.

The beetle-type diagram in Figure 11.2 provides a very simple illustration of some of the major stressors on water, groundwater, and soil resources responsible for the unsustainable outcome of the nonliving renewable natural resources. Other than the use, misuse, abuse, etc., by industry and humans shown in the top left-hand side of the illustration, most of the stressors and their impacts have been identified and discussed in the previous chapters. Some of the key actions needed to drive the renewable natural resources of

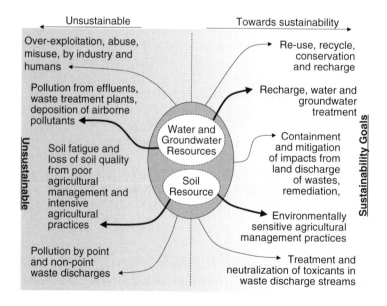

FIGURE 11.2

Some of the major stressors on groundwater, water, and soil resources responsible for the unsustainable state of these nonliving renewable natural resources (shown on the left). A sample of the types of actions needed to permit these renewable natural resources to meet sustainability goals and objectives.

water, groundwater, and soil toward sustainability goals and objectives are shown on the right-hand side of the illustration. The major impact of pollutants on both water and soil resources is degradation of the quality of these resources. In the case of soils, for example, degradation of soil quality will lead directly to loss of productivity and lower agricultural yields. For water, degradation of water quality leads directly to loss of drinking water status at the upper end of water usage, to relegation for agricultural and industrial use, and finally to its being rendered nonusable.

Pollutants are chemical stressors that can severely affect water, groundwater, and soil quality. These include land-based nonpoint source pollutants such as herbicides, pesticides, fungicides, etc., and point source pollutants from effluents, waste treatment plants, and liquid discharges as wastes and spills from industrial plants (e.g., heavy metals and organic chemicals). Some of the methods and procedures for mitigating the major impacts from pollutant stressors, together with treatment and remediation options, have been discussed in the previous chapter. Undoubtedly, there will be more advanced and sophisticated methods developed in the very near future to meet the challenges posed by these stressors. Unsustainable renewable natural resources become nonrenewable natural resources — a condition that should not be allowed to happen.

11.2.2 Geoenvironment and Management of Renewable Resources

From a geoenvironmental perspective, the actions shown on both sides of Figure 11.2 are some of the more significant actions that detract from or lead to conservation and recovery of nonliving renewable resources. The arrows that emanate from the total elliptical shell indicate actions affecting the total land environment, i.e., both soil and water–groundwater resources. Those arrows emanating specifically from the top or bottom resources (water–groundwater resources and soil resources) indicate actions affecting those specific resources. Not all the pertinent or necessary actions for all of these are shown. Only the more significant ones are depicted in the diagram.

11.2.2.1 Unsustainable Actions

Many of the impacts due to the unsustainable actions shown on the left-hand side of Figure 11.2 have been discussed in the previous chapters and have been referred to in Section 11.1.1. The points of note include: (1) Overexploitation, abuse, and misuse of water and soil resources by humans are actions requiring corrective measures that extend beyond the geoenvironmental sphere. (2) Pollution from effluents, waste treatment plants, and deposition of airborne pollutants is both a point and a nonpoint source of contaminants that degrades water quality and creates contaminated ground. The extent and seriousness of health threats from the pollutants and the degree of degradation of both water and land resources are functions of both the quantity/concentration and toxicity of the discharges and atmospheric depositions on land. Runoffs over land surface and industrial chemical spills can severely pollute receiving waters and groundwaters.

The tainted water problem in the town of Walkerton, Ontario, Canada, that led to the deaths of seven residents in the period from May to the end of July 2000 (referred to in Chapter 3) is testimony to the effects of runoff and infiltration of polluted recharge rainwater. Reported evidence indicated that heavy rains on May 12 washed bacteria from cattle manure into Walkerton's shallow town well, and that town residents were exposed to *Escherichia coli* over the next few days. As reported, more than 500 residents reported *E. coli* symptoms and a further 150 residents sought hospital treatment (http://camillasenior.homestead.com/Walkerton_Chronology-2004.pdf).

Another example of note is the Love Canal problem that surfaced in the 1970s. As is well known, this is probably the first well-publicized hazardous waste dumping site in North America (September 2005, http://onlineethics.org/environment/lcanal/index.html). Prior to this period, awareness of the problems and seriousness of indiscriminate dumping of toxic materials was not appreciated by the general public, and it was claimed that the Love Canal site was the recipient of such hazardous materials for a period of at least 20 years. Tests conducted by the New York Department of Environmental Concern showed severe pollution of ground and waters in the area, resulting in the declaration of a state of emergency by the governor of the state of New York and the closing of schools and relocation of several families.

Many other examples of water and ground contamination from inadvertent and deliberate dumping can be cited. There are other cases of pollution, however, that are indirectly caused by man-made activities. A good example of this is leaching of exposed sulfide ores and rocks (Chapter 5), and waste piles in landfills will also produce polluted recharge water. Other causes of pollution of water resources have been discussed in Chapter 3.

Soil fatigue and loss of soil quality can occur from natural causes such as those leading to aridification and desertification. Long periods of rainfall deprivation leading to aridification and finally desertification are conditions of nature. Desertification can also result from a prolonged process of degradation of a once-productive soil. The root causes for desertification are deemed to be a complex mix of various degradative actions. The present concern is soil fatigue and loss of soil quality from human activities leading to poor forest and agricultural management, and intensive forestry and agricultural practice are conditions that will eventually render the soil useless for production of plants and crops. In respect to agriculture, as we have noted in Chapter 2, *soil quality* is a determinant of the capability of a soil to sustain plant and animal life and their productivity, and any diminution of soil quality will have an impact on its capability to provide the various functions, such as plant and animal life support and forestry and woodland productivity, and will undoubtedly result in the loss of biological activity and biodiversity and depletion of nutrients in the soil.

We should note that soil quality as a measure of the functionality or capability of a soil is not confined exclusively to the agricultural usage. Soil can also serve other kinds of functions. These include (1) containment and management of wastes and waste streams, (2) resource material for production of building blocks, and (3) sub-base support for structures and facilities. The determinants for soil quality for these types of functions will differ from the classic definition, which was developed for agricultural use. This is discussed further in Section 11.3.

11.2.2.2 Toward Sustainability

For sustainability management of the land environment and particularly of the water and soil resources, a necessary requirement is for recharge materials and processes to be devoid of pollutants and other detrimental and degrading agents. This is particularly acute for reuse and recycle of process water. At all times, the quality of water and soil needs to be maintained and even improved. For this to occur, we need to establish water and soil quality indices and further establish baseline values for these indices. These indices will require analyses involving indicators — both status and material indicators. A more detailed discussion of these and the quality indices will be given in a later section in this chapter.

11.2.2.2.1 Recharge of Water Resources

The sources of natural recharge of receiving waters and groundwater are direct precipitation (rain, snow, hail, and sleet) and snowmelt delivered as percolation and infiltration.

The chemistry of precipitation that defines the quality of the precipitation is a function of the nature, chemistry, and concentration of airborne particulates through which precipitation occurs. As noted in Chapter 10, the noxious substances in the atmosphere derived from man-made activities include NO_x (nitrogen oxides), SO_2 (sulfur dioxide), CO (carbon monoxide), Pb and other metals (such as Al, As, Cu, Fe, La, Mg, Mn, Na, Sb, V, and Zn) as airborne particulates, volatile organic chemicals (VOCs), aromatic hydrocarbons (such as benzene, toluene, and xylene), and polycyclic aromatic hydrocarbons (PAHs) (such as anthracene and naphthalene). The same holds true for the chemistry of the snowpack that serves as the storage for recharge as snowmelt. Thus, for example, Nanus et al. (2003) report that high-elevation areas in the Rocky Mountains annually receive large amounts of precipitation, most of which accumulates in a seasonal snowpack. They maintain that all the accumulated atmospheric deposition is delivered in a very short period of time to the ground and receiving waters during spring snowmelt. The presence of the noxious gases, together with other airborne particulates, ensures that the pH of rainfall onto the ground surface will be acidic. Spatial variations in atmospheric deposition of acid solutes are the result of precipitation amount in combination with concentration, and that deposition does not necessarily reflect variations in concentration alone (Nanus et al., 2003).

Deposition of airborne particulates with rainfall will also ensure that these will be carried with the surface runoff and also with infiltrating water. The other causes of pollution of precipitation recharge water include (1) runoffs from polluted land surfaces, as might be found on agricultural lands, and (2) infiltration into subsurface through land surface polluted with pesticides, fungicides, other surface wastes, organic debris, heaps, leach piles, sulfide rock piles, etc., as illustrated in Figure 11.1. The evidence shows that in regions where urbanization, industrialization, and exploitation are present, it is difficult to find precipitation recharge devoid of pollutants and airborne pollutants. Furthermore, in these regions, it is also difficult to rule out pollution of the receiving waters and groundwaters from contaminated runoffs and infiltration. For regions remote from the effects of industrialization and urbanization, and also far remote from airborne pollutants, one would have better chances of finding uncontaminated recharge precipitation. Treatment of polluted or contaminated recharge precipitation is not generally practical since it is more than likely that pollution already exists in the water and land receptors in urbanized and industrialized regions. Instead, passive treatment using natural processes, together with aggressive treatment of extracted water, is used to provide safe drinking water (see Chapter 3).

11.2.2.2.2 *Improvement of Soil Quality for a Sustainable Soil Resource*

Soil is an important resource material. It contains most of the nutrients required for plant growth and is rich with microorganisms. Besides being the most critical medium for agricultural food production, as well as production of other kinds of crops and trees, such as cotton and palm trees, it is also a very important tool for management of wastes and waste discharges in the ground, as seen in the previous chapter. It serves as a dynamic resource not only for production of food and raw materials, but also for the soil microorganisms contained in the soil. These microorganisms not only play an important role in the natural bioremediation of harmful organic chemicals in the ground, but also participate intimately in the recycling of carbon, nitrogen, phosphorus, and other elements in the soil. In essence, they are significant contributors to the control or management of greenhouse gases, water flow in soils, soil quality, and, through all of these, the life support systems for humankind.

It is recognized that loss of nutrients, loss of biodiversity, loss of soil organics, salinization, acidification, and degradation of physical, chemical, and biological properties of soil occur with time — through leaching processes, intensive agricultural practice, erosion,

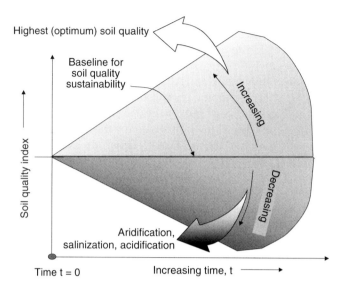

FIGURE 11.3

Increasing (enhanced) and decreasing (diminished) soil quality in relation to time. The soil quality index (SQI) is a composite index determined on the basis of measures of achievement of levels required by prescribed physical, chemical, and biological indicators. Note that different SQIs are needed, depending on whether one is concerned with agricultural production or, for example, the use of soil for waste management.

and overuse and poor land management. Natural or man-assisted recharge of soil as a resource material, i.e., recharge of soil quality, is required if sustainability of soil quality is to be achieved. We consider *recharge of soil quality* to consist of any or all of the types of physical, chemical, and biological amendments, methods, and processes that serve to increase soil quality. Figure 11.3 shows a schematic description of soil quality changes with time. The ordinate shown in the diagram represents soil quality index (SQI). The SQI is a composite index that incorporates analyses that include consideration of the physical, chemical, and biological indicators relating to the soil resource application under review. Thus, for example, the specific component indicators for the physical, chemical, and biological indicators leading to calculations of the SQI for waste management would differ from those obtained for agricultural production or for forestry. A more detailed discussion of the SQI and the various indicators will be found in a later section in this chapter.

Artificial recharge, i.e., recharge through human intervention, involves the use of soil conditioners, fertilizers, added nutrients, biological agents, and good agricultural land management practice. The term *soil amendments* is used as a general catch-all term to include all the preceding items and processes. These soil amendments, when introduced into a tired soil, are designed to improve the soil qualities through improvement of the physical, chemical, and biological properties of the soil — to allow for better use of the soil as a resource material. Improvements in soil permeability, infiltration, water retention, nutrient-holding capacity, and soil structure — on top of the addition of nutrients and fertilizers — constitute the artificial soil recharge process.

11.2.2.3   Protection of Soil and Water Resources

The impacts of significant consequence on the soil and water resources in the geoenvironment are due to physical disturbances of the land environment and direct or indirect contamination of the soil and water resources, as shown in Figure 11.2. The various factors, conditions, and circumstances wherein impacts are generated have been discussed in the previous chapters. The three different categories for protection of the soil and water resources in the land environment are:

- Category 1: Direct and specific protective measures to ensure no degradation of soil and water qualities. This category of actions and measures assumes that the soil and water qualities are at acceptable levels, and that with proper management, they will be sustainable.

- Category 2: Measures and actions to mitigate and minimize detrimental impact to both soil and water qualities. We assume that the impacts are managed to the extent that their effects do not degrade both soil and water qualities, and also do not pose health threats to humans and other biotic receptors.

- Category 3: Application of treatment and remediation technologies to return soil and water qualities to levels acceptable for use. This category applies to situations where soil or water quality has degraded to the state that treatment and remediation are required to return it to the level of quality required for use.

The actions included in the various measures undertaken for the three categories of protection and management of soil and water resources range from passive to aggressive. The physical and chemical buffering properties of the soil are central to the effectiveness of the passive protection technique. At the other end of the scale are physical protection barriers such as liner–barrier systems that prevent the migration of leachates and pollutants, and treatment and remediation techniques that require aggressive physical, chemical, and even biotechnological intervention. Application of any of these techniques depends on (1) the nature and scope of the perceived threat, (2) the resource being threatened and its functions or use, (3) the predicted intensity and type of damage done to the resource by the impact, (4) the extent of resource protection needed, and (5) the economic impact. In Figure 11.1, for example, assuming the absence of noxious gases and airborne particles, the perceived source of threat is represented by the pesticides, insecticides, and other surface pollutants that will move toward the river and also infiltrate into the ground during periods of precipitation. The resources being threatened are the river, surface layer soil, and unconfined aquifer. Assuming that the river and the unconfined aquifer serve as drinking water sources, and further assuming that the soil is an agricultural soil resource, the need and extent of protection required for these resources will be evident.

11.3 Water and Soil Quality Indicators

The discussion in this section extends the discussion on indicators in Section 9.2 of Chapter 9. Figure 9.1 in Chapter 9 depicts the role of indicators in the situation created by precipitation falling through airborne noxious gases and particulates. The water and soil quality indicators identified as monitoring targets include both system status and material performance — or material property–status types. Water quality indicators and soil quality indicators are essentially material property–status indicators. They are meant to indicate the quality of the material (water or soil). The quality of the material under consideration or analysis is established with specific reference to its intended function. In regard to soil quality, for example, we have seen from the previous section that the classic definition developed for agricultural use needs to be broadened to encompass the use of soils for various other purposes — from waste management to building supplies and construction.

This is also true for water quality indicators. The range of usage starts from the top, with drinking water standards setting the height of the water quality bar. At the low end of water usage would be water for agricultural purposes and other similar functions. Indicators for all the various functions of both water and soil would vary in both form (type) and detail.

There are several levels of specificity (i.e., levels of detail) in the prescription of water quality and soil quality indicators. These depend on (1) the intended function and management goals, as for example drinking water usage or irrigation purposes, (2) the ability to obtain all the necessary data sets, (3) the available and applicable remedial or corrective technological capabilities, (4) the scale and risk tolerance, and (5) economic factors. Perhaps the overriding factors in all of these are *management goals* and *risk tolerance*.

11.3.1   Quality and Index

In Figure 11.3 and in the previous section, we talked about soil quality index (SQI) as a measure of soil quality. Similar to the different intended functions for water and soil, determination of SQI and water quality index (WQI) will also depend on many of the same factors described in the preceding paragraphs. Development of indices requires full consideration of the many different properties and influences that ultimately combine to produce the material status. Since this is a dynamic process dependent on applications or processes applied to the soil, internal soil reaction rates, and elapsed time, the indices will also vary in accord with circumstances and time. Quantification of SQI and WQI permits one to arrive at determinations that show whether the material, and finally the system itself, will be sustaining. Taking the SQI as an example and referring to Figure 11.3, when calculations show that the SQI at any one particular time is greater than the baseline value, we will have increasing soil quality, and we can be assured of the sustainability of the function served by the soil. Evaluation and quantification of SQI is application or function specific; i.e., they depend on management goals for the material (water and soil).

11.3.1.1   *Example of SQI Development*

To illustrate the procedures that one would follow to evaluate and determine the appropriate SQI, we will use the role of soil as a resource material for management of the impacts from contaminant discharge into the ground. We recall from the previous chapter that the basic properties contributing to the development of the assimilative capability of soils are physical, chemical, and biological. From this starting point, determination of what pertinent attributes are significant and measurable is required. Furthermore, it needs to be determined whether or how these attributes vary with circumstances specific to the problem at hand, i.e., functions or use of the soil. Figure 11.4 is a schematic illustration of the physical, chemical, and biological properties that are considered to be significant in the development of the assimilative capability of the soil. If one were to compare the kinds of attribute data sets with the information given in Figure 9.19 in Chapter 9, it would be immediately evident that many of the basic interactions developed between contaminants and pollutants have been incorporated in the measured attributes. We use the term *attribute* in the discussion to mean the property or characteristic being measured.

The data obtained from tests and other kinds of measurements (field and laboratory) of the physical, chemical, and biological attributes can be used as (1) input to compare with individual attribute indicators, thus leading to immediate comparison of the sustaining capability of each individual attribute, or (2) input to statistical and analytical models

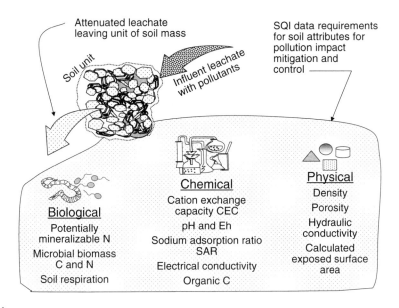

FIGURE 11.4
Soil properties pertinent to use of soil as a management tool for mitigation of impact from contaminants discharge in the ground. Data from these attributes serve as input to determination of the pollution mitigation soil quality index (SQI).

developed to produce a *lumped* (combined or total) index parameter. Prescription of individual attribute indicators is performed once again on the basis of the intended function of the soil. Take, for example, the set of attributes in the "Physical Properties" list in Figure 11.4. Density, porosity, calculated exposed surface area, and hydraulic conductivity have been chosen as the set of pertinent attributes. Consider two specific applications for the soil: (1) use as a permeable reactive barrier (PRB) material, as in Figure 10.5 in Chapter 10, and (2) use as an engineered clay barrier (ECB) in the liner system, shown in Figure 1.10 in Chapter 1. The primary controlling property in both the PRB and ECB applications is the hydraulic conductivity. For the PRB, one permits the transporting fluid to penetrate the PRB at a rate that allows for partitioning and transformation processes to occur. Residence time in the PRB is paramount. This is controlled by an appropriate soil permeability and thickness of the PRB. In general, one might want to design a wall thickness in conjunction with a Darcy coefficient of permeability k in the range of 10^{-5} to 10^{-7} cm/s, depending on the partitioning processes envisaged. In the case of the ECB application, a k value of considerably less than 10^{-7} cm/s is generally sought. The prescription of k indicators for desired objectives can now be obtained. The prescription of attribute indicators and their applications can be seen in Figure 11.5 for the example of hydraulic conductivity. Given that hydraulic conductivity (as characterized by the Darcy k value) is a direct function of density and porosity, i.e., $k = f(\gamma, n)$, where γ refers to soil density and n refers to porosity, the *weighting* of data for and n becomes important. Application of weighting factors in such situations is to a large extent based on knowledge of previous behavior.

Determination of weighting factors to be used for all the data sets relating to the physical, chemical, and biological attributes can be a challenging task. Much depends on the experience and knowledge of the analyst. The results of the weighted data are used in a deterministic model that is designed to produce a lumped index known as the quality index. As stressed previously, the quality index will have a prefix that denotes the function

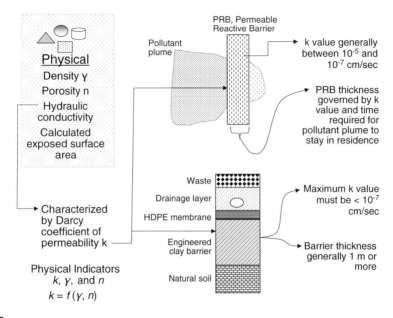

FIGURE 11.5
Prescription of hydraulic conductivity indicators based on intended function of soil. Since the *k* value determined is a function of density and porosity, the weighting factor for *k* is considerably larger than those for density and porosity. The calculated exposed surface area is a parameter of interest and may be neglected. The thicknesses of the PRB and ECB are determined by situation-specific conditions. The 1-m or more thickness for the ECB is quite common for containment of municipal solid waste landfills.

of the material, as for example *drinking water quality index* and *pollution mitigation soil quality index*.

11.3.1.2   Water Quality Index

Research and development of water quality indices (WQIs) have not received the same level of attention as SQI in soil science. Instead, attention has been focused more on the establishment of national standards. Because of the very direct relationship between the quality of drinking water and human health, drinking water quality standards are set by the regulatory bodies for most, if not all, of the countries in the world. As discussed in Section 3.3.1 in Chapter 3, the main parameters usually monitored for drinking water are biological oxygen demand (BOD), color, turbidity, N, P, suspended solids, odor, heavy metals, VOCs, pesticides, bacterial level (such as coliform-forming units, or CFUs), and perhaps other microorganisms. We should note that infectious diseases caused by pathogenic bacteria, viruses, and protozoa in drinking water are by far the greatest health threat.

   While these national standards set the basis for the water quality for the various countries, production of a national drinking WQI for each and every country will require considerable effort in developing the relationships and weighting functions. To a large extent, this is because of the risk–benefit approach adopted by many responsible authorities in articulating quantitative values for the parameters chosen for monitoring. The risk–benefit consideration is not economically driven — at least not directly. Rather, it is driven by the need to provide acceptable drinking water to most people without endangering public health. Setting standards that may be considered "too stringent" may on the one hand be prudent and safe, but on the other hand may make drinking water unavailable to a large percentage of the population, especially in regions of water deprivation. Because of these kinds of factors, development of drinking WQIs becomes more than a challenge.

11.4 Sustainability Attempts

We will look at some case studies in this section. These are cases demonstrating either sustainable practices and the implementation of methods to evaluate the sustainability of a project, or process with regards to the geoenvironment. At least one case study is presented for each of the sectors urbanization, resource exploitation, food production, industrial development, and marine environment.

11.4.1 Rehabilitation of Airport Land

Between 150,000 and 350,000 m^3 of soil in the contaminated site at a former Norwegian airport needed to be remediated (Ellefsen et al., 2001). A method was used to incorporate environmental effects into an evaluation of different remedial options. The environmental costs and benefits were determined, which became part of the decision assessment. One of the main environmental targets was the reuse of the treated soil for landscaping. Asphalt and concrete would also be reused. To perform the assessment, a model developed by the Danish National Railway Agency and the Danish State Railways was utilized. A life cycle approach for the remediation was used. Consumption of materials, fuel and energy, effects of noise, odor and other annoyances to humans, and emissions to air, soil, and water were calculated.

The site (including soil and groundwater) was contaminated with diesel and heating oil between 3.5 and 5 m below the surface due to leaking storage tanks and runoff water. Free-phase oil was also found. Two remedial options were chosen and compared: (1) excavation followed by biological treatment and (2) *in situ* treatment by biosparging and removal of six tanks.

Energy consumption for the excavation option was found to be five times higher than for the *in situ* procedure. Additionally, emissions of greenhouse gases were estimated to be three times more than with the *in situ* option. For the *in situ* option, electricity consumption was considered to be the main environmental cost of the biosparging process. However, as hydroelectricity is the major source of electrical power, negligible amounts of CO_2 were emitted. Material consumption was equal in both cases, but differed in origin. Iron and manganese were required for excavation machines, while nickel and copper were used for the air injection pipes and electrical materials. Biosparging was thus selected as the remedial option. The area is used for parking and storage facilities and will be used in the future for housing and parks. Other environmental assessments in the future will be performed using this approach.

A follow-up report by Ellefsen et al. (2005) indicated that the cleanup was completed in 2003. The area will become a green area with reserves for nature. Another challenge of the area was the management of the asphalt and sub-base contamination from PAHs from an old runway. Instead of transporting the waste to a hazardous waste facility, bitumen at a level of 3% was added to 20,000 tonnes of the soil via a cold mix process. The stabilized mixture was then used as a road foundation in the area. Leaching tests with water and road salt indicated that the material was appropriate for reuse. Another 80,000 tonnes of the contaminated soil were used without stabilization in the same road, while 60,000 m^3

were used for other road construction and 80,000 tonnes for new terrain construction. In total, 200,000 m³ of PAH-contaminated soil were used, with only 15,000 m³ requiring hazardous waste disposal.

11.4.1.1   Sustainability Indicators: Observations and Comments

There are several indicators that can be used to determine whether the remediation project meets the aims or principles of *sustainability*. These include:

- Land use: The results show that if one uses the contaminated land as a starting point, the original plan for remediation and rehabilitation of the land to permit usage as parking and storage facilities is a step upward, i.e., better land use. The subsequent report indicating use of the rehabilitated land as green space is a positive step toward sustainability goals. We need to note that the land use indicators here are not in reference to the initial airport land use. Because of the new intended green space land use, the sustainability indicators can now be cast in terms of "return to nature" indicators.

- Energy utilization: Conservative energy use as a target for remediation procedures does not always produce results that will support complete site remediation. Comparing two specific remediation procedures for energy use is a good procedure in minimizing depletion of energy resources, especially nonrenewable energy resources. Since the energy resource to be used for both remediation and rehabilitation procedures is hydroelectric based, and assuming that this is fully renewable, the sustainability feature here can be viewed more as a conservation measure. The energy indicators are referenced specifically to the remediation–rehabilitation processes, and not to the production of hydroelectricity.

- Noxious emissions: The use of hydroelectricity as the source of power has essentially limited noxious emissions. Since the impact of CO_2 discharges has been minimized with the type of energy used, one will need to accept that emissions indicators for full sustainability cannot be realistically set. A set of realistic parameters and values for emissions indicators needs to be prescribed.

- Nonrenewable and renewable resource materials: The metals used are nonrenewable resources. While reuse of the asphalt pavement material and the underlying and contiguous contaminated soil for road construction shows a positive approach to the principles of sustainability, there is a requirement for monitoring to ensure that these materials do not present future contamination problems to the immediate environment. As with the situation of emissions, a set of appropriate parameters and values for material and system status indicators is required.

The use of recycled products in construction of roads has also been recently demonstrated in a project in Finland designed to show sensitivity to sustainability objectives in construction of roads (Lahtinen et al., 2005). New types of road construction materials based on the industrial by-products, fly ash, and fiber ash were evaluated in new roads. Fiber ash was evaluated in light traffic paths and for the widening of safety lanes. The pilot construction took place in 2002 and 2003. Monitoring of the road performance is under way until the end of 2005. The fly ash was obtained from the incineration of bark, peat, and sludge, and the fiber ash is a fiber sludge from the paper industry, with fly ash and cement binder. Up to now, the results of using the recycled materials in the road construction are positive both technically and economically and provide a potential way of saving virgin, nonrenewable resources.

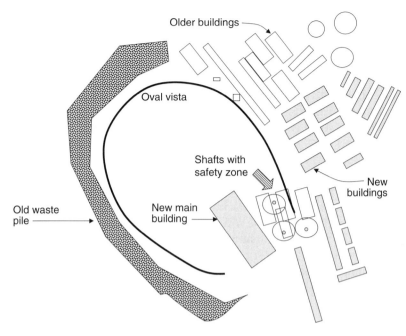

FIGURE 11.6
Redeveloped Mont Cenis site. New buildings are in gray and older buildings are in white. The three abandoned shafts (encircled with a safety zone) are used to recover energy. (Adapted from EMC, *Mont-Cenis*, report of the Entwicklungsgesellschaft, Mont-Cenis, Herne D, 1998.)

11.4.2 Sustainable Mining Land Conversion

Mount Cenis was established as a coal mine near Herne in the Ruhr District of Germany in 1871 (Genske, 2003). Subsequent coal washing and coking facilities were then built. The mine was one of the largest in the area, but it was closed in 1978 due to the coal and steel crisis in Europe. At the site, 26 ha of land were contaminated. There was subsidence, acid mine drainage, and many underground structures. However, in 1990, a large project was conceptualized to remediate and reuse the land for companies and enterprises. The main features of the project, as shown in Figure 11.6 (EMC, 1998), included:

- An academy for the Ministry of Interior (the largest building)
- Various public service buildings, such as a meeting hall, civic administration buildings, and a library
- Shops and services for a shopping mall that already existed
- Two hundred fifty housing units
- A recreation park

Wood was chosen for the structures due to resource efficiency, as detailed in a report for the Club of Rome of 1995 (Weizäker et al., 1997). Forests close to the construction site were chosen as the source of the wood to reduce transportation requirements. Concrete buildings were chosen to reduce climate control requirements. Energy savings of 23% for heating enabled an 18% reduction of the CO_2 emissions. Approximately 10,000 m² of solar cells covered the roof of the wood and glass structure for the academy, hotel, living quarters, and sports facilities. Energy consumption was reduced to 32 kW-h per year. The power plant on the roof provides 1 MW of power, more than twice that needed by the center (EMC, 1998).

Approximately 120 million m^3 of methane is generated from the abandoned mines. The general practice of burning the methane releases approximately 8 million tonnes of CO_2. Therefore, it was decided to capture the gas containing 60% methane. This was converted to 2 million kW-h of electricity and 3 million kW-h of heat for the nearby buildings. A supplementary natural gas plant (1800 kW-h) and a hot water storage tank were constructed to ensure adequate energy and heat due to the fluctuating nature of the methane production. As a backup, a connection was made to the municipal energy system. This also was used for discharge of excess energy. Rainwater was collected from buildings for use in toilets, washing solar panels, and watering gardens.

Infiltration of the water was allowed only where the soil was low in contamination. Excavation of the contaminated soil was not performed since this fills landfills and transfers the problem to another place. The soil was instead placed on clay liners or membranes to prevent leaching of the contaminants to the groundwater. Herb gardens were grown on top of the contaminated land. Gravel and sand filters were placed above the liners to collect the precipitation. The entire project cost 110 million euros and was shared between the community and private investors.

11.4.2.1   *Sustainability Indicators: Observations and Comments*

As with Section 11.4.1, there are several indicators that can be examined to evaluate whether sustainability or the path toward the goals of sustainability has been taken:

- Land use: Increased land use capability has been achieved with the remediation–rehabilitation scheme. As with the land use case in Section 11.4.1, the land use indicators chosen are in specific reference to the initial condition prior to remediation and rehabilitation, and not before mining.

- Energy sources: The multisource energy input, from solar to methane gas capture and reuse, coupled with more efficient climate control in the buildings, shows good attention to energy conservation. This agrees with the requirement for reduction in depletion rate of nonrenewable energy resources as a goal toward *conservation for sustainability.*

- Water use: As with the conservation strategy for energy, utilization and reuse of water show good accord with water sustainability indicators.

- Remediation of contaminated land: The concern for not dumping contaminated soil into another landfill as a reason for placing the contaminated soil on secure membranes and left on-site is a responsible attitude. One assumes that the appropriate requirements for (1) monitoring the contaminated soil facility, (2) prescribing indicators for the safety–recovery status of the contaminated soil facility, on the assumption that the contaminated soil will be remediated through intrinsic remediation processes, and (3) developing appropriate risk management procedures were employed.

11.4.3   Agriculture Sustainability Study

A study was performed by Hoag et al. (1998) to evaluate soil quality for corn production. The objective of the study was to evaluate the sustainability of corn production that used both renewable and nonrenewable resources. The study could also be applied to forestry. Sustainability literature, economic theory, and an empirical model of production were linked. The three questions examined included (1) the impact of the definition of sustainability, (2) the relationship of U.S. soil conservation policies and sustainability objectives, and (3) the effect of substitution, reversibility, and uncertainty on optimal soil use.

The three definitions of sustainability retained for the study included (1) sustainability as constant consumption (referred to as weak sustainability) (Hartwick, 1978; Solow, 1974), (2) sustainability as a constant stock of natural resources (referred to a strong sustainability) (Pearce and Atkinson, 1995), and (3) sustainability for intergenerational equity (WCED, 1987). The two indices used for determination of soil quality included the quality of the soils for producing crops and nitrate leaching. The index for soil quality was based on a model by Pierce et al. (1983). Soil productivity (PI) was determined as

$$PI = \sum_{i=1}^{r} (SAWC_i \times SBD_i \times SPH_i \times WF_i) \tag{11.1}$$

where *SAWC* is the sufficiency of available water capacity, *SBD* is the sufficiency of bulk density, *SPH* is the sufficiency of pH, *WF* is the weighting factor for each *i*th horizon, and *r* is the number of 10 cm horizons in the root depth.

Weather, physical, and chemical inputs (e.g., tillage, fertilizers, chemicals, water) and soil quality determine yield for agricultural production. A dynamic model was then developed to evaluate sustainability in production based on Pierce et al. (1983), Hoag (1998), and others. The yield of the product is a function of soil quality (SQ), soil with the aid of physical inputs (tillage), and soil with neutral inputs (soil nitrogen, nitrogen fertilizer application, and sprayed pesticides). SQ is a function of available water capacity, bulk density, pH, and soil organic matter (SOM). Soil nitrogen is a function of soil nitrogen level, the amount of nitrogen applied, tillage level, and nitrogen taken up and leached. Leaching is a function of the soil nitrogen, nitrogen application, tillage, precipitation, and crop uptake. The producer thus must maximize production, which depends on SQ and the environmental by-products.

Data were collected in Minnesota, Iowa, and Missouri. The simulation model Environmental Policy Integrated Climate (EPIC) was used to evaluate the productivity, soil resource degradation, water quality, and effects of input levels and practices. Nine scenarios were simulated.

$$SQ = \sum_{i=1}^{r} (SAWC_i \times SBD_i \times SPH_i \times SOMC_i \times WF_i) \tag{11.2}$$

where *SAWC* is the sufficiency of available water capacity, *SBD* is the sufficiency of bulk density, *SPH* is the sufficiency of pH, *SOMC* is the soil organic matter content, *WF* is the weighting factor for each *i*th horizon, and *r* is the number of 10 cm horizons in the rooting depth. *SQ* was calculated within one year as one summation of the factors at each horizon. The General Algebraic Model (GAMS)/MINSO approach (Brooke et al., 1992) was used for optimization.

It was found that the ability to meet sustainability goals depends on soil type and the definition of sustainability. In some cases, soil can be managed many ways, but in others, some trade-offs are required. Lower SQs can require complex solutions for management. From Figure 11.7, it can be seen that soil conservation over a long period of time (approximately 100 years) can assist susceptible soils, whereas a soil that is of poor quality to start with cannot be saved even by conservation practices. Soil sustainability can be easily reached by stable soils.

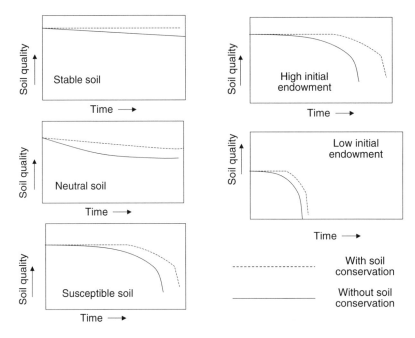

FIGURE 11.7
Comparison of soil quality (SQ) over time for different soils with and without soil conservation. (Adapted from Hoag, D. et al., *Sustainability and Resource Assessments: A Case Study of Soil Resources in the United States*, EPA/600/R-98/038, U.S. EPA Office of Research and Development, Washington, DC, 1998.)

Many of the assumptions should be further evaluated in the future. These include items such as (1) time frame, (2) definitions used, (3) technological advancements, (4) determination of how costs could be shared by society and the producer, and (5) consideration of improvement to the soil quality index so that other components, such as contaminants, can be incorporated.

Another case study was performed for arable farming systems using dynamic balance studies for heavy meals (Moolenaar and Lexmond, 2000). Inputs consisted of fertilizers (N, P, and K), animal manure, organic additives sewage sludge and compost, and atmospheric deposition, as shown in Figure 11.8. Outputs are the crops, soil erosion, and leaching losses.

An experimental farm in Nagele, the Netherlands, was chosen for the study. A conventional arable system with a 4-year crop rotation (seed potato, sugar beet, chicory, onion, winter wheat, and spring barley) was used with mineral fertilizers (CAFS-MF), and another one with organic fertilizers (CAFS-OF). The integrated system (IAFS) included carrot instead of chicory, while the ecological system (EAFS) was a 6-year rotation of seed potato, spring wheat, celery and onion, spring barley, and carrot and oats. The ecological system also used solid goat and cattle manure, while the IAFS used a combination of organic and mineral fertilizers. Heavy metal balances were calculated in all systems. Soil samples came from the top 30 cm of soil. The static balances are shown in Figure 11.9. Dynamic balances were then calculated. For the conventional system, Cd values were found to exceed the Dutch standard. Crops such as carrots, sugar beets, potatoes, and onions have a significant effect on Cd uptake. The CAFS-MF system also showed high accumulation of Cd due to high fertilizer inputs. The results indicated that crop selection and fertilizer type have significant impacts on heavy metal accumulation. However, it was noted that decreasing Cd inputs increased Cu and Zn inputs. In the conventional system,

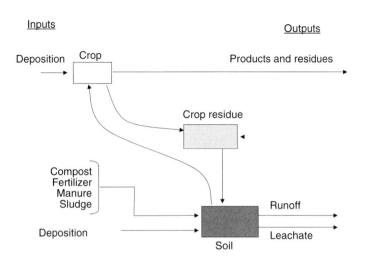

FIGURE 11.8
Inputs and outputs into an arable agricultural system. (Adapted from Moolenaar, S.W. and Lexmond, Th.M., in *Heavy Metals: A Problem Solved? Methods and Models to Evaluate Policy Strategies for Heavy Metals*, van der Voet, E. et al., Eds., Kluwer Academic Publishers, Dordrecht, the Netherlands, 2000, pp. 47–64.)

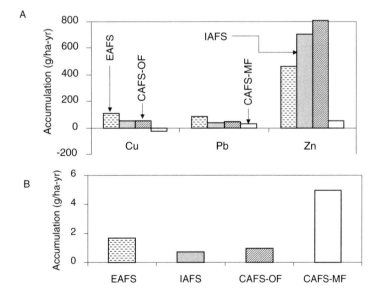

FIGURE 11.9
Comparison of the accumulation of (A) Cu, Pb, and Zn and (B) Cd in an ecological (EAFS), integrated (IAFS), and conventional system with mineral fertilizer (CAFS-MF) and in a conventional system with organic fertilizer (CAFS-OF). (Data from Moolenaar, S.W. and Lexmond, Th.M., in *Heavy Metals: A Problem Solved? Methods and Models to Evaluate Policy Strategies for Heavy Metals*, van der Voet, E. et al., Eds., Kluwer Academic Publishers, Dordrecht, the Netherlands, 2000, pp. 47–64.)

copper depletion occurred. While the integrated system was found to be best for Cd, this was not the case for Cu, Pb, and Zn. This dynamic balance approach enabled comparison of different agricultural systems for optimizing heavy metal accumulation. For example, triple superphosphate application increased Cd accumulation, and copper and zinc inputs increased during manure application.

11.4.3.1   Sustainability Indicators: Observations and Comments

The use of *soil quality* as an indicator for sustainability in agricultural production has a history of almost 30 years. The use of soil quality as a tool for assessing the health of a soil with particular reference to agricultural purposes was first discussed in the late 1970s (Warkentin and Fletcher, 1977). The term *soil health*, which was (and may still be) used by farmers, refers to the functional capability of the soil to support crops and other plants. With the quantification procedures indicated in this section, it is seen that a structured effort is being made in agriculture to meet the goals and requirements for sustainable agriculture. The production of quantification techniques for soil quality provides the means for comparing the dynamic state of the agricultural soil, and offers the opportunity to develop the methodology for determination of soil quality indices. As with the other soil quality indices developed for the various soil functions discussed in Section 11.3.1, these agricultural soil quality indices are the necessary constituents of the indicators for agricultural soil sustainability.

11.4.4   Petroleum Oil Well Redevelopment

The Damson Oil Site in California, near Venice Beach, is an abandoned oil well that stopped production in 1989 (CCLR, 2000). Damson deconstructed the facilities in 1991. However, after the oil wells were capped, the company filed for bankruptcy, leaving soils contaminated with hydrocarbons, sumps with oil and sludge from the extraction process, vaults with oil, and several miles of pipeline. Further contamination occurred as a result of deliberate dumping of debris by passersby. Since it was deemed too expensive to restore the site to a sandy beach, it was decided to establish an in-line skating facility. The oil site was to be capped with concrete, and the remediation costs and improvements made would enable the facility to be economically viable. The plan consisted of (1) an environmental site assessment, (2) waste removal for all surface soils, liquid wastes, and sludges, (3) a demolition plan for the pipeline and other structures and tanks, (4) a construction plan for a skating facility, (5) possible restoration of the beach, (6) establishment of other facilities, and (7) negotiation of a risk management plan with the regional board for remediation objectives and standards. By 2000, environmental assessments and cost estimates were completed and the construction of the boardwalks was initiated. Funding was obtained from the Brownfield Task Force of Los Angeles. The project is ongoing.

Another former oil refinery was converted to a business and recreational opportunity in Casper, WY (Applegate et al., 2005). The refinery had operated since the early 1900s but closed in 1991 due to environmental liabilities. The cost of the site remediation was estimated at $350 million. Various risk assessments were undertaken. To protect the river and remediate the groundwater, a horizontal wall for air sparging and venting was designed and installed, in addition to a sheet-pile barrier wall. Pipes were also removed to eliminate the pollution source. Final remediation strategies included (1) removal of sediment from the lake and (2) removal of tanks, pipes, concrete, and other material from the refinery area. Cleaning of the groundwater involved oil recovery, sparging, venting,

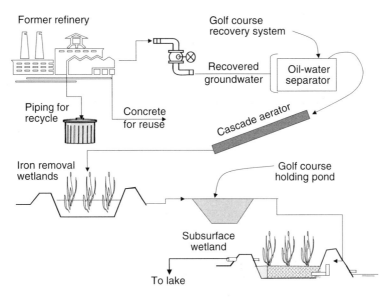

FIGURE 11.10

Golf course water treatment system for recovered groundwater from former refinery and tank farm. (Adapted from Applegate, D. et al., *Civ. Eng.*, 75, 44–49, 2005.)

phytoremediation, and monitored natural attenuation in the refinery and tank farm areas. All concrete (nearly 272,000 tonnes) that was removed was crushed for reuse at the site, and most of the pipes were sent to recyclers. Some examples of concrete reuse include (1) use as drainage in the water treatment system, (2) construction of a barrier that prevented animals from entering the waste depository near the lake, and (3) construction of roads. It is estimated that oil recovery at the site will take approximately 25 years — on the basis of analysis of the mobility of the oil strongly adsorbed to the alluvium. As a golf course was also constructed at the site, oil recovery wells had to be designed so that they would not be placed in the fairways and greens.

Water at the site was also to be reused. Therefore, a system was set up that included management of the storm water, irrigation of the golf course, and pumping of the water into the lake for the migratory birds. The schematic of the water management system is shown in Figure 11.10 and can handle between 1890 to 5670 l/min. It is mainly hidden and integrated in the golf course. A kayaking course will also be placed in the river. This former industrial site was opened to the public in June 2005.

11.4.4.1 Sustainability Indicators: Observations and Comments

From a land use sustainability perspective, the ongoing project in the first example, together with remediation of the contaminated site in the second example and use of recycled materials, it appears that improved land use has been obtained. The remediation–rehabilitation plan in the first example provides opportunities for the prescription of indicators for sustainability. It appears that site restoration was performed in a fashion that will return the site to conditions and usage beyond initial sandy beach conditions. From a land use standpoint, the remediation–rehabilitation scheme is a positive step. One presumes, however, with the initial environmental site assessment, that a proper accounting has been given to avoidance of negative impacts from the development and operation of all the new facilities. A prescription of *facilities–operation indicators* would be useful.

11.4.5 Mining and Sustainability

The Sullivan Mine in Kimberley, Australia, was discovered in 1892 (Teckcominco, 2001). While the community was previously dependent on a single industry (mining), since 1990, the community became more diversified and the area is now a resort destination. To determine the viability of the community, a set of indicators was developed based on the following guidelines: (1) Canadian Mining Association Guidelines for Sustainable Development, (2) Australian Minerals Industry, (3) the Global Reporting Initiative, and (4) the World Business Council for Sustainable Development. These indicators were divided into economic, environment, and social categories for several components. The environment indicators for sustainability were (1) ecosystem health based on soil erosion and species diversity, (2) the availability of natural resources such as minerals, (3) the effect of the company on the ecological amenity, and (4) sustainability. The features included for the environment sector were a tailings pond reclamation and a drainage water treatment plant. Reuse, recycling, and safe handling of all products are practiced. The Mark Creek has now been carefully landscaped with a golf course. A Sullivan Mine Public Liaison Committee was formed to ensure public participation in the process.

Another example of the practice of sustainability (Stanton-Hicks, 2001) can be found in an iron industry at another site in Australia (Pilbara). A theoretical model (CSIRO, as shown in Figure 11.11) has been incorporated. Mining was initiated in the region in the 1960s with large-scale technology and inland mining operations. A number of programs have been established since 1990 to train and work with the indigenous employees and contractors in the area. The life cycle cost approach is being used, with eventual rehabilitation of the site to be considered at an early stage. Fugitive dust is a particular issue for mine dumps. In addition to using wind block and conveyor coverings, the appropriate amount of water must be added to ensure that dust does not become airborne.

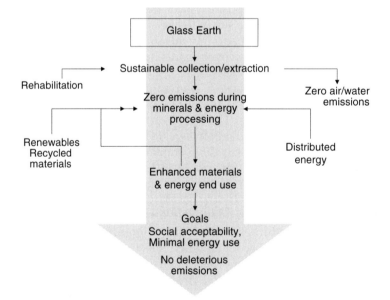

FIGURE 11.11
Concept of the CSIRO model. (Adapted from Stanton-Hicks, E., *Sustainability and the Iron Ore Industry in the Pilbara: A Regional Perspective*, Sustainability and Mining in Western Australia, Institute for Sustainability and Technology Policy, Murdoch University, Perth, 2001.)

Hamersley Iron has developed the HIsmelt Technology, which enables the use of more feedstocks, and subsequently more recycling and high-quality recycled products from the high-phosphorus iron ore. Phosphorus is no longer problematic due to the reduction process, which replaces the conventional oxidation one. Heat energy can be recaptured, and scrubbing reduces the release of toxins and particulates in addition to greenhouse gases. The process is under development. The area still has many challenges, including ensuring the maintenance of the small ecosystems (desert, mountain, and coastal) in the areas.

11.4.5.1 Sustainability Indicators: Observations and Comments

The aspects of sustainability include:

- Maximization of economic, ecological, and sociocultural efficiencies through integration
- Extension of mine life through new techniques and development of new markets
- Land use agreements to benefit economic and socioeconomic sustainability
- Pastoral management practices to improve environmental management and cooperation with land management agencies

11.4.6 Organic Urban Waste Management in Europe

In Europe more than 40% of all waste is biologically treated via composting or anaerobic digestion (Barth, 2005). In the European Union (EU), this amounts to a potential organic waste of around 50 million tonnes of food and garden residues. Forty-two percent of this waste can easily be source separated. The composts can serve as an organic fertilizer or soil improver. Open windrows are the main technology for European countries. Another technology, anaerobic digestion, on the other hand, is divided into 46% dry digestion and 49% wet digestion, with many installations treating more than 40,000 tonnes per year of waste. Commercial and food waste in Germany and Austria is increasing, as well as mixed municipal waste in Spain. Denmark digests household biowaste in agricultural digestion sites. The aim is also to produce organic fertilizers and soil improvers.

Various policies with the EU Commission also will promote biological treatment directly or indirectly (Figure 11.12). The landfill directive states that fermentable waste must be reduced by 25% by 2006 and by 65% by 2016. This will improve landfill structures and reduce biogas production in landfills. The directive on biowaste is expected by 2005. Its aim is to promote recycling of biowaste and to promote certified compost that prevents contamination. The European Soil Protection Strategy promotes the protection of soil by mitigating the three threats to soil, including decreased organic matter, soil contamination, and erosion due to desertification. The addition of composts and residues of organic matter to soil can clearly enhance organic matter and restore the agronomical and microbiological properties of soil. In addition, the addition of these types of organic matter to the soil can mitigate climate change by providing a carbon sink in the soil.

The European Climate Change Program (ECCP) promotes the buildup of carbon in the soil by organic fertilizers, which can thus serve as a carbon sink (via the process of carbon sequestration) for up to 2 gigatonnes of carbon per year. This is compared to the 8 gigatonnes of carbon per year that is emitted in to the atmosphere. Therefore, the practice of biological waste treatment of separated wastes that is developing in Europe is clearly more sustainable than landfilling solid waste.

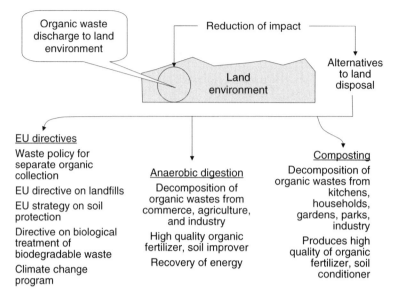

FIGURE 11.12
European initiatives for managing organic waste. (Adapted from Barth, J., From Waste to Valuable Product: 15 Years of Successful Experiences with Recycling of Organic Waste in Europe, paper presented at AMERICANA 2005, Montreal, April 6–8, 2005.)

11.4.7   Sediment Reuse: Orion Project, Port of New York and New Jersey

Sediments must often be removed by dredging to maintain waterways and ports. Approximately 5 to 10% of these sediments are contaminated (Urban Harbors Institute, 2000). Management of these materials must be planned carefully with environmental protection and economic viability. The Port of New York and New Jersey has to dredge approximately 4 million m^3 of sediment every year. Since one third of the sediments do not meet standards, alternate uses are required. More than 1.2 million m^3 have been used as foundation fill for a parking lot. The sediments were dredged, transported to a screening facility, and then pumped on to the shore for mixing with cement kiln dust to improve compressive strength. The mixture was placed on a 24 ha lot, and asphalt was used to cover the fill. No virgin material was required for the lot foundation.

11.5   Concluding Remarks: Toward Geoenvironmental Sustainability

It is clear that as long as depletion of the nonrenewable natural resources contained within the geoenvironment occurs, sustainability of the geoenvironment cannot be attained. When one adds the burden of natural catastrophic disasters and their consequences, together with physical, chemical, and biological impacts to the geoenvironment from the various stressors described in the previous chapters, it becomes all the more evident that geoenvironmental sustainability is an impossible dream. One has two simple choices: (1) to concede that the sustainability of the geoenvironmental resources that provide society with its life support systems cannot be realized and prepare to face the inevitable, or (2) to correct those detrimental elements that can be corrected and find substitutes and alternatives to replace the depleting geoenvironmental resources. The material in this book is a first step in a long series of required steps in adoption of the second choice.

It has been argued that if the global population were to be reduced to some small limiting size, say in the order of just over a billion people, and if replacements or substitutes for the nonrenewable resources could be found, sustainability could be achieved. Working on the assumption that this will not likely happen, we have chosen to address the problem of protection of the geoenvironmental base that provides society with its life support systems. The need to protect the environment, and especially the natural resources that provide the basis for the sustenance and well-being of society, is eminently clear. The subject addressed in this book is a difficult one, not only from the viewpoint of the basic science–engineering relationships involved in ameliorating adverse impacts on the geoenvironment, but as much or more so from the fact that many crucial elements contributing to the generation of these same impacts could not be properly addressed. This is a fact and a realization that many of these elements either were not within the purview of this book (especially the critical subject of biological diversity) or were elements that were dictated by forces influenced by business, public awareness, and political will. Prominent among these are (1) social-economic factors and business–industrial attitudes and relationships, (2) public attitudes, awareness, sensitivity, and commitment, and (3) political awareness and will.

In adopting the second choice, we have focused on the importance of the geoenvironment as a resource base for (1) provision of the required sustenance of the human population and (2) production of energy and goods. We have attempted to develop a better understanding of the stressors on the geoenvironment and to lay emphasis on the need to better manage the geoenvironmental natural renewable and nonrenewable resources. Again, the absence of discussion relating to the direct primary sources of stressors, such as the decision makers responsible for the upstream and downstream industries, means that this book can only provide the geoenvironmental perspective on results of the main impacts resulting from these stressors. The oceans and the coastal marine environments are also significant resource bases, and are essential components of the life support for the human population and must not be neglected.

Outside of the calamitous natural events in the very recent years, in the form of earthquakes, hurricanes, landslides, floods, etc., that have caused death and severe distress to countless numbers of unfortunate humans, it is seen that pollution of air, land, and water resources is the greatest anthropogenically derived threat to the human population and the geoenvironment. The various discards, spills and loss of materials (chemicals, etc.), and discharges of wastes, either in liquid form or as solids, are common to all types of human activities within (1) the built urban environment, (2) mineral and hydrocarbon exploitation, (3) the agricultural ecosystem, and (4) industries. Some of these activities include wastewater discharges, use of nonrenewable resources as energy input and also as raw materials for the industries, injection wells, leachates from landfills and surface stockpiles, open dumps, illegal dumping, underground storage tanks, pipelines, irrigation practices, gaseous and noxious particulate airborne emissions, production wells, use of pesticides and herbicides, urban runoff, and mining activities. These pose significant threats to the land environment and the receiving waters, and to the inhabitants of these environments.

The degree of environmental impact due to pollutants in a contaminated ground site is dependent on (1) the nature and distribution of the pollutants, (2) the various physical, geological, and environmental features of the site, and (3) existent land use. Through management and education, the sources of pollution must be controlled to maintain water quality and supply for future generations. Environmental management, including various remediation and impact avoidance tools, has been developed so that technology can be used to implement the replacement of the nonrenewable resources that are being depleted. Mitigation and management of pollutants in the subsoil should seek to reduce and eliminate the presence of pollutants in the soil. Engineering the natural attenuation capability of soils, through enhancements of the attenuation capability with geochemical, biological, and nutrient aids, will provide greater management options. Considerable attention needs to be paid to many of these issues by researchers, policy makers, and other professionals to alleviate the stresses to the geosphere and seek sustainability and ways for society to live in harmony with the environment now and in the future.

References

Applegate, D., Degner, M., Deschamp, J., and Haverl, S., (2005), Highly refined, *Civ. Eng.*, 75:44–49.

Barth, J., (2005), From Waste to Valuable Product: 15 Years of Successful Experiences with Recycling of Organic Waste in Europe, paper presented at AMERICANA 2005, Montreal, April 6–8.

Brooke, A., Kendrick, D., and Meeraus, A., (1992), *GAMS: A User's Guide*, Scientific Press, Redwood City, CA.

California Center for Land Recycling (CCLR), (2000), *Brownfield Redevelopment Case Studies*, CCLR, San Francisco, 47 pp.

DEFRA, (2000), National Waste Production Survey, Environmental Agency Water UK, London. www.environment-agency.gov.uk/environment/statistics/waste.

Ellefsen, V., Westby, T., and Andersen, L., (2001), Sustainability: The Environmental Element: Case Study 1, paper presented at the Clarinet Final Conference, Sustainable Management of Contaminated Land, Proceedings, Vienna, June 21–22.

Ellefsen, V., Westby, T., and Systad, R.A., (2005), Remediation of contaminated soil at Fornebu Airport-Norway, stabilization and re-use of PAH-contaminated soil, in *Proceedings of the 16th ICSMGE*, Osaka, Japan, pp. 2365–2370.

EMC, (1998), *Mont-Cenis*, report of the Entwicklungsgesellschaft, Mont-Cenis, Herne D, 44 pp.

Genske, D.D., (2003), *Urban Land: Degradation, Investigation, Remediation*, Springer-Verlag, Berlin, 331 pp.

Graham, E., (1989), Oilspeak, Common Sense and Soft Science, *Audubon*, September, pp. 102–111.

Haigh, M.J., (1978), *Evaluation of Slopes on Artificial Landforms, Blaenavon, UK*, Research Paper 183, Department of Geography, University of Chicago.

Hartwick, J., (1978), Substitution among exhaustible resources and intergenerational equity, *Rev. Econ. Stud.*, XLV-2:347–354.

Hoag, D., (1998), The intertemporal impact of soil erosion on non-uniform soil profiles: a new direction in analyzing erosion impacts, *Agric. Syst.*, 56:415–429.

Hoag, D., Hughes Popp, J., and Hyatt, E., (1998), *Sustainability and Resource Assessments: A Case Study of Soil Resources in the United States*, EPA/600/R-98/038, U.S. EPA Office of Research and Development, Washington, DC (also available online at http://www.epa.gov/ncea/soil.htm).

Lahtinen, P.O., Maijala, A., and Kolkka, S., (2005), Environmentally friendly systems to renovate secondary roads, Life Environment Project: Kukkia Circlet, LIFE02 ENV/FIN/000329, in *Proceedings of the 16th ICSMGE*, Osaka, Japan, pp. 2407–2410.

McLean, I. and Johnes, M., (2000), *Aberfan: disasters and government*, Welsh Academic Press, Cardiff, 274 pp.

Moolenaar, S.W. and Lexmond, Th.M., (2000), Application of dynamic balances in agriculture, in *Heavy Metals: A Problem Solved? Methods and Models to Evaluate Policy Strategies for Heavy Metals*, van der Voet, E., Guinee, J.B., and Udo de Haes, H.A. (Eds.), Kluwer Academic Publishers, Dordrecht, the Netherlands, pp. 47–64.

Nanus, L., Campbell, D.H., Ingersoll, G.P., Clow, D.W., and Mast, M.A., (2003), Atmospheric deposition maps for the Rocky Mountains, *J. Atmos. Environ.*, 37:4881–4892.

Payne, J.R. and Phillips, C.R., (1985), *Petroleum Spills in the Marine Environment*, Lewis, Chelsea, MI, 148 pp.

Pierce, D. and Atkinson, G., (1995), Measuring sustainable development, in *The Handbook of Environment Economics*, Bromley, D. (Ed.), Blackwell Publishers, Cambridge, MA, pp. 166–181.

Pierce, F.J., Larson, W.E., Dowdy, R.H., and Graham, W.A.P., (1983), Productivity of soils: assessing long term changes due to erosion, *J. Soil Water Conserv.*, 38:39–44.

Solow, R., (1974), The economics of resources or the resources of economics, *Am. Econ. Rev.*, 64:1–13.

Stanton-Hicks, E., (2001), *Sustainability and the Iron Ore Industry in the Pilbara: A Regional Perspective*, Sustainability and Mining in Western Australia, Institute for Sustainability and Technology Policy, Murdoch University, Perth.

Tanacredi, J.T., (1999), Nature of pollution, in *Encyclopedia of Environmental Science*, Alexander, D.E. and Fairbridge, R.W. (Eds.), Kluwer Academic Publishers, Dordrecht, the Netherlands, pp. 482–485.

Teckcominco, (2001), The Sullivan Mine: A Case Study on Mining and Sustainability, paper presented at the Mineral Councils of Australia Environmental Workshop, Canberra, October.

Tirion, H.B., (1999), Agricultural impact on environment, in *Encyclopedia of Environmental Science*, Alexander, D.E. and Fairbridge, R.W. (Eds.), Kluwer Academic Publishers, Dordrecht, the Netherlands, pp. 10–13.

Urban Harbors Institute, (2000), *Green Ports: Environmental Management and Technology at U.S. Ports*, EPA 825706-01-0, U.S. EPA, Washington, DC, March.

U.S. Congress OTA, (1989), *Partnerships under Pressure, Managing Low Level Radioactive Waste*, OTA-0-426, Office of Technology Assessment, U.S. Government Printing Office, Washington, DC, November.

U.S. EPA, (1990), *Disposal Tips for Home Health Care, Solid Waste and Emergency Response*, EPA 625-689-024, U.S. EPA, Washington, DC, January.

U.S. EPA, (1995a), *Profile of the Fabricated Metal Products Industry*, Office of Compliance Sector Notebook Project, EPA 310-R-95-007, U.S. EPA Office of Compliance, Washington, DC, September.

U.S. EPA, (1995b), *Profile of the Petroleum Refining Industry*, Office of Compliance Sector Notebook Project, EPA 310-R-95-013, U.S. EPA Office of Compliance, Washington, DC, September.

U.S. EPA, (1995c), *Profile of the Stone, Clay, Glass and Concrete Products Industry*, EPA 310-R-95-017, Office of Compliance Sector Notebook Project, U.S. EPA Office of Compliance, Washington, DC, September.

U.S. EPA, (2002), *Profile of the Pulp and Paper Industry*, 2nd ed., EPA 310-R-02-002, Office of Compliance Sector Notebook Project, U.S. EPA Office of Compliance, Washington, DC, November.

Warkentin, B.P. and Fletcher, H.F., (1977), Soil quality for intensive agriculture, in *Proceedings of the International Seminar on Soil Environment and Fertility Management in Intensive Agriculture, Society, Science, Soil and Manure*, National Institute of Agricultural Science, Tokyo, pp. 594–598.

Washington State Department of Ecology, (1989), *Solid and Hazardous Waste Program*, report to the Legislature, Washington State Infectious Waste Project and Attachments, Olympia, WA, December 30.

Weizäker, E.U. von, Lovins, A.B., and Hunter Lovins, L., (1997), *Factor Four: Doubling Wealth Halving Resource Use*, Earthscan, London, 322 pp.

Wilson, A., (1993), Cement and Concrete: Environmental Consideration, *Environmental Building News*, Vol. 2, March/April, http://www.buildinggreen.com/features/cem/cementconc.html.

World Commission on Environment and Development (WCED), (1987), *Our Common Future*, Oxford University Press, Oxford, 398 pp.

Index

A